72 Structure and Bonding

Bioinorganic Chemistry

With contributions by
F. A. Armstrong L. Banci I. Bertini
S. R. Cooper C. Luchinat D. M. P. Mingos
S. C. Rawle L. Zhenyang

With 58 Figures and 14 Tables

Springer-Verlag
Berlin Heidelberg New York
London Paris Tokyo Hong Kong

ISBN 3-540-51574-7 Springer-Verlag Berlin Heidelberg New York
ISBN 0-387-51574-7 Springer-Verlag New York Berlin Heidelberg

Typesetting: Macmillan India Ltd, Bangalore-25; Printing: Colordruck, Berlin; Bookbinding: Lüderitz & Bauer, Berlin
2151/3140-5 4 3 2 1 0 – Printed on acid-free paper

Table of Contents

Crown Thioether Chemistry

Stephen R. Cooper* and Simon C. Rawle

Inorganic Chemistry Laboratory, University of Oxford, Oxford OX1 3QR, United Kingdom

The synthetic, structural, and coordination chemistry of crown thioethers with both transition and p-block metal ions is reviewed comprehensively through December 1988. Emphasis falls upon the electronic structures and redox properties induced in metal ions by coordination to crown thioethers. Examples include stabilization of mononuclear Rh(II), Pt(III), and low-spin octahedral Co(II). A subsidiary theme concerns the influence of ligand conformation in determining both the binding efficacy and the qualitative coordination chemistry associated with a given crown thioether. The review concludes with a view toward potential future applications of crown thioethers in catalysis, in sequestration or biological delivery of heavy metal ions, and in fundamental studies directed toward rational design of ligands.

1 Introduction

1.1 Scope of the Review

This review covers the synthesis, conformation, and coordination chemistry of crown thioethers, with particular emphasis on 1,4,7-trithiacyclononane (9S3), 1,5,9-trithiacyclododecane (12S3), 1,4,8,11-tetrathiacyclotetradecane (14S4), and 1,4,7,10,13,16-hexathiacyclooctadecane (18S6) (Fig. 1). It incorporates results on other crown thioethers such as 1,4,7,10-tetrathiacyclododecane (12S4), 1,4,7,10,13-pentathiacyclopentadecane (15S5), and 1,5,9,13,17,21-hexathiacyclotetracosane (24S6). The review discusses the molecular and electronic structures of the complexes, including their kinetic and redox properties, as well as the structural features of the free ligands. Coverage of the literature is complete through December 1988.

We confine attention to macrocyclic polythioethers that contain at least three sulfur atoms within the macrocyclic ring and have at least two methylene units between S atoms (i.e., excluding dithioacetals). The review discusses neither sulfur-containing cyclophanes nor their metal complexes.

For several reasons we further restrict consideration to those compounds that comprise solely thioethers as donor groups. First, complexes of such compounds most clearly evince the electronic consequences of thioether coordination. Second, homogeneity of donor group simplifies analysis of conformation. Such studies provide the trends and general principles for interpretation of the coordination chemistry of mixed donor macrocycles. Third, the surge of research effort in all-thioether macrocycles has yielded sufficient information to permit detailed comparisons between complexes of closely related ligands; inclusion of mixed-donor macrocycles contributes little to such discussion.

We name the crown thioethers non-systematically by an extension of the crown nomenclature introduced by Pedersen [1]. Thus 1,4,7,10,13,16-hexathiacyclooctadecane is called hexathia-18-crown-6. This name is often further abbreviated to

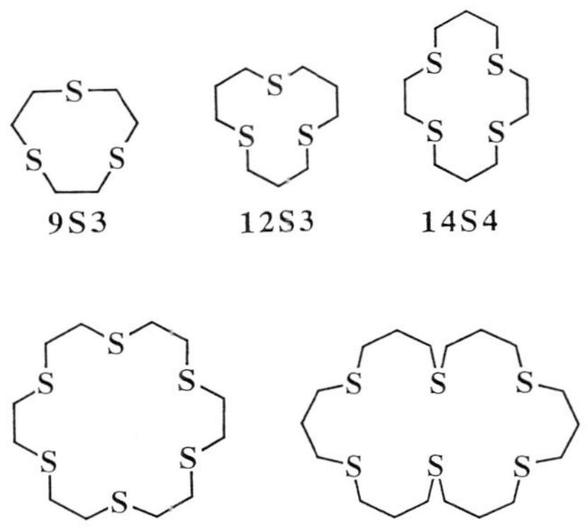

9S3 12S3 14S4 18S6 24S6

Fig. 1. Crown thioethers discussed in this review

18S6, where the numbers denote the ring size and number of sulfur atoms, respectively. This nomenclature rarely leads to ambiguity since most crown thioethers used as ligands feature the most symmetric possible disposition of donor atoms.

1.2 Motivation

Over the last 15 years reviews have dealt separately with the coordinative properties of thioethers [2], the synthesis of macrocyclic sulfides [3], and with macrocyclic compounds in general, including macrocyclic thioethers [4]. Since then, the growing use of crown thioethers has generated the need for a review confining itself to this more circumscribed area. Interest in these ligands has burgeoned in recent years, fuelled by four considerations: 1) the possible analogy between the coordination chemistry of thioethers and phosphines, 2) the relevance of thioether coordination to the blue copper proteins, 3) synthetic improvements that made crown thioethers readily available, and 4) the increasing availability of X-ray diffraction facilities, a vital tool for this field.

The first of these considerations, the potential parallel between thioethers and phosphines, suggests that thioethers might have extensive and industrially useful coordination chemistry. This possibility spurred earlier efforts to examine the complexes of macrocyclic thioethers [5, 6].

Another source of interest came from biochemistry. Research on the blue copper proteins revealed unusual electronic properties (redox potential and kinetics, EPR and optical behavior) that were suspected of arising from interaction of the copper ion with a thioether group from methionine [7]. While crystallographic studies established a weak interaction (Cu ··· S 2.9 Å) [8, 9, 10], its influence on the electronic properties of the Cu site is now considered questionable. Nevertheless, the controversy regarding the blue copper proteins, like the analogy to phosphines, served to focus attention on the broad issue of how thioether coordination affects the electronic structure of transition metal ions. Homoleptic thioether complexes provide the best way of assessing these consequences, since no other groups obscure the effect of thioether coordination.

Despite this interest in crown thioethers, arduous synthetic routes to the ligands impeded extensive investigation of their chemistry until recently. However, advances in synthetic methodology in the last five years has opened the door to work on the coordination chemistry of these ligands. This is particularly true of 9S3, the first synthesis of which proceeded in such low yield (0.04%) as to preclude further study [11].

Fruitful exploration of crown thioether coordination chemistry also had to await the routine availability of X-ray diffraction facilities. The paramagnetism of many crown thioether complexes vitiates the utility of NMR, while uninformative charge transfer bands dominate their optical spectra. Hence X-ray diffraction has proven indispensable to the development of crown thioether chemistry; it provides one of the few ways of determining the ligand denticity, as well as the coordination geometry and stereochemistry at the metal. More fundamentally, however, the issues raised by these complexes often focus on metrical features and ligand conformations. These can only be studied by diffraction methods. Analysis of compression or dilation of the metal ion coordination sphere (i.e. M–L distances), of the ligand conformation, and of coordination mode all demand structural characterization.

Crown thioethers prompt attempts to impose poly(thioether) coordination upon metal ions, even those for which few thioether complexes were previously known. As a class, simple thioethers (e.g. Me_2S) are not particularly good ligands for both electronic and steric reasons. Thioethers exhibit weaker σ-donor and π-acceptor ability than, for example, phosphines. Consequently they also show less binding affinity. In addition, the appreciable bulk of their terminal alkyl groups further hinders complexation. This latter factor is partially circumvented in acyclic polythioethers, and such ligands display respectable chelate effects.

Complexes of crown thioethers are usually considerably more stable even than those of comparable acyclic ligands. This enhanced stability (the macrocyclic effect [12], although not all complexes of macrocyclic thioethers manifest one [13, 14]) makes synthetically tractable the often otherwise impossible imposition of the oligo(thioether) environment. For this reason, the growing attention given to crown thioethers has spurred a blossoming of thioether coordination chemistry generally. These macrocyclic ligands thereby provide entry into new chemistry, since corresponding complexes of simple thioethers often defy synthesis [15]. Indeed, their use has yielded the first examples of homoleptic thioether complexes (i.e., those in which the coordination sphere comprises solely thioethers) of numerous elements. The development of such excellent ligands as 9S3 also encourages use of thioethers as coligands in, for example, catalytic applications and cluster chemistry.

In addition, crown-type ligands can also alter the properties of a metal ion by, e.g. constricting or dilating its coordination sphere (the macrocyclic constriction effect [16, 17, 18]). Both through imposition of an unusual environment – comprising predominantly or exclusively thioether coordination – and through manipulation of that environment, coordination complexes of crown thioethers often exhibit unusual properties and reactivities. Crown thioether complexes are an ideal system in which to study the effect on optical and redox properties of geometric deformations of the coordination sphere.

The development of both organometallic and bioinorganic chemistry focused attention on the reactivity of metal complexes. This arose in two contexts: (i) practical industrial applications and (ii) fundamental comprehension of how, e.g. metalloenzymes and metal-containing redox proteins function. Such interest naturally spurs attempts to control the reactivity of metal ions. Experimentally accessible parameters include coordination number, stereochemistry, redox potential, and steric accessibility of the metal ion. These properties of metal ions may, in principle, be manipulated by ligand design.

Crown thioethers such as 9S3 and 18S6 commonly enforce six-coordination even in some cases where lower coordination numbers might have been expected. In addition, through use of different crown thioethers it is possible to study how compression/dilation of the $M(SR_2)_6$ coordination sphere affects, e.g. redox potentials and hyperfine splittings. Thus crown thioethers provide an excellent means not only of imposing a high-symmetry homoleptic thioether environment, but also one that is well-suited to study the effect of systematic geometric deformations on the properties of metal ions.

In summary, while a considerable number of thioether complexes have been studied, their low stabilities have frustrated attempts at a systematic survey of thioether coordination chemistry. The use of crown thioethers solves this stability problem. Incorporation of the thioether groups into a macrocycle often – but not always [13] – greatly increases the stability of the resulting complexes compared to those of corresponding linear multidentate thioethers, and particularly those of dimethyl sulfide. In contrast to thioethers, phosphines presently constitute a ligand class of central importance in the chemistry of low-valent transition metals, which in turn have found extensive applications in homogeneous catalysis. It is important to bear in mind, however, that only 25 years ago phosphines were widely considered to have sparse coordination chemistry. In some respects the coordination chemistry of thioethers is now in a similar state of development to that of phosphines before their importance came to be appreciated, initially largely through the work of Chatt and coworkers.

1.3 π-Acidity, d-orbital Participation, and Charge Neutralization

Coordination to thioethers typically stabilizes the lower oxidation states of metal ions, and, where relevant, the lower spin states as well. These recurring themes derive largely from two characteristics of thioether ligands: their π-acidity, and their failure to neutralize positive charge effectively.

Several lines of evidence impute appreciable π-acidity to thioethers. Infrared studies of substituted metal carbonyl complexes show that trans thioethers increase the CO stretching frequency more than pyridine or an aliphatic amines, but less than phosphines [2, 19, 20, 21]. To the extent that such changes derive from differences in π-backbonding, these results indicate π-acidity intermediate between that of amines and that of phosphines.

Magnetic properties further support π-acidity. Quantitatively, g values reflect π-effects because deviations from g_e (the free electron value) arise from unquenched orbital angular momentum (from circulation of the t_{2g} electrons). Delocalization of *d* electrons into π* levels on the ligand diminishes the unquenched orbital angular momentum (as measured by k, the orbital reduction factor), and thereby Δg. For example, in copper(II) complexes of thioether ligands, g values deviate from 2 by little compared to those of harder ligands such as OH_2 or NH_3 [22].

Stabilization of low-spin states indicates π-interaction in a more dramatic fashion. Delocalization of t_{2g} electron density into ligand orbitals of π* symmetry (with respect to the M–S bond) diminishes electron–electron repulsion and thereby reduces the spin pairing energy. As a consequence, complexes of thioethers, like those of phosphines, typically assume the low-spin state.

Nephelauxetic ("cloud expanding") effects of thioethers point to the same conclusion. Like g values, nephelauxetic ratios (β, where $\beta = B_{complex}/B_{free\ ion}$, and B is the second Racah parameter) measure the delocalizing effect of the ligands. Thus smaller nephelauxetic ratios indicate greater delocalization. In this respect thioethers consistently exceed aquo and amine ligands. For example, β for $[Ni(SR_2)_6]^{2+}$ complexes averages approximately 0.7; in corresponding amine complexes $\beta \geqslant 0.9$ [23]. The difference in β manifests the greater ability of thioethers to delocalize metal d electron density, a reflection of π-acceptance. This occurs despite the presence of a putative residual lone pair (sp^3 hybrid) on a coordinated thioether.

Reactivity trends also reflect the π-acidity of thioethers. For example, *cis*-$[Ru(14S4)(NO_2)_2]$ resists thermal decomposition to the corresponding nitrosyl [24]. This inertness contrasts with the reactivity of the corresponding amine complex, *cis*-$[Ru(14N4)(NO_2)_2]$, which decomposes readily. Similar observations obtain for decomposition of *cis*-$[RuL(N_3)_2]$ to *cis*-$[RuL(N_2)_2]$ (L = 14S4, 14N4). This difference arises from delocalization of t_{2g} electron density onto the thioether ligands, with concomitant reduction in reactivity of the coordinated $-NO_2$ and $-N_3$ groups [24].

Photoelectron spectroscopy (PES) provides further evidence for π-acidity. The ionization energies obtained from PES directly reflect π-delocalization, since the energy of the HOMO (t_{2g}) depends critically on the overlap with ligand π* orbitals. PES studies of $[Cr(CO)_5L]$ (L = SMe_2, PEt_3) indicate that thioethers and phosphines place roughly comparable amounts of electronic charge on the metal; both do so more than CO. This is consistent with corresponding degrees of π-acidity [25]. This conclusion agrees with that obtained from a structural study of $[Cr(CO)_4(dto)]$ (where dto = 3,6-dithiaoctane). Comparison of M–C distances in $[Cr(CO)_4(L)]$ (L = dto and 1,2-bis(diphenylphosphino)ethane) suggested that the thioether ligand exert slightly less π-acidity [26].

Finally, redox potentials of thioether complexes also indicate the π-acidity of the ligands. Coordination to thioethers invariably raises redox potentials relative to those of analogous aquo or amine complexes. Thus electrochemical studies of $[M(bpy)_2L_2]^{n+}$ complexes (M = Ru, Os) with L = amine, thioether, and phosphine ligands) [27] show that thioethers exceed amines in stabilizing the low (n = 2) oxidation state, but fall short of phosphines. Part of this increment stems from the limited ability of thioethers to neutralize charge through σ-donation. Nevertheless, π-acceptance probably contributes significantly as well.

Historically, the π-acidity of second-row donor ligands has been attributed to the participation of *d* orbitals in backbonding from the metal to the ligand. This explanation has attracted considerable skepticism; calculations indiate that the *d* orbitals lie too high in energy to interact significantly with metal ions [28, 29]. Metal-phosphorus and P–C (or P–O) bond lengths in redox pairs suggest that the C–P σ* bond (which has π-symmetry with respect to the M–P bond) of phosphines plays a central role in π-acidity [30]. This explanation neatly reconciles the experimental observations of M–P π-interaction (as evidenced by trends in, e.g., ν_{co} of substituted metal carbonyls) with the theoretical results. It also dovetails with the observed increase in C–P distance, the magnitude of which correlates with the usual measures of π-acidity. The same argument clearly obtains for thioethers also.

Charge neutralization – or rather the lack of it – also plays an important role in thioether coordination chemistry. Because of their low σ-donor ability, thioethers generally fail to displace anions from the coordination spheres of metals.

Instead, cations tenaciously retain coordinated anions (as in, e.g. $[(NbCl_5)_2(14S4)]$ [31, 32, 33], $[CuCl_2(12S3_2)]$ [34], and $[(14S4)(HgCl_2)_2]$ [35]). Coordination of "non-coordinating" anions such as $CF_3SO_3^-$ and ClO_4^- occurs with surprising frequency in thioether coordination chemistry. This role becomes even more important in complexes of crown thioethers, where thioethers dominate the coordination sphere. Rorabacher and his coworkers [36, 37] have documented the considerable affinity (even in aqueous solution) of Cu(II) crown thioether complexes for such seemingly innocuous anions as ClO_4^-, BF_4^-, and $CF_3SO_3^-$. Consistent with the charge neutralization argument, perchlorate interacts more strongly with Cu(II)-thioether complexes than it does with the Cu(II) aquo ion [36].

Charge neutralization also affects the redox properties of thioether complexes. It contributes to the marked stabilization of lower oxidation states found in all cases. Thioether complexes of metal ions in high oxidation states may approach the boundaries of the electroneutrality principle. Apart from any π-acidity of the ligands, simple electrostatic considerations suggest that poor charge neutralization by the ligands disfavors higher oxidation states.

These arguments raise the question of whether π-effects need be considered at all. Clearly, however, charge neutralization fails to account for the stabilization of low spin Fe(II) and Co(II) ions in their crown thioether complexes. Similarly, it does not account for the small Δg values of thioether complexes, or the high nephelauxetic influence of the ligands. Thus, in summary, π-acceptance and the accumulation of cationic charge combine to generate the characteristic electronic properties of thioether complexes. Last, molecular orbital calculations support the ability of thioethers to serve as π-acceptors [38, 39].

1.4 History

Crown thioether chemistry dates from 1886, when Mansfeld reported the synthesis of 9S3 [40]. To determine whether ring sizes greater than six could be prepared (!) he allowed ethylene bromide to react with sodium sulfide. From this reaction he isolated a product that differed in properties from *p*-dithiane; he suggested it might be 9S3 (Table 4). A similar procedure with 1,3-dibromopropane led to a compound tentatively proposed to be 12S3.

In 1920 Ray published the first of a series of papers on 9S3 [41, 42, 43]. He reported that preparation of ethanedithiol (by reaction of ethylene bromide with potassium hydrosulfide) leaves behind 9S3 after distillation (Table 4). In contemporaneous work Bennett [44] and coworkers [45] used mixed melting point and cryoscopic molecular weight determination to show that Ray's product was in fact *p*-dithiane, not 9S3. In a similarly convincing fashion they also disproved Mansfeld's earlier claim for 9S3. Tucker and Reid [46] were also unable to repeat Ray's work.

Meadow and Reid [47] subsequently (1934) found that reaction of dithiolates with dihalides in ethanol produces small amounts ($< 2\%$) of cyclic compounds in addition to copious quantities of polymer (Table 4). Using this route they prepared the first samples of 18S6 and 16S4. Meadow and Reid also cast doubt on Ray's work on cyclic polythioethers.

Investigation of crown thioethers then lay dormant for 35 years, to be revived in 1969 by Rosen and Busch, who studied the coordination chemistry of 14S4 [48]. They prepared the ligand in 7.5% yield [49] (subsequently increased to 55% through use of high dilution methods [50]) from the reaction of 1,4,8,11-tetrathiaundecane with 1,3-dibromopropane. Analogous synthetic schemes afforded 12S3, 12S4, and 13S4 in yields of 3, 4, and 16% (Table 4) [51].

In the same year Black and McLean reported preliminary work [52] (subsequently followed by a full paper [53]) on the synthesis of 18S6 in 31% yield from the reaction sodium 3-thiapentane-1,5-dithiolate with ethylene bromide in EtOH (Table 4). This surprisingly high yield has not been repeated; instead two laboratories have independently obtained an 8% yield from this reaction [54].

Advances in synthetic methodology, however, now afford crown thioethers in high yield. Ochrymowycz and coworkers laid the basis for this development with their pioneering work on the synthesis of crown thioethers [55]. Their exploratory synthetic work revealed that reaction of sodium 3-thiapentane-1,5-dithiolate with bis(2-chloroethyl)sulfide (mustard gas) gives 18S6 in 32.8% yield (Table 4).

In 1980 Buter and Kellogg further improved yields by introducing the use of cesium carbonate to mediate formation of macrocyclic rings [56, 57]. Application of this procedure to thiacrown synthesis greatly diminishes the extent of polymer formation, with concomitant increase in yield of the desired macrocyclic products. Use of Cs_2CO_3 — but not, e.g. Na_2CO_3 or K_2CO_3 — promotes the high dilution cyclization of α, ω-dithiols with α, ω-dihalides to give crown thioethers in high yield (typically $\geqslant 75\%$). As a result of these various synthetic improvements, crown thioethers are now readily available (Table 4) [58, 59].

The success of Buter and Kellogg's procedure focuses attention on how cesium carbonate fosters the formation of macrocyclic rings. Template effects clearly play no role: even reactants that lack potential donor groups for Cs^+ readily yield macrocyclic products. For example, reaction of 1,10-dimercaptodecane with 1,5-dibromopentane gives the cyclic dithia compound (1,7-dithiacycloheptadecane) in 90% yield [57]. Recent ^{133}Cs NMR work [60] confirms the earlier suggestion [61] that the Cs^+ ion promotes cyclization by ion-pairing only weakly (if at all) with RS^-. This weak ion pairing (which follows from the low charge/radius ratio of Cs^+) generates exceptionally nucleophilic thiolate anions. High reactivity ensures low concentrations of unreacted starting material, and thereby fosters high dilution conditions. This in turn favors the desired intramolecular reaction.

Despite the very early claims for its preparation, 9S3 eluded synthesis until 1977, when Ochrymowycz and coworkers prepared it in heartbreakingly low yield (0.04%) from the reaction of ethylene chloride with sodium 3-thiapentane-1,5-dithiolate in EtOH (Table 4) [11]. Subsequently, Glass and coworkers [62] improved the yield to 4.4% through use of $BzMe_3N^-$ ^-OMe to deprotonate 3-thiapentane-1,5-dithiol for reaction with ethylene chloride (Table 4). Application of the cesium carbonate procedure (for the same reactants in DMF) affords 9S3 in 50% yield [59]. Sublimation of ligand from the crude reaction mixture greatly facilitates the isolation of what historically has been the most elusive of all the crown thioethers. In another approach, Sellmann and Zapf [63, 64] recently published an ingenious synthesis of 9S3 based upon the mediation of a $Mo(CO)_3$ template. Reaction of $[Mo(CO)_3((SCH_2CH_2)_2S)_3]$ (prepared from $[Mo(CO)_3(MeCN)_3]$ and 3-thia-pentane-1,5-dithiol) with ethylene bromide couples the two thiolate termini to yield 9S3 in 60% yield. This sequence is in principle catalytic, since addition of more dithiol liberates the 9S3 and regenerates the starting material. Another method was used by Fujihara et al. [65], who synthesized functionalized propyl-linked crown thioethers from a dithiospiropentane (a masked $(MeSCH_2)_2C(CH_2I)_2$) equivalent.

Unlike the corresponding oxa-crowns, crown thiothers with fused benzo rings have seen relatively little work. Klar and coworkers [66] prepared the hexamethoxy derivative of tribenzo-9S3 by a modified Adams-Ferretti reaction between 1,2-dibromo-4,5-dimethoxybenzene with 1,2-dimercapto-4,5-dimethoxybenzene (Table 4). A variant of this procedure also affords the corresponding 12S4 analogue with four fused benzo units [67].

Sellmann and coworkers prepared dibenzo-hexathia-18-crown-6 on a metal template [68, 69]. Two stepwise alkylations of $[Fe(\text{benzene-1,2-dithiolate})_2(CO)_2]$ with $S(CH_2CH_2Br)_2$ in THF (the first of which gives $[Fe(\text{2,3,11,12-dibenzo-1,4,7,10,13-pentathiatridecane})(CO)]^{2-}$), followed by treatment with HCl (to extract the Fe(II) ion) gives dibenzo-18S6 in 68% yield. A similar procedure gives dibenzo-15S5 in 20% yield.

1.5 Properties of Crown Thioethers

Crown thioethers occur as colorless crystalline solids, odor-free when pure. Their lack of odor results from their low vapor pressure; of the common ligands only 9S3 sublimes readily. Most crown thioethers melt near 100 °C; 12S4 is unusual in melting at over 200 °C. They dissolve readily in acetone, dimethylformamide, dimethylsulfoxide, ethylacetate, dichloromethane, chloroform, toluene, dioxane, and THF, to a lesser extent in diethyl ether, and to a still lesser extent in pentane, methanol, or water. They can be purified most easily by recrystallization from, e.g. hexane/acetone or chloroform; in the case of 9S3 vacuum sublimation provides an especially convenient means of purification. The purification of crown thioethers by high-performance liquid chromatography [70], and their analysis by mass spectrometry has been described [71].

As ligands crown thioethers present both advantages and disadvantages. Their advantages include their ready availability (owing to facile synthetic routes; they are also now sold commercially [72]). More importantly, certain of them – especially 9S3 – coordinate avidly to a wide variety of transition and main group metal ions. Consequently they should find wide use as auxiliary ligands in synthetic inorganic chemistry. In addition, the free ligands lack chirality, unlike corresponding cyclic tertiary phosphines. Last, crown thioethers do not participate in protic equilibria, and do not undergo oxidation by air (cf. alkyl phosphines) under normal conditions.

Their disadvantages center mostly on their relatively low σ-donor ability. Accordingly, they generally fail to displace anions from metal ion coordination spheres. This property limits use of, e.g., metal halides as starting materials. (Abstraction of halide from the metal ion with, e.g., Ag(I), must precede addition of the crown thioether, since Ag(I) itself has great affinity for thioether coordination.) In addition, like phosphines, thioethers react with strong oxidants (e.g., MnO_4^-, H_2O_2, HNO_3), yielding sulfoxides, sulfones, or even sulfonic acids. For this reason crystallization of their complexes as perchlorate salts poses a *severe explosion hazard* [73, 74, 75]; $CF_3SO_3^-$ or BF_4^- counterions usually crystallize well, and provide safe alternatives.

1.6 Synthesis of Complexes

As indicated above, because of both their inability to neutralize positive charge and their weak σ-donor properties, thioethers usually fail to displace anions from the coordination spheres of metal ions. In addition, they often compete poorly with good donor solvents – most notably water. Most syntheses have followed the synthetic methods developed by Rosen and Busch [49, 51]. Useful solvents include CH_3NO_2, CH_3CN, acetone, and ethanol and non-oxidizing acidic media such as Ac_2O. Metal-solvato complexes (e.g. $[Ru(Me_2SO)_6]^{2+}$, $[Ni(EtOH)_6]^{2+}$) of poorly cordinating anions (e.g. BF_4^-, $CF_3SO_3^-$, picrate) have proven the most generally useful starting materials. Once prepared, some crown thioether complexes (e.g., $[Ru(9S3)_2]^{2+}$) either thermodynamically or kinetically resist attack by water.

2 Ligand Conformations; Implications for Binding

Conformational analysis of crown thioethers builds upon the pioneering of Dale and coworkers [76] on analogous oxocrowns. Ligand strain expresses itself primarily in torsional angles; the small force constants for torsion lead to shallow potential wells compared with, e.g. those for bond length or bond angle changes. Dale [77] pointed out that by trading strain into torsional changes, molecules optimize their conformations in the energetically cheapest way. Deviations of torsional angles from the optimum values of 60° (*gauche*) or 180° (*anti*) therefore provide a sensitive measure of the strain inherent in the molecule.

For oxocrowns, information on conformation comes from both X-ray diffraction and ^{13}C NMR studies. Unfortunately, for thiacrowns ^{13}C NMR provides little insight into ligand conformation [76, 78]. Consequently X-ray diffraction has played a crucial role in the study of thiacrown conformations. Reasoning from such solid state data to behavior in solution must, of course, be conducted with caution.

Comparison of oxa- and thia-crowns reveals some interesting contrasts. In the free state, most crown thioethers adopt peculiar "inside-out" conformations in which the S atoms point out the ring. In their important work on 14S4, DeSimone and Glick [79] termed this arrangement of hetero atoms "exodentate", as opposed to the endodentate orientation commonly found in oxa- and aza-crowns. This difference in ring conformation originates in the conformational preferences of the constituent bonds, and it has important ramifications for coordination chemistry. Exodentate S atoms necessitate extensive conformational rearrangement before chelation can occur; in effect, the ligand must turn "right-side-in", with a corresponding enthalpic cost. As a consequence, the greater the number of exodentate S atoms, the weaker the tendency to chelation and the stronger the tendency to bridge two or more metal ions.

Thiacrowns commonly manifest these exodentate conformations because C–S–C–C units (i.e., C–S bonds) slightly prefer gauche placement, whereas S–C–C–S fragments strongly prefer the *anti* conformation. [80]. These preferences originate at least in part from the differences in 1,4-interactions at C–S and C–C bonds (Fig. 2). In ethyl-linked crown thioethers, $(SCH_2CH_2)_n$, n > 3), the conformational preferences of both C–S and C–C bonds act in concert to generate "bracket" S–C–C–S–C–C–S units, in which the central S atom marks a corner (Fig. 2). Such bracket units form the fundamental building block of crown thioethers. For example, 12S4 results from fusion of two such units at the S atoms.

2.1 *Ethyl-linked Crown Thioethers, i.e. $(SCH_2CH_2)n$*

2.1.1 *9S3 (n = 3)*

In the solid state, 9S3 crystallizes with imposed C_3-symmetry (Fig. 3) [81]. Unlike other crown thioethers, all of its S atoms point into the central cavity. This anomaly presumably results from the severe ring (Baeyer) strain associated with a nine-membered ring. The resulting endodentate conformation closely resembles that required for chelation to a metal ion. Coordination slightly diminishes the torsional angles at the C–C bonds; in the free ligand, higher values of these torsional angles minimize S · · · S repulsions [82]. In fact, the correspondence between the conformation of the ligand in the free and bound state probably accounts for the extraordinary ligating ability of 9S3.

Fusion of three aromatic residues to 9S3, as in $(MeO)_6$-tribenzo-9S3, necessarily constrains the macrocycle conformationally. Klar and coworkers identified three likely conformations – termed "crown", "saddle", and "pseudo-saddle" – defined by the position of the phenylene moieties with respect to the plane of the three S atoms [83]. In the crown conformation, all three phenylene units lies on the same side of the S_3 plane. In the saddle conformation, one phenylene ring lies in the S_3

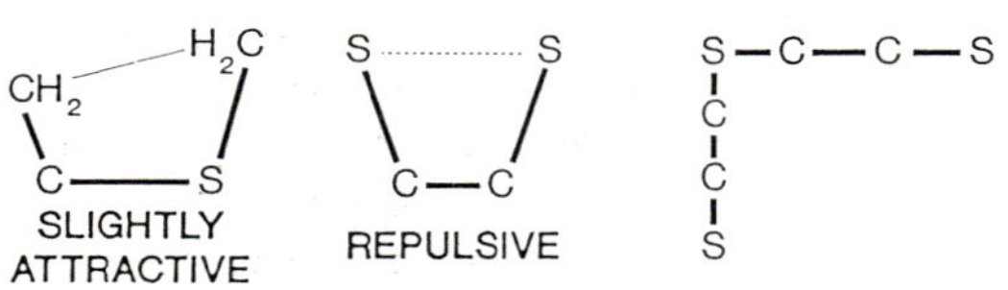

Fig. 2. 1,4 interactions at C–C–S–C and S–C–C–S units in *gauche* placement (left); a bracket unit, the recurring structural motif of ethyl-linked thiacrowns

plane and the other two lie on opposite sides of it. Last, the pseudo saddle conformation differs from the saddle form in a "flattening" to give a twofold axis relating two phenylene units. X-ray crystallography shows that in the solid state $(MeO)_6$-tribenzo-9S3 assumes the saddle conformation [84] (cf. the corresponding O_3 macrocycle, which adopts the pseudo-saddle conformation) [83].

2.1.2 12S4 ($n = 4$)

In the solid state, 12S4 adopts a quadrangular structure (derived from the parent cycloalkane) with all four S atoms exodentate (Fig. 3) [80, 85]. In principle, two conformations for 12S4 are consistent with the quadrangular structure of the parent cycloalkane. In the conformation actually adopted the S atoms assume corner positions (cf. 12O4); thus chelation to a metal ion necessitates a particularly radical – and therefore energetically costly – conformational rearrangement.

As in the 9S3 case, fusion of aromatic residues to each ethylene linkage grossly perturbs the ring conformation. In $(MeO)_8$-tetrabenzo-12S4 two phenylene rings lie in the S_4 plane; the others lie perpendicular to it on opposite sides, to yield a centrosymmetric structure [67]. Unlike unsubstituted 12S4, where all C–S bonds exhibit *gauche* placement, in the benzo$_4$-12S4 analogue half of the C–S bonds adopt anti placement.

2.1.3 15S5 ($n = 5$)

15S5 [80] adopts a structure derived from that of 12S4. The structure of 15S5 (Fig. 3; Table 1) results from the "prying open" of 12S4 at one corner and the insertion of another CH_2CH_2S unit. Examination of the resulting structure clearly reveals its relation to 12S4. In another perspective 15S5 results from fusion of two

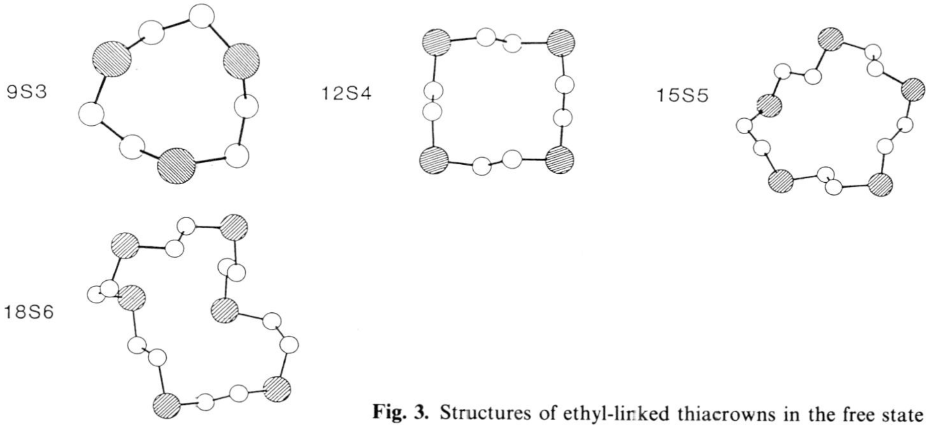

Fig. 3. Structures of ethyl-linked thiacrowns in the free state

bracket units at one end, and their linkage at the other end by a CH_2CH_2S sequence. This leads to an exodentate orientation for all five S atoms.

2.1.4 *18S6 (n = 6)*

Like 15S5, 18S6 [80, 86] can be derived from 12S4 (Fig. 3; Table 1). Union of two bracket units through CH_2CH_2S linkages results in the observed structure. Unlike 12S4 and 15S5, however, 18S6 does not have solely exodentate S atoms; two of the six adopt endodentate orientations. Nevertheless, all the C–S bonds are *gauche*, and four of the six S–C–C–S units are *anti*, in agreement with the conformational approach based on 1,4-interactions.

Inclusion of two fused benzo groups causes dibenzo-18S6 to differ conformationally from the unsubstituted parent compound [69]. All six donor atoms – indeed, the entire ligand except for the benzo groups – lies essentially in a plane (Table 1). The benzo groups lie centrosymmetrically disposed on opposite sides of the S_6 plane at an angle of 83° (cf. dibenzo-18O6, 124°). The two non-benzo S atoms adopt *anti* placements about their bonds to the adjacent carbon atoms; they also assume an exodentate orientation with respect to the crown cavity.

2.2 *Crown Thioethers Containing Propyl Linkages*

The conformational preferences discussed above express themselves more clearly in ethyl-linked crown thioethers than in analogous propyl-linked ones. In fact, propyl-linked crown thioethers have thus far resisted simple predictive generalization. In part this failure reflects the paucity of structural data from which a pattern might emerge. In any case the conformational preferences of propyl-linked ligands will be less clearcut than those of ethyl-linked ones. To two types of linkages (C–C–S–C and S–C–C–S) with cooperating conformational influences, propyl-linked crowns add a third potentially conflicting one (C–C–C–S). Any conflicts that arise may well preclude simple analysis or generalization. In particular, completely propyl-linked crown thioethers lack the impetus of strongly anti preferring S–C–C–S units in determination of their conformations. These cases are determined solely by weak *gauche* preferences, which may easily be overturned by, e.g., packing interactions.

2.2.1 *14S4*

Like 12S4, 14S4 adopts a quadrangular structure derived from the parent hydrocarbon, cyclotetradecane. The four S atoms occur at corners of the quadrangle (i.e., in exodentate orientation) (Fig. 4; Table 1) [79]. This structural feature doubtless contributes to the lack of a macrocyclic effect for 14S4 noted by Smith and Margerum [13]. The structure of 14S4 conceptually derives from fusion of two $SCH_2CH_2SCH_2CH_2CH_2S$ "homo-bracket" units. ("Homo-bracket" implies inclu-

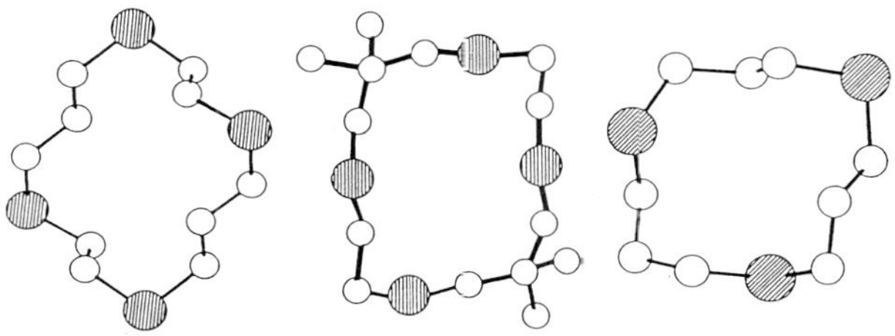

Fig. 4. Structures of free 14S4, 6,6,13,13-Me_4-14S4, and 12S3

Table 1. Crown Thioethers: Crystal Structure Determinations

Complex	Ideal. geom.[a]	M–S, Å	Remarks	Ref.
		9S3		
$[Mn(9S3)(CO)_3]^+$	O_h	2.338(5) 2.321(3) 2.327(4) 2.341(4) 2.314(4) 2.321(4)	Mn–C_{ave} 1.824 Å	[94]
$[Fe(9S3)_2]^{2+}$	O_h	2.251(1) 2.241(1) 2.259(1)	l.s.	[96]
$[Fe(9S3)(9S3(O))]^{2+}$	O_h	2.260(1) 2.250(1) cation 1 2.263(1) 2.258(1) cation 2	disord.; Fe–S=O est. 2.16 Å;	[74]
$[Co(9S3)_2]^{2+}$	O_h	2.356(6) 2.240(7) 2.367(5)	l.s.	[62. 102]
$[Ni(9S3)_2]^{2+}$	O_h	2.377(1) 2.380(1) 2.400(1)		[62]
$[Cu(9S3)_2]^{2+}$	O_h	2.419(3) 2.426(3) 2.459(3)		[62]
$[Zn(9S3)_2]^{2+}$	O_h	2.491(3) 2.497(3) 2.494(3)		[119]
$[Ru(9S3)_2]^{2+}$	O_h	2.344(1) 2.338(1) 2.336(1)		[120]
	O_h	2.344(1) 2.331(1) 2.339(1)		[120]
	O_h	2.327(1) 2.336(1) 2.333(1)		[122]
$[Pd(9S3)_2]^{2+}$	D_{4h}	2.333(2) 2.318(2)(eq) 2.957(2)(ax)	anhyd.; quasi-O_h	[106, 136]
	D_{4h}	2.309(1) 2.314(1)(eq) 3.005(1)(ax)	hyd.; quasi-O_h	[106]
cis-$[PdBr_2(9S3)]$	D_{4h}	2.275(2) 2.257(2)(eq) 3.125(1)(ax)		[106]
cis-$[PdCl_2(9S3)]$	D_{4h}	2.267(2) 2.2456(2)	Pd–Cl_{ave} 2.33	[137]
$[Pt(9S3)_2]^{2+}$	C_{4v}	2.25–2.30(eq) 2.88–2.93(ax)		[107]
$[PtMe_3(9S3)]^+$	O_h	2.411(3) 2.405(3) 2.405(3)	Pt–C_{ave} 2.08 Å	[139]
$[Hg(9S3)_2]^{2+}$	O_h	2.723(2) 2.713(2) 2.649(2)		[141]
$[Pb(9S3)_2]^{2+}$	D_{4d}	3.129(5) 3.084(4) 3.015(2)	Pb–$OClO_3$ 2.72(2)	[119]
$[Co(9S3)_2]^{3+}$	O_h	2.258(1) 2.253(1) 2.249(1)		[102]
$[Cu_2(9S3)_3]^{2+}$	T_d	2.231(1) 2.302(2) 2.325(2) 2.323(1) 2.244(1) 2.336(1) 2.302(1) 2.329(1)		[117]
$[Ag(9S3)_2]^+$	O_h	2.753(1) 2.727(2) 2.696(2)		[114–116]
$[Ag_3(9S3)_3]^{3+}$	T_d	2.724(2) 2.595(4) 2.613(4)	2.480(2)	[115]
$[Ag(9S3)Cl]$	T_d	2.618(1) 2.599(1) 2.598(1)		[116]
$[Cu(9S3)I]$	T_d	2.331(1) 2.343(1) 2.329(1)	Cu–I 2.490(1)	[115]
$[Rh(9S3)_2]^{3+}$	O_h	2.331(2) 2.345(3) 2.348(3)		[128–131]
$[Pd(9S3)_2]^{3+}$	O_h	2.545(2) 2.356(1) 2.369(2)		[132]

(Continued)

Table 1. (continued)

Complex	Ideal. geom.[a]	M–S, Å	Remarks	Ref.
$[Rh(9S3)(COD)]^+$	C_s	2.452(1) 2.448(1) 2.308(1) 2.108(2) 2.197(2) 2.217(2)	Rh–C 2.098(2)	[130]
$[Mo(9S3)(CO)_3]$	O_h	2.512(6) 2.504(6) 2.543(7)		[82]
$[AuCl(9S3)]_x$	D_{4h}	2.270(3)	Au–Cl 2.267(3) Au···Au 3.310	[141]
$[Ru(benzo\text{-}9S3)(CO)Br_2]$	O_h	2.300(4) 2.306(4) 2.426(4)	$Ru–Br_{ave}$ 2.54 Ru–C 2.023(20)	[126]
$[PtCl_2(benzo_3\text{-}9S3)]$	C_{4v}	2.246(5) 2.245(5) 2.859(5)	Pt–Cl 2.33	[140]
$[Rh(NO_3)_3(benzo_3\text{-}9S3)]$	O_h	2.2274(3)	Rh–O 2.064(9)	[134]
$[CuBr(benzo_3\text{-}9S3)]$	T_d	2.322(5) 2.332(5) 2.323(4)	Cu–Br 2.311(3)	[118]
		18S6		
$[Ni(18S6)]^{2+}$	O_h	2.397(1) 2.389(1) 2.377(1)		[23, 148]
$[Co(18S6)]^{2+}$	O_h	2.251(1) 2.479(1) 2.292(1)	l.s.; J–T dist.	[98, 99]
$[Cu(18S6)]^{2+}$	D_{4h}	2.323(1) 2.402(1) 2.635(1)	J–T dist.	[93]
$[Cu(18S6)]^+$	T_d	2.253(2) 2.358(2) 2.360(2)		[93]
$[Cu_2(MeCN)_2(18S6)]^{2+}$	T_d	2.32 2.34 233 Cu–N 1.94 Å		[150]
$[Pt(18S6)]^{2+}$	D_{4h}	2.298(2) 2.295(2)(eq) 3.380(3)(ax)	2 semi-coord S	[108]
$[Pd(18S6)]^{2+}$	D_{4h}	2.311(1) 2.307(2)(eq) 3.273(2)(ax)	2 semi-coord S	[108]
$[AgBr(18S6)]^+$	T_d	2.514(1) 2.571(1) 2.636(1)	Ag–Br 2.636(1)	[116]
$[CuCl(18S6)]^+$	T_d	2.296(2) 2.331(1) 2.376(2)	Cu–Cl 2.306(2)	[153]
$[(Cp^*RhCl)_2(18S6)]$	C_{3v}	2.377(1) 2.365(1)	Rh–Cl 2.387(1) $Rh–C_{ave}$ 2.173	[157]
		20S6		
$[(Rh(COD))_2(20S6)]^{2+}$	O_h	2.462(1) 2.482(1) 2.320(1)		[133]
		24S6		
$[Ni(24S6)]^{2+}$	O_h	2.413(1) 2.444(1) 2.437(1)		[159]
		12S3		
$[Ni(12S3)_2]^{2+}$	O_h	2.409(1) 2.420(2) 2.434(2)		[23]
$[Ru(12S3)_2]^{2+}$	O_h	2.3772(4) 2.3676(4) 2.3736(4)		[121]
$[Rh(12S3)_2]^{3+}$	O_h	2.363(4) 2.355(3) 2.360(3) 2.356(4) 2.350(4) 2.354(4)		[143]
$[Ag(12S3)(O_3SCF_3)]\}_x$	T_d	2.463(2) 2.477(2) 2.621(2)	Ag–O 2.482(6)	[116]
$\{[CuCl_2(12S3)]\}_x$	D_{4h}	2.447(1) 3.050(1)	exo; Cu–Cl 2.20 bridging Cu	[34]
$[Cu_2(12S3)_3]^{2+}$	T_d	2.275(6) 2.281(5) 2.281(5) 2.347(4) 2.246(5) 2.278(4) 2.258(5) 2.343(4)	12S3-bridged	[117]
		15S5		
$[Cu(15S5)]^{2+}$	C_{4v}	2.331(2) 2.315(2) 2.289(2) 2.338(2) 2.398(2)	Cu 0.41 Å above S_4 plane	[191]
$[Cu(15S5)]^+$	T_d	2.338(5) 2.243(5) 2.317(5)		[191]

Table 1. (continued)

Complex	Ideal. geom.[a]	M–S, Å	Remarks	Ref.
		14S4		
$[Ni(14S4)]^{2+}$	D_{4h}	2.177(1) 2.175(1)	*anti*	[166]
$[Cu(14S4)]^{2+}$	D_{4h}	2.308(1) 2.297(1)	*anti* Cu–$OClO_3$ 2.65	[89, 152, 171]
$\{[Cu(14S4)]^{+}\}_x$	T_d	2.260(4) 2.338(4) 2.327(4) 2.342(3)		[151, 152]
$[(NbCl_5)_2(14S4)]$	O_h	2.713(8)	Nb–Cl_{ave} 2.30	[31–33]
$[Ru(14S4)Cl_2]$	C_{2v}	2.262(1) 2.333(1)	folded; Ru–Cl 2.471(1)	[163]
$\{[Rh(14S4)]\}_2^{2+}$	D_{4h}	2.261(3) 2.264(6) 2.261(3) 2.264(6)	*syn* Rh···Rh 3.314	[182]
$[Hg(14S4)(OH_2)]^{2+}$	C_{4v}	2.58(4) 2.51(5) 2.60(5) 2.71(4)	*syn*; Hg–O 2.35 Hg 0.48 Å > S_4	[35]
$[(HgCl_2)_2(14S4)]^{2+}$	T_d	2.580(2) 2.699(2) 2.407(3) 2.419(3)	exo L; Hg–Cl 2.42, 2.41	[35]
$[HgI_2(14S4)]^{2+}$	T_d	2.75	exo; Hg–I 2.65 2.67	[156]
$[Pd(14S4)]^{2+}$	D_{4h}	2.23–2.33	*syn*	[129]
		Miscellaneous		
$[Cu(12S4)(OH_2)]^{2+}$	C_{4v}	2.34(1) 2.30(1) 2.37(1) 2.32(1)	Cu 0.58 Å above S_4 Cu–O 2.11(2)	[89]
$[AlMe_3(12S4)]$	D_{3h}	2.718(3) 3.052(3)	Al–C_{ave} 1.95	[85]
$[CuCl((MeO)_8$ benzo$_4$-12S4)]	C_{4v}	2.395(5) 2.511(5) 2.644(7) 2.722(7)	Cu–Cl 2.181(5)	[190]
$[Cu(13S4)(OH_2)]^{2+}$	C_{4v}	2.334(4) 2.333(1)	*syn*; Cu 0.38 Å above S_4	[89]
		2.310(5) 2.330(5)	Cu–O 2.14(1) (mol 2)	
		2.321(4) 2.333(4)	*syn*; Cu 0.37 Å above S_4	[89]
		2.213(5) 2.322(5)	Cu–O 2.16(2)	
$[Cu(15S4)(OClO_3)_2]^{2+}$	D_{4h}	2.323(3) 2.313(3)	Cu–O 2.53(1)	[89]
$[Ag(benzo\text{-}15S4)]_2^{2+}$	C_{4v}	—	dimeric; CN = 5	[206]
		16S4		
$[Cu(16S4)(OClO_3)_2]^{2+}$	D_{4h}	2.331(1) 2.387(2)	Cu–O 2.482(5)	[89]
sym-$[Mo_2(SH)_2$-$(16S4)_2]^{2+}$	O_h	2.320(1) 2.483(2) 2.537(2) 2.461(2)	Mo–SH 2.471(2) Mo–SR_2–Mo	[185]
[(OEt)Mo(16S4)–O–Mo(O)(16S4)]$^{3+}$	O_h	EtOMO–S_{ave} 2.48 O=Mo–S_{ave} 2.46	Mo–O_{br} 1.76 Mo–O_{br} 2.14 Mo=O 1.76	[185, 186]
$[MoO(SH)(16S4)]^{+}$	O_h	2.48 ave	Mo–SH 2.49 Mo=O 1.67	[185]
$[Hg(16S4)(OClO_3)$–$(O_2ClO_2)_2)]$	Cl[b]	2.563(5) 2.587(6) 2.605(6) 2.663(5) 2.76(2)	Hg–O(C_{3v}) 2.76 Hg–O(C_{2v}) 3.08, 3.26	[189]
$[Mo(CO)_2(Me_8$–16S4)]$^{+}$	O_h	2.434(2) 2.439(2) 2.432(2) 2.438(2)	*trans* isomer; Mo–C 1.99	[187]

(Continued)

Table 1. (continued)

Complex	Ideal. geom.[a]	M–S, Å	Remarks	Ref.
$[Mo(N)_2(Me_8\text{–}16S4)]^+$	O_h	2.424(2) 2.428(2) 2/419(2) 2.424(2)	*trans* isomer;	[188]
		Free Ligands		
9S3		3 endo S		[81]
12S4		4 exo S, at corners		[80, 85]
15S5		5 exo		[80]
18S6		2 endo, 4 exo S		[80, 86]
14S4		4 exo S, at corners		[79]
12S3		1 S at corner		[34]
$(MeO)_6benzo_3$-9S3		saddle form		[84]
$(MeO)_8benzo_4$-12S4		four exo		[67]

[a] Note: the idealized geometry refers to the symmetry of the coordination sphere, irrespective of the identity of the donor atoms comprising it.
[b] This complex approaches D_{4h} symmetry if the asymmetrically bound bidentate and monodentate ClO_4^- groups in the axial positions are ignored.

sion of an extra methylene group in one side; in this case the extra methylene group does not disturb the S–C–C–S–C–C–S bracket motifs.

2.2.2 *Me₄14S4*

As discussed above, the preferred conformations at C–S and C–O bonds differ in part because of their different 1,4-interactions, Replacement of one $-CH_2-$ with $-CMe_2-$ should generate 1,4-repulsions at *gauche* C–S bonds analogous to those experienced by *gauche* C–O bonds. Synthesis and X-ray diffraction studies of 6,6,13,13-tetramethyl-14S4 [87] bear out this expectation (Fig. 4; Table 1) [86]. Introduction of the methyl groups grossly changes the conformation from that of the parent ligand to one resembling that of the oxa analogue, 14O4. This conformation more closely resembles that required for ion binding, with the S atoms approaching endodentate orientation.

2.2.3 *12S3*

Structurally, 12S3 sets in opposition the conflicting priorities of a quadrangular structure on one hand and *gauche* placement at the C–S bonds on the other. A quadrangular structure (like that of 12S4 and the parent alkane, cyclododecane) would require two *anti* C–S bonds. On the other hand, *gauche* placement at all of the C–S bonds necessitates a non-quadrangular structure. Crystallography reveals that 12S3 adopts the former (Fig. 4; Table 1) [34], consistent with relaxation of the *gauche* preference of C–S bonds in propyl-linked thiacrowns in the absence of the cooperative anti preference of any S–C–C–S units. Examination of molecular models suggests that a structure with all C–S bonds in *gauche* conformations would suffer intolerable H · · · H contacts.

2.3 Conclusions

In crown thioethers the heteroatoms often adopt exodentate orientations, in contrast with the endodentate heteroatoms of analogous oxa- and aza-crowns. This reversal comes at least in part from the difference in C–E bond length (E = S vs. E = O, NH). Its impact on the stability of crown thioether complexes qualitatively determines which crown thioethers have extensive coordination chemistry. In addition, the reversal of conformational preferences between thia- and first-row (N, O) crowns vitiates simple reasoning from one to the other. For example, comparison of 14N4 with the seemingly analogous 14S4 exchanges a ligand aided by conformational effects for one hindered by them (quite apart from the electronic differences between the donor groups).

3 Coordination Chemistry [88]

A recurrent theme in the coordination chemistry of crown thioethers concerns the interplay between conformational preferences of the ligands and their coordinative behavior. In particular, the structures of complexes result from a compromise between the conformational preferences of the ligands and the electronic requirements of the metal ion. Crown thioethers such as 12S4 show a diminished propensity for chelation because of the exodentate orientation of the S atoms in the free ligand. Exodentate structures reflect the antipathy of most crown thioethers to chelation. As a consequence, complexes with incomplete chelation by the ligand form a substantial fraction of this review.

Indeed, exodentate rings typify free thiacrowns. This conformational feature considerably influences their coordination chemistry. Because of it many thiacrowns display low affinity for chelation of a single metal ion. Instead they bind in a mono- or bidentate fashion, which often leads to dimeric or oligomeric structures. Examples include $[(NbCl_5)_2(14S4)]$ [31–33], $[CuCl_2(12S3)]_x$ [34], $[Ag(O_3SCF_3)–(12S3)]_x$ [116], $[Cu(L)(OClO_3)_2]$ (L = 15S4 [89], 16S4 [89]), and $[(HgCl_2)_2(14S4)]$ [35], each of which is discussed later.

These compounds share two common characteristics. First, in conformation the bound ligand differs little from the free form; chelation through all of the S atoms, on the other hand, would require a gross change [90]. Second, each complex retains at least one anion in the coordination sphere. Owing to their weak σ-donor properties thioethers generally fail to displace strongly held counterions such as halides. However, even such poorly coordinating anions as triflate, perchlorate [36], and tetrafluoroborate [37] interact strongly with thioether complexes – sometimes in preference to additional thioether ligands. Association with such a variety of poorly coordinating anions (of varying ability to form bonds) suggests that electrostatic interaction dominates bond formation in this association.

This observation in turn highlights the importance of charge neutralization in thioether coordination chemistry. Thioethers fail to neutralize charge on metal ions very well. Poor charge neutralization contributes to the high redox potentials typically found for thioether complexes. Charge neutralization assumes even greater importance in solvents that solvate anions poorly (e.g. CH_3CN and CH_3NO_2 [91], both of which are commonly used in crown thioether chemistry), or low dielectric constant.

Even in water, however, charge separation is a problem. Rorabacher et al. have shown that Cu(II) thioether complexes associate significantly with ClO_4^- and other anions in aqueous solution [36, 37]. Furthermore, ions such as perchlorate interact more strongly with Cu(II) thioether complexes than they do with the aquo ion itself [36]. This buttresses the suggestion that Cu(II) coordinated to thioethers apparently has a greater positive charge than it does coordinated to water, which parallels the much higher redox potential.

Such coordination with "non-coordinating" anions appears not only in solution but also in the crystal structures of crown thioether complexes. Although perchlorate has long been considered the archetypal non-coordinating anion, ClO_4^- coordination occurs surprisingly often among thioether complexes. The Cu(II) complexes of 14S4, 15S4, and 16S4 [89] all show axially coordinate ClO_4^-, as does the bis(2,5-dithiahexane) complex of Co(II) [92]. This recurring observation doubtless reflects the failure of thioethers to neutralize positive charge on the metal ion either electrostatically through possessing a negative charge itself, or covalently through σ- or π-electron donation.

Charge neutralization is not the whole story, however. Several lines of evidence show that thioethers do exert considerable π-acidity. First, CO stretching frequencies in substituted metal carbonyls increase on introduction of thioethers. Second, complexes of thioethers show smaller g anisotropy than do those of other, harder ligands such as amines and water [93]. This diminution of Δg reflects greater quenching of orbital angular momentum by thioethers, presumably by greater acceptance of metal t_{2g} electron density. Third, despite their relatively weak ligand field strength, thioethers often form low-spin complexes. As in their influence on g anisotropy, they do this by accepting t_{2g} electron density into orbitals of π^*-symmetry (with respect to the metal center). By delocalizing the t_{2g} electron density thioethers reduce electron–electron interactions and thereby reduce the spin pairing energy.

Thus the electronic consequences of thioether coordination result from the interplay of both charge neutralization as well as π-acidity. The magnitude of these effects depends on the number of thioethers in the coordination sphere. This is particularly important for charge neutralization effects, which clearly will be most pronounced in the absence of anionic auxiliary ligands.

3.1 Tridentate Crown Thioethers

3.1.1 9S3 – First-row Metals

Among crown thioethers 9S3 is unique in retaining its conformation on binding to a trigonal face of a metal. In effect, the enthalpic price for arranging the donor atoms for coordination is paid in the synthesis of the ligand. This retention of conformation confers unique stability on chelating 9S3 complexes. Owing to its powerful chelating ability 9S3 provides a unique opportunity to study the effects of polythioether coordination on a wide variety of metal ions. Consequently, despite being until recently the most difficult crown thioether to synthesize, 9S3 is now the most extensively studied.

3.1.1.1 Chromium

Comparison of the optical spectrum of $[Cr(9S3)_2]^{3+}$ with those of other Cr(III) complexes shows that 9S3 generates a smaller ligand field splitting than comparable amine complexes (e.g., $[Cr(9N3)_2]^{3+}$) [94]. This compound represents one of the first homoleptic thioether complexes of Cr; it was prepared by heating $Cr(ClO_4)_3$ with 9S3 in the absence of a solvent. This extremely hazardous procedure cannot be recommended. The compound was characterized by elemental analysis and optical spectroscopy (Table 3).

Reaction of $CrCl_3 \cdot 6H_2O$ (in MeCN) or $[CrCl_3(THF)_3]$ (in dioxane) with 9S3 gives $[Cr(9S3)Cl_3]$ as sparingly soluble blue–violet microcrystals [94]. Stirring of this complex with triflic acid results in replacement of chloride with triflate. Infrared spectroscopy establishes coordination of the triflate residues in the blue–green $[Cr(9S3)(triflate)_3]$ complex, but no structural data are available.

3.1.1.2 Manganese

Reaction of $[Mn(CO)_5X]$ (X = Cl, Br, I) with 9S3 in DMF gives the facially-coordinated $[Mn(CO)_3(9S3)]X$ as air- and water-stable yellow crystals (Eq. 1a) [95]. The complex crystallizes with three crystallographically independent cations per unit cell; each cation has mirror symmetry. Carbon atoms in the 9S3 chain disorder to accommodate the mirror symmetry. Kinetic studies show that displacement of CO is zeroth-order in 9S3, consistent with a limiting dissociative mechanism in which loss of CO *cis* to X is the rate determining step. Upon treatment with 5 M HCl $[Mn(CO)_3(9S3)]^+$ expels one CO to give $[Mn(CO)_2(9S3)Cl]$ (Eq. 1b), while treatment with $NOBF_4$ oxidatively removes one CO to afford $[Mn(CO)_2(OH_2)(9S3)]$ (Eq. 1c).

$$[Mn(CO)_5X] + 9S3 \xrightarrow{DMF} [Mn(CO)_3(9S3)]^+ \tag{1a}$$

$$(X = Cl, Br, I)$$

$$[Mn(CO)_3(9S3)]^+ \xrightarrow{5\ M\ HCl} [Mn(CO)_2(9S3)Cl] \tag{1b}$$

$$[Mn(CO)_3(9S3)]^+ + NOBF_4 \longrightarrow [Mn(CO)_2(9S3)Cl] \tag{1c}$$

3.1.1.3 Iron

Iron(III) perchlorate with excess 9S3 (which serves as reductant as well as ligand; Eq. 2a) readily affords the pink cation $[Fe(9S3)_2]^{2+}$ (which can also be made from

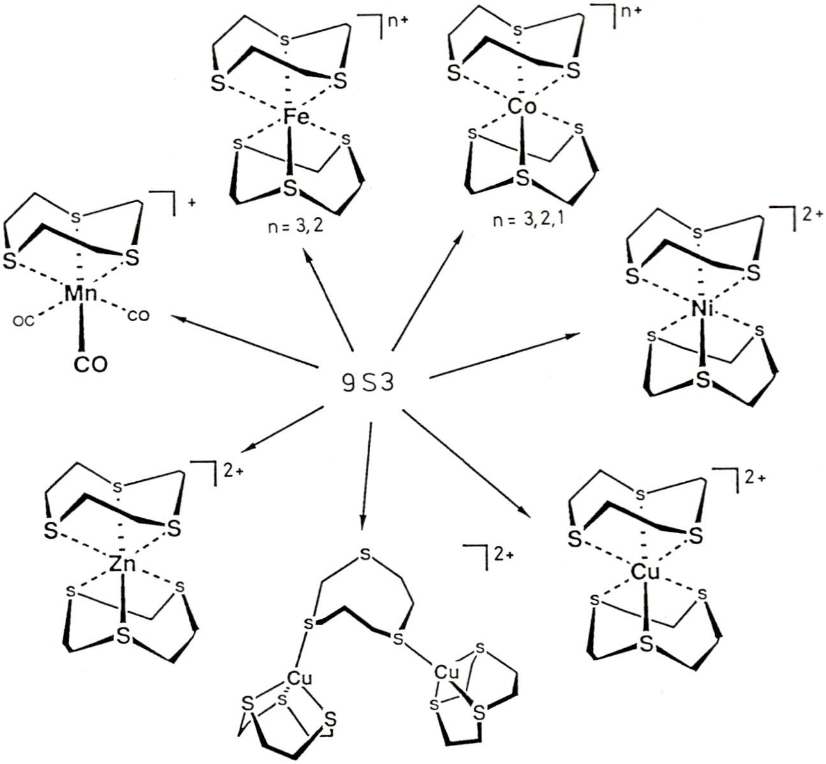

Fig. 5. First-row complexes of 9S3

$Fe(ClO_4)_2 \cdot 6H_2O$) (Fig. 5) [74, 96]. Surprisingly, in view of the relatively weak ligand field strength of thioethers [97], this octahedral complex contains a low-spin FeS_6 core (Table 5).

$$[Fe(OH_2)_6]^{3+} + \text{excess } 9S3 \xrightarrow{\text{MeOH}} [Fe(9S3)_2]^{2+} \quad (2a)$$

$$[Fe(9S3)_2]^{2+} + \text{oxidant} \xrightarrow{\text{MeOH}} [Fe(9S3)_2]^{3+} \quad (2b)$$

(oxidant = electrode, PbO_2)

$$[Fe(9S3)_2]^{2+} + S_2O_8^{2-} \longrightarrow [Fe(9S3)(9S3(O))]^{2+} \quad (2c)$$

$$[Fe(OH_2)_6]Cl_3 + 9S3 \xrightarrow{\text{MeOH}} [Fe(9S3)Cl_3] \quad (2d)$$

Electrochemically $[Fe(9S3)_2]^{2+}$ shows a reversible one-electron oxidation at a highly oxidizing potential (Table 2). Oxidation of $[Fe(9S3)_2]^{2+}$ either by PbO_2

Table 2. Crown Thioether Complexes: Redox Potentials

Couple	E_f/V	Remarks	Ref.
		9S3	
$[Fe(9S3)_2]^{3+/2+}$	+ 0.98	vs $Fc^{+/0}$; rev; MeCN	[96]
$[Co(9S3)_2]^{3+/2+}$	− 0.013	vs $Fc^{+/0}$; rev; MeCN	[96, 102]
	+ 0.680	vs SHE; rev; $MeCN_2$	[99]
	+ 0.573	vs SHE; rev; MeCN	[100]
$[Co(9S3)_2]^{2+/+}$	− 0.86	vs $Fc^{+/0}$; rev; MeCN	[96, 102]
	− 0.21	vs SHE; rev; $MeNO_2$	[99]
	− 0.292	vs SHE; rev; MeCN	[100]
$[Ni(9S3)_2]^{3+/2+}$	+ 0.97	vs $Fc^{+/0}$; quasi-rev; MeCN	[96]
$[Cu(9S3)_2]^{2+/+}$	+ 0.61	vs SCE; rev; CH_3NO_2	[93]
$[Ru(9S3)_2]^{3+/2+}$	+ 1.99	vs SHE; quasi-rev; MeCN	[120, 121]
	+ 1.41	vs $Fc^{+/0}$; quasi-rev; MeCN	[106]
$[Pd(9S3)_2]^{3+/2+}$	+ 0.605	vs $Fc^{+/0}$; quasi-rev; MeCN	[132, 136]
$[Pt(9S3)_2]^{3+/2+}$	+ 0.388	vs $Fc^{+/0}$; quasi-rev; MeCN	[107]
$[Ag(9S3)_2]^{2+/+}$	+ 1.10	vs SCE; quasi-rev; CH_3NO_2	[114, 116]
$[Rh(9S3)_2]^{3+/2+}$	− 0.309	vs SCE; rev; CH_3NO_2	[128]
	− 0.71	vs $Fc^{+/0}$; MeCN	[131]
$[Rh(9S3)_2]^{2+/+}$	− 0.721	vs SCE; quasi-rev; CH_3NO_2	[128, 131]
	− 1.08	vs $Fc^{+/0}$; MeCN	[131]
		18S6	
$[Co(18S6)]^{3+/2+}$	+ 0.844	vs SHE; rev; CH_3NO_2	[98, 99]
$[Cu(18S6)]^{2+/+}$	+ 0.72	vs SCE; rev; $MeNO_2$	[93]
		12S3	
$[Ru(12S3)_2]^{3+/2+}$	+ 1.81	vs SHE; quasi-rev; MeCN	[7]
$[Rh(12S3)_2]^{3+/2+}$	− 0.177	vs SCE; quasi-rev; CH_3NO_2	[130]
$[Rh(12S3)_2]^{2+/+}$	− 0.420	vs SCE; irrev; CH_3NO_2	[130]
$[Cu(12S3)_n]^{2+/+}$	+ 0.789	vs SHE; 80% aq. MeOH	[143]
		14S4	
$[Cu(14S4)]^{2+/+}$	+ 0.689	vs SHE; 80% MeOH/20% H_2O	[143]
	+ 0.60	H_2O	[170]
cis-$[Ru(14S4)Cl_2]^{+/0}$	+ 0.852	vs SHE; aq. HCl; rev.	[24]
cis-$[Ru(14S4)Cl_2]^{+/0}$	+ 0.230	vs $Ag^{+/0}$; MeCN; rev.	[24]
cis-$[Ru(14S4)(N_3)_2]^{-/0}$	+ 0.104	vs $Ag^{+/0}$; MeCN; rev.	[24]
cis-$[Ru(14S4)Br_2]^{+/0}$	+ 0.304	vs $Ag^{+/0}$; MeCN; rev.	[24]
cis-$[Ru(14S4)(NCS)_2]^{+/0}$	+ 0.522	vs $Ag^{+/0}$; MeCN; quasirev;	[24]
cis-$[Ru(14S4)(NO_2)_2]^{+/0}$	≈ + 0.868	vs $Ag^{+/0}$; MeCN; irrev.	[24]
cis-$[Rh(14S4)Cl_2]^{3+/+}$	− 0.72	vs Ag^+/AgCl; MeCN	[6]
		Miscellaneous	
$[Cu(12S4)]^{2+/+}$	+ 0.723	vs SHE; 80% aq. MeOH	[143]
	+ 0.64	vs SHE; H_2O	[170]
$[Cu(13S4)]^{2+/+}$	+ 0.674	vs SHE; 80% aq. MeOH	[143]
	+ 0.52	vs SHE; H_2O	[143]
$[Cu(15S4)]^{2+/+}$	+ 0.785	vs SHE; 80% aq. MeOH	[143]
	+ 0.64	vs SHE; H_2O	[143]
$[Cu(16S4)]^{2+/+}$	+ 0.798	vs SHE; 80% aq. MeOH	[143]
	+ 0.69	vs SHE; H_2O	[143]

(Continued)

Table 2. (continued)

Couple	E_f/V	Remarks	Ref.
$[Cu(15S5)]^{2+/+}$	+ 0.855	vs SHE; 80% aq. MeOH	[143]
	+ 0.70	vs SHE; H_2O	[170]
$[Cu(20S6)]^{2+/+}$	+ 0.805	vs SHE; 80% aq. MeOH	[143]
$[Cu(21S6)]^{2+/+}$	+ 0.852	vs SHE; 80% aq. MeOH	[143]
cis-$[Ru(13S4)Cl_2]^{+/0}$	+ 0.918	vs SHE; aq. HCl; rev.	[24]
cis-$[Ru(13S4)(N_3)_2]^{+/0}$	+ 0.130	vs $Ag^{+/0}$; MeCN; rev.	[24]
cis-$[Ru(13S4)Cl_2]^{+/0}$	+ 0.274	vs $Ag^{+/0}$; MeCN; rev.	[24]
cis-$[Ru(13S4)Br_2]^{+/0}$	+ 0.334	vs $Ag^{+/0}$; MeCN; rev.	[24]
cis-$[Ru(13S4)(NCS)_2]^{+/0}$	≈ + 0.334	vs $Ag^{+/0}$; MeCN; irrev.	[24]
$[Mo_2(SH)_2(16S4)_2]^{2+}$	− 0.95	vs SCE; MeCN; quasi-rev.	[185]
$[Mo_2O_2(16S4)_2(OEt)]^{3+}$	≈ − 0.7	vs SCE; MeCN; irrev.	[185]
$[MoO(SH)(16S4)]^{+}$	≈ − 0.7	vs SCE; MeCN; irrev.	[185]
trans-$[MoBr_2(Me_8\text{-}16S4)]^{3+/2+}$	+ 1.21	vs SCE; MeCN; quasi-rev.	[187]
trans-$[MoBr_2(Me_8\text{-}16S4)]^{2+/+}$	− 0.15	vs SCE; MeCN; quasi-rev.	[187]
trans-$[MoCl_2(Me_8\text{-}16S4)]^{3+/2+}$	+ 1.19	vs SCE; MeCN; quasi-rev.	[187]
trans-$[MoCl_2(Me_8\text{-}16S4)]^{2+/+}$	− 0.28	vs SCE; MeCN; quasi-rev.	[187]
$[Co(24S6)]^{3+/2+}$	+ 0.894	vs SHE; quasi-rev; $MeNO_2$	[c]

[c] Rawle, S. C.; Cooper, S. R.; unpublished work

or by electrochemical means affords the highly reactive, low-spin (Table 5) green Fe(III) complex (Eq. 2b). Mössbauer results [74] establish the metal (rather than the ligand) as the site of this oxidation. They further show that the low-spin t_{2g} [5] ion undergoes appreciable static Jahn-Teller distortion at room temperature.

In curious contrast, however, oxidation with $S_2O_8^{2-}$ instead of PbO_2 results in oxygen atom transfer to a thioether group rather than oxidation of the metal ion (Eq. 2c) (cf. the analogous reaction of $S_2O_8^{2-}$ with $[Co(9S3)_2]^{2+}$, which yields the Co(III) complex [74]). The product of this reaction, $[Fe(9S3)(9S3O)]^{2+}$, contains an S-bonded sulfoxide and five thioether groups coodinated to a low-spin Fe(II) [74]. Despite crystallographic disorder, structural methods clearly show that oxidation to the sulfoxide shortens the Fe–S distance (by 0.09 Å; Table 2) [74]. Replacement of 9S3 with its monosulfoxide increases E_f (Fe(III/II)) considerably.

3.1.1.4 Cobalt

Treatment of $[Co(OH_2)_6]$ $(BF_4)_2$ with 9S3 in absolute ethanol affords $[Co(9S3)_2]$ $(BF_4)_2$ as purple crystals (Eq. 3a). Structural work revealed short Co–S bonds [62] that were suggested [98], and subsequently shown [96, 99, 100], to arise from a low-spin configuration at the Co(II) ion. In agreement with this low-spin d^7 formulation, $[Co(9S3)_2]^{2+}$ (^2E_g in O_h) exhibits a substantial axial Jahn-Teller compression (0.12 Å). This compression contrasts with the more usual Jahn-Teller axial elongation found in $[Co(ttn)_2]^{2+}$ [99], and $[Co(18S6)]^{2+}$ [98, 99]. These complexes are among the few proven examples of low-spin octahedral

Co(II) complexes [101].

$$[Co(OH_2)_6]^{2+} + 2\,9S3 \xrightarrow{EtOH} [Co(9S3)_2]^{2+} \quad (3a)$$

$$[Co(9S3)_2]^{2+} + \text{oxidant} \xrightarrow{H_2O \text{ or } MeCN} [Co(9S3)_2]^{3+} \quad (3b)$$

(oxidant = electrode, $S_2O_8^{2-}$, $NOBF_4$)

$$[Co(9S3)_2]^{2+} + e^- \xrightarrow{MeCN} [Co(9S3)_2]^{+} \quad (3c)$$

In $[Co(9S3)_2]^{2+}$ 9S3 not only stabilizes the low-spin state, but also the low (2+) oxidation state (with respect to 3+). Electrochemistry of $[Co(9S3)_2]^{2+}$ reveals oxidation to the Co(III) form at an unusually high potential [99, 100]. Surprisingly, a second reversible wave corresponding to the Co(II/I) couple also occurs [99, 100, 102]. Comparison of the cobalt(II) complexes of 9S3 and 9N3 underscores the stabilization of the lower oxidation state by the crown thioether. Although these two ligands exert comparable ligand field strength, E_f for $[Co(L)_2]^{n+}$ (n = 3, 2; L = 9S3) exceeds that for the L = 9N3 [103] complex by over 800 mV. In practical terms, the air-sensitivity of the amine complex contrasts with the need for powerful oxidants (e.g. Br_2) to oxidize the thioether analogue.

EPR spectroscopy suggests that part of this stabilization arises from π-delocalization. g values in $[Co(9S3)_2]^{2+}$ (Table 5) deviate little from g_e, a reflection of appreciable delocalization of metal electron density onto the ligands [22, 99].

Oxidation of $[Co(9S3)_2]^{2+}$ by either electrochemical or chemical ($S_2O_8^{2-}$ or $NOBF_4$) means gives the orange Co(III) complex (Eq. 3b) [102]. This cation contains an octahedral CoS_6 core with Co–S_{ave} = 2.25 Å (Table 1), in agreement with distances found in the isoelectronic $[Fe(9S3)_2]^{2+}$ cation (Fe–S_{ave} = 2.25 Å). Thus neither the difference in charge nor the marked decrease in ionic radius (0.75 vs 0.68 Å for low-spin Fe(II) and Co(III), respectively) [104] significantly influence the metal–thioether bond lengths. Perhaps a greater covalent contribution for the Fe(II) complex swamps both charge and ionic radius considerations.

Both the high potential of the $[Co(9S3)_2]^{n+}$ (n = 3, 2) couple [96, 99, 102], and its electrochemical reversibility derive in large measure from the π-acidity of the hexakis-thioether donor set. Electron transfer entails no spin state change ($t_{2g}^6 \rightarrow t_{2g}^6 e_g^1$) between the two low-spin species. Two factors should promote rapid electron transfer. First, the extensive delocalization of metal electron density over the ligands (shown by the EPR data) increases accessibility to outer sphere electron transfer. Second, the minimal reorganization of the metal ion coordination spheres (inferred from the crystal structures) implies a small Franck-Condon barrier. Consistent with this discussion and the electrochemical results, stopped flow measurements show that $[Co(9S3)_2]^{n+}$ (n = 3,2) exhibits much more rapid electron self-exchange (10^5-fold) than corresponding aza compounds ($1.3 \times 10^4\ M^{-1}s^{-1}$ vs $0.19\ M^{-1}s^{-1}$). Subsequent remeasurement of rate by NMR methods yielded a value of $1.6 \times 10^5\ M^{-1}s^{-1}$ [105].

$[Co(9S3)_2]^{2+}$ also undergoes an intriguing reduction (electrochemically reversible) to a green Co(I) species (Eq. 3c; Table 2). So far this complex has eluded isolation and structural characterization. Possible structural models of this d^8 species include the isoelectronic cations $[Ni(9S3)_2]^{2+}$ (octahedral) [62], $[Pd(9S3)_2]^{2+}$ (quasi-octahedral) [106], $[Pt(9S3)_2]^{2+}$ (square pyramidal) [107] or $[M(18S6)]^{2+}$ (M = Pd, Pt) (square planar) [108].

3.1.1.5 Nickel

Interaction of $[Ni(OH_2)_6]^{2+}$ with 9S3 in ethanol affords the octahedral brick red $[Ni(9S3)_2]^{2+}$ cation (Eq. 4a) [62]. Nickel–sulfur distances (Table 1) [62] fall well short (0.06 Å) of the sum of ionic radii [2, 104]. Such short Ni–S distances (compared to Ni(II) complexes of acyclic thioethers) [109, 110] stem from compression induced by the macrocyclic nature of the ligand (the macrocyclic constriction effect) [16, 17, 18].

$$[Ni(OH_2)_6]^{2+} + 2\,9S3 \longrightarrow [Ni(9S3)_2]^{2+} \quad (4a)$$

$$[Ni(9S3)_2]^{2+} \longrightarrow [Ni(9S3)_2]^{3+} + e^- \quad (4b)$$

Nickel(II) provides an ideal case for evaluation of the ligand field properties of thioethers. In the high-spin [96] (Table 5) $[Ni(9S3)_2]^{2+}$ cation 9S3 exerts a ligand field strength comparable with that of 9N3 (Δ(9S3) 1265 cm^{-1} [96]; Δ(9N3) 1250 cm^{-1} [111]; Table 3). Consistent with the greater π-acidity of thioethers, the two ligands differ more widely in nephelauxetic ratio, which measures the "cloud expanding" capability of the ligand (β(9S3) 0.66; β(9N3) 0.90) [111]. In agreement with the structural results, the macrocyclic constriction effect [16, 17, 18] increments the effective ligand field by approximately 10% (compared with other hexakis(thioether) complexes). The cyclic triamine ligand 9N3 shows similar behavior [112].

Electrochemical investigation of $[Ni(9S3)_2]^{2+}$ shows a reversible one-electron oxidation process at a high redox potential, (Eq. 4b) [96]. The nature of this oxidized species – in particular the location of the electron hole – has not yet been elucidated.

3.1.1.6 Copper

Addition of $[Cu(OH_2)_6](BF_4)_2$ to a solution of 9S3 in ethanol gives the bis(9S3) complex of Cu(II) (Eq. 5a). Unlike the corresponding Co(II) complex, the brown $[Cu(9S3)_2]^{2+}$ (d^9; ^2E_g in O_h) cation shows minimal Jahn-Teller distortion (Cu–S range over 0.04 Å) [62]. Nevertheless, the sum of Cu–S bond lengths compares closely with that of $[Cu(18S6)]^{2+}$, a case that shows clear-cut Jahn-Teller elongation [93]. This lack of apparent distortion could arise from the operation of: 1) dynamic Jahn-Teller distortion, or 2) static Jahn-Teller distortion coupled with a positional disorder of the cations. On either a time- or space-average (respectively) no distortion would be apparent. (A third possibility, that 9S3 lacks sufficient

Table 3. Crown Thioether Complexes: Optical Data

Complex	Color[a]	$\lambda_{max}(\varepsilon)$	Comments	Ref.
		9S3		
		nm(M^{-1} cm^{-1})		
$[Cr(9S3)_2]^{3+}$	pink	509 685	diffuse reflect.	[94]
$[Cr(9S3)(triflate)_3]$	bl-gr	715(sh) 635 475 457(sh)	diffuse reflect.	[94]
$[Cr(9S3)Cl_3]$	bl-vi	778(sh) 734(sh) 690 516(sh) 493	diffuse reflect.	[94]
$[Mn(CO)_3(9S3)]^+$	yel	346(2400)	acetone	[94]
$[Fe(9S3)_2]^{2+}$	vi	523(53) 395(52)		[96]
$[Co(9S3)_2]^{2+}$	vi	730(11) 480(92) 338(6600) 264(6500)		[62, 102]
$[Ni(9S3)_2]^{2+}$	red	784(27) 527(26) 325(14000)		[62]
$[Ru(9S3)_2]^{2+}$	yel	338(172) 292(195)		[120, 122]
$[Pd(9S3)_2]^{2+}$	gr	615(55)		[106, 136]
$[Pd(9S3)_2]^{3+}$	or	475(3000)		[136]
$[Pt(9S3)_2]^{2+}$	yel	432(95)		[107]
$[Pt(9S3)_2]^{3+}$	or	401(3500)		[107]
$[Co(9S3)_2]^{3+}$	or	458(100) 333(89)		[102]
$[Fe(9S3)_2]^{3+}$	gr	634(700) 587(sh) 498(sh) 458(520) 340(7600)		[74]
$[Fe(9S3)(9S3(O))]^{2+}$	or-br	465(123) 374(96) 268(16000)		[74]
		18S6		
$[Ni(18S6)]^{2+}$	pink	815(23) 520(27)		[148]
		24S6		
$[Ni(24S6)]^{2+}$	bl-gr	905(100) 590(70)		[23]
		12S3		
$[Ni(12S3)_2]^{2+}$	bl	890(25) 570(34)		[23, 51]
$[Ru(12S3)_2]^{2+}$	yel	348(210) 312(290)		[121]
		15S5		
$[Cu(15S5)]^{2+}$	bl	414(6200) 565($\approx$ 2000)		[170]
		14S4		
$[Ni(14S4)]^{2+}$	red	495(263) 417(97.5)	CH_3NO_2	[49]
$[Ni(14S4)Cl_2]^{2+}$	red	1080(sh, 25) 940(48) 610(28)	*trans*?; CH_3NO_2	[49]
$[Ni(14S4)Br_2]^{2+}$	red	1110(sh, 16) 940(48) 610(53)	CH_3NO_2; *trans*?	[49]
$[Ni(14S4)I_2]^{2+}$	red	700(sh, 58) 540(315)	CH_3NO_2; in equil	[49]
$[Ni(14S4)(NCS)_2)]^{2+}$	red	1010(sh, 34) 915(54) 570(28)	CH_3NO_2; *trans*?	[49]
$[Cu(14S4)]^{2+}$	bl	390(8200) 570(1900)		[170]
cis-$[Co(14S4)Cl_2]^+$	red-br	533(654) 420 340	CH_3NO_2	[50]
cis-$[Co(14S4)Br_2]^+$	red-br	550(640) 310	CH_3NO_2	[50]
trans-$[Co(14S4)I_2]^+$	bl	640(820) 470(3420)	CH_3NO_2	[50]

(Continued)

Table 3. (continued)

Complex	Color[a]	$\lambda_{max}(\varepsilon)$	Comments	Ref.
cis-$[Co(14S4)(NCS)_2]^+$	red-br	540(840) 420	CH_3NO_2	[50]
cis-$[Co(14S4)(ox)_2]^+$	red	528(579)	CH_3NO_2	[50]
cis-$[Co(14S4)(NO_2)_2]^+$	or	470(814)	CH_3NO_2	[50]
cis-$[Rh(14S4)Cl_2]^+$	yel	350(2270) 320(sh) 252(26950)	H_2O	[50]
cis-$[Rh(14S4)Br_2]^+$	yel	370(2180) 245(24150)	H_2O	[50]
cis-$[Rh(14S4)I_2]^+$	yel	405(2460) 320(8500) 245(22550)	H_2O	[50]
cis-$[Ru(14S4)Cl_2]$	yel	430(sh. 125) 360(1010) 298(1090) 246(sh, 3550)	CH_3CN	[180]
cis-$[Ru(14S4)Cl_2]^+$	red	560(932) 485(sh, 740) 424(1350) 350(sh, 1620) 323(2050)	HCl	[180]
		16S4		
$[Cu(16S4)(OClO_3)_2]$	bl	440(6100) 603(800)		[170]
$[Mo(SH)_2(16S4)_2]^{2+}$	or	470(3700) 285(21600)	bridging 16S4	[185]
$[Mo_2O_2(16S4)_2(OEt)]^{3+}$	bl	610(6950) 240(16800)	μ-O and Mo=O	[185]
$[MoO(SH)(16S4)]^+$	red-br	625(200) 505(400)	Mo–SH, Mo=O	[185]
fac-$[MoCl_3(Me_8\text{-}16S4)]$	or	453(160) 290(2800) 750(25)	$CHCl_3$	[187]
$[MoBr_2(Me_8\text{-}16S4)]$	or	320(5900) 495(45)	$CHCl_3$	[187]
$[MoCl_2(Me_8\text{-}16S4)]$	yel	334(4200) 457(33)	$CHCl_3$	[187]
		Miscellaneous		
$[Cu(12S4)(OH_2)]^{2+}$	bl	387(6900) 675(2000)		[170]
$[Cu(13S4)(OH_2)]^{2+}$	bl	390(6100) 625(1800)		[170]
$[Cu(15S4)(OClO_3)_2]$	bl	414(8000) 565(1100)		[170]
cis-$[Rh(benzo\text{-}15S4)Cl_2]^+$	yel	350(1935) 255(20300)	benzo-15S4; H_2O	[50]
$[Ni(benzo\text{-}15S4)]^{2+}$	red	510(273) 450(sh, 142)	benzo-15S4; CH_3NO_2	[50]
cis-$[Ru(benzo\text{-}15S4)Cl_2]$	yel	440(sh, 72) 362(984) 284(sh, 1660)	CH_3CN	[180]
$[Ni_2(28S8)]^{4+}$	red	502(210) 440	low-spin; sq. pl.; $MeNO_2$	[192]
$[Ni_2(NCS)_4(28S8)]$	bl-gr	1010(44) 910(57) 590(47)	high-spin; $MeNO_2$	[192]

[a] bl = blue; vi = violet; yel = yellow; gr = green; or = orange; br = brown

flexibility to permit distortion, can be discounted in light of the 0.12 Å distortion found in $[Co(9S3)_2]^{2+}$ [62].) Low temperature EPR measurements [22, 113] reveal that the distortion can be frozen out, and thereby establish the presence of dynamic Jahn-Teller distortion at room temperature.

$$[Cu(OH_2)_6]^{2+} + 2\,9S3 \xrightarrow{\text{EtOH}} [Cu(9S3)_2]^{2+} \tag{5a}$$

$$[Cu(9S3)_2]^{2+} + e^- \xrightarrow{\text{MeCN}} [Cu(9S3)_2]^+ \tag{5b}$$

$$[Cu(MeCN)_4](BF_4) + 9S3 \xrightarrow{\text{MeOH}} [Cu_2(9S3)_3]^{2+} \tag{5c}$$

$$[Cu(MeCN)_4](ClO_4) + 9S3 \xrightarrow{\text{MeOH or MeCN}} [Cu_3(9S3)_3]^{2+} \tag{5d}$$

$$CuI + 9S3 \xrightarrow{\text{MeOH}} [CuI(9S3)] \tag{5e}$$

Electrochemical studies on $[Cu(9S3)_2]^{2+}$ show reversible reduction to the colorless Cu(I) complex (Table 2; Eq. 5b) [93, 96]. Three obvious structures can be envisaged for $[Cu(9S3)_2]^+$, which has so far eluded structural characterization. Like the Cu(II) complex, $[Cu(9S3)_2]^{2+}$ could be six-coordinate ([3 + 3] coordination). This unlikely (c.f. $[Cu(18S6)]^+$ below) possibility cannot be dismissed in light of six-coordination in the congeneric complex $[Ag(9S3)_2]^+$ [114, 115, 116]. Alternatively, tetrahedral coordination could arise from slippage of one 9S3 ring with respect to the other (tridentate) one to coordinate in a monodentate fashion ([3 + 1] coordination). Tetrahedral coordination could also be effected by slippage of the two rings such that each coordinates in bidentate fashion ([2 + 2] coordination).

A case can be made for each of the three possibilities, [3 + 3], [3 + 1] and [2 + 2] coordination. Investigation of other group Ib complexes of 9S3 has recently revealed examples of [3 + 3] coordination ($[Ag(9S3)_2]^-$ [114–116]) while the dimeric $[Cu_2(9S3)_3]^{2+}$ exhibits [3 + 1] coordination (Fig. 5; Table 1) [117]. The closest approach to [2 + 2] coordination occurs in $[Pd(9S3)_2]^{2+}$ [106], $[PdBr_2(9S3)]$ [106] and $[Pt(9S3)_2]^{2+}$ [107], each of which contains an effectively bidentate 9S3. On balance, [3 + 1] coordination appears most likely in view of $[Cu_2(9S3)_3]^{2+}$, in which two {Cu(9S3)} units, each containing a tridentate unit, coordinates to one S atom from a third 9S3 (Fig. 6; Table 1) [116]. This complex contains two Cu moieties each displaying [3 + 1] coordination; it results from reaction of $[Cu(MeCN)_4](BF_4)$ with two equivalents of 9S3 in MeOH (Eq. 5c); use of the hexafluorophosphate salt under identical conditions yields instead $[Cu(9S3)_2](PF_6)$ [116]. Structurally the coordination sphere around Cu consists of three S atoms from a chelating 9S3, and a fourth, considerably shorter bond to the capping S atom from the bridging 9S3 (Table 1). A similar reaction carried out instead with the perchlorate salt of $[Cu(MeCN)_4]^+$ yields $[Cu_n(9S3)_n]^{n+}$ [115]. While the extent of oligomerization is unknown, this compound probably has a structure like that of the congeneric $[Ag_3(9S3)_3]^{3+}$ cation [115].

Reaction of 9S3 with CuI yields [(9S3)CuI] (Eq. 5e), in which 9S3 chelates Cu(I) to yield a complex of tetrahedral microsymmetry (Fig. 5; Table 1) [115]. The limited bite of 9S3 enforces a trigonal distortion of the CuS_3I coordination sphere that decreases $\angle$S–Cu–S (all $\approx 90°$) and increases $\angle$S–Cu–I (all $\geqslant 118°$) from tetrahedral values. Similar behavior occurs in the analogous tribenzo 9S3 ligand $(MeO)_6$-tribenzo-9S3 (L). In [CuBr(L)] [118] the macrocycle adopts the "crown" conformation to give a molecule with a conical appearance. The presence of the three fused benzo rings causes some of the torsional angles at comparable single bonds to differ substantially from those in 9S3 itself.

3.1.1.7 Zinc

As a class a ion, Zn(II) normally prefers oxygen coordination over sulfur. In addition, Zn(II) often adopts tetrahedral rather than octahedral coordination. Nevertheless, in acetonitrile solution $Zn(ClO_4)_2$ reacts with 2 equivalents of 9S3 to give the colorless centrosymmetric $[Zn(9S3)_2]^{2+}$ cation as its perchlorate salt (Fig. 5; Table 1) [119]. This octahedral complex contains the first example of a structurally characterized ZnS_6 core; most Zn complexes of sulfur-based ligands assume tetrahedral geometry. In common with other first-row complexes of this ligand, the limited bite of 9S3 necessitates a slight trigonal elongation of the coordination sphere, which increases non-chelating S–Zn–S angles over chelating ones.

$$[Zn(OH_2)_6]^{2+} + 2\,9S3 \xrightarrow{MeCN} [Zn(9S3)_2]^{2+} \tag{6}$$

3.1.2 9S3 – Second-row Metals

3.1.2.1 Molybdenum

Reaction of $[Mo(CO)_3(MeCN)_3]$ with 9S3 gives $[Mo(CO)_3(9S3)]$ as pale yellow crystals (Eq. 7) [82]. Structurally the complex consists of the expected octahedral coordination sphere (twist angle 26.5(9)°) in which 9S3 coordinates facially (Fig. 6; Table 1). Molybdenum–carbon distances fall 0.1 Å short of those in $[Mo(CO)_6]$, which reflects the weaker π-acceptor capability of thioethers than CO. Similar indications come from ν_{CO}, which decreases by 200 cm^{-1} on going from the hexacarbonyl to $[Mo(CO)_3(9S3)]$. Conformationally coordination primarily changes torsional angles at the S–C–C–S units (from 58° in free 9S3 to 48° in the complex). This change in torsional angle directs the lone pairs on the sulfur atoms to a common locus on the idealized threefold axis of the ligand [82].

$$[Mo(CO)_3(MeCN)_3] + 9S3 \xrightarrow{MeCN} [Mo(CO)_3(9S3)] \tag{7}$$

3.1.2.2 Ruthenium

Two synthetic methods yield the oxidatively and hydrolytically robust cation $[Ru(9S3)_2]^{2+}$ [120, 121, 122]. Crystallography reveals a regular octahedral cation (as expected for low-spin d^6) (Fig. 6; Table 1) in which 9S3 adopts its usual conformation.

This complex not only resists oxidation, but also hydrolysis. Despite the relatively weak coordination generally offered by thioethers, $[Ru(9S3)_2]^{2+}$ can be recrystallized from boiling water. While the resistance to hydrolysis certainly derives from kinetic as well as possibly thermodynamic factors, the alacrity with which Ru(II) coordinates to a variety of thioethers suggests that particular affinity attends this interaction. Electrochemical studies support this generalization. The highly oxidizing potential of the quasi-reversible $[Ru(9S3)_2]^{3+/2+}$ couple (Table 2) reflects strong stabilization of Ru(II) with respect to Ru(III). As in the congeneric iron system, this oxidation could be either predominantly metal- or ligand-centered. Preparative electrochemical oxidation of $[Ru(9S3)_2]^{2+}$ yields a bright blue solution [123], but no compound has yet been isolated. These (III/II) potentials exceed those in comparable amine complexes (e.g. $[Ru(9N3)_2]^{n+}$ [124] by approximately 1.5 V. This comparison is particularly apt in light of the similarity in ligand field splitting between the two classes of compounds, and it illustrates the ability of thioethers to stabilize lower oxidation states.

$$[Ru(DMSO)_6]^{2+} + 2\,9S3 \xrightarrow{MeOH} [Ru(9S3)_2]^{2+} \tag{8a}$$

$$\text{“}RuCl_3 \cdot xH_2O\text{”} \xrightarrow[MeOH]{Ag+} \xrightarrow{\text{excess } 9S3} [Ru(9S3)_2]^{2+} \tag{8b}$$

$$[RuCl_2(\text{benzene})]_2 + \text{excess } 9S3 \xrightarrow{MeOH} [Ru(9S3)_2]^{2+} \tag{8c}$$

$$[Ru(9S3)_2]^{2+} \longrightarrow [Ru(9S3)_2]^{3+} + e^- \tag{8d}$$

$$\text{“}RuCl_3 \cdot xH_2O\text{”} + 9S3 \xrightarrow{Me_2SO} [Ru(9S3)(Me_2SO)_2Cl]^+ \tag{8e}$$

$$[Ru(Me_2SO)_4Cl_2] + 9S3 \xrightarrow{MeNO_2} [Ru(9S3)(Me_2SO)Cl_2] \tag{8f}$$

$$[Ru(L)(CO)] + S(CH_2CH_2Br)_2 \xrightarrow{DMF} [Ru(\text{benzo-}9S3)(CO)Br_2] \tag{8g}$$

$$[Ru(L)(CO)] + BrCH_2CH_2Br \xrightarrow{DMF} [Ru(\text{benzo-}9S3)(CO)Br_2] \tag{8h}$$

Treatment of $RuCl_3 \cdot 2H_2O$ with 9S3 in Me_2SO, followed by $NaPF_6$, affords $[Ru(9S3)(Me_2SO)_2Cl]\,(PF_6)$ (Eq. 8e) [122], while use of $[Ru(Me_2SO)_4Cl_2]$ and $MeNO_2$ gives instead $[Ru(9S3)(Me_2SO)Cl_2]$ (Eq. 8f) [125].

Another type of synthetic approach was taken by Sellmann and his coworkers [126]. Extending their work on template cyclizations, they prepared [Ru(benzo-9S3)(CO)Br_2] by reaction in hot DMF of [Ru(L)(CO)] (L = 2,3,11,12-dibenzo-1,4,7,10,13-pentathiatridecane dianion) either with $S(CH_2CH_2Br)_2$ (a mustard gas analogue) (Eq. 8g) or with ethylene bromide (Eq. 8h). During this reaction C–S bond fission occurs, presumably by alkylation of the thioether linkage followed by cleavage of the sulfonium salt (a mechanism first proposed by Bennett [127]). Coordination to the metal ion apparently favors formation of the nine-membered ring in preference to the 18-membered one. Use of the aprotic highly ionizing DMF, particularly as elevated temperatures, favors intermediate formation of sulfonium salts over simple alkylation.

Structurally [Ru(benzo-9S3)(CO)Br_2] consists of a facially coordinated benzo-9S3 with the benzo group approximately perpendicular to the S_3 plane (Table 1). In its conformation the 9S3 ring closely resembles analogous complexes of the unsubstituted ligand. Owing to the trans influence of CO the unique Ru–S_{trans} distance exceeds the other two by over 0.1 Å (Table 1).

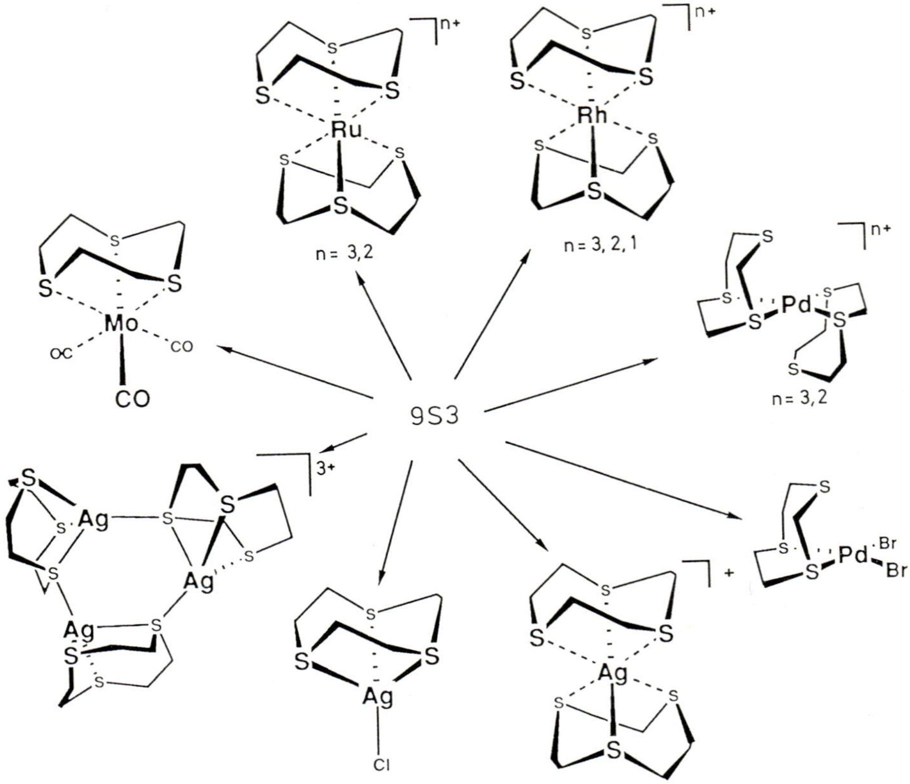

Fig. 6. Second row complexes of 9S3

3.1.2.3 Rhodium

Reaction of $RhCl_3 \cdot x\,H_2O$ with silver triflate in MeOH followed by 9S3 affords $[Rh(9S3)_2]^{3+}$ (Eq. 10a), in which a Rh(III) ion nestles between two tridentate 9S3 rings ([3 + 3] coordination) (Fig. 6; Table 1) [128, 129, 130, 131]. The RhS_6 coordination sphere closely approaches octahedral symmetry, as expected from the high ligand field stabilization energy associated with a low-spin d^6 ion. Metal–sulfur distances compare closely with those in the isoelectronic $[Ru(9S3)_2]^{2+}$ cation (Table 1). Thus, as in the congeneric Co(III) and Fe(II) complexes, the difference in charge exerts minimal effect on the coordination spheres.

$$RhCl_3 \cdot xH_2O \xrightarrow[\text{MeOH}]{Ag^+} \xrightarrow{9S3} [Rh(9S3)_2]^{3+} \tag{10a}$$

$$[Rh(9S3)_2]^{3+} + e^- \xrightarrow{MeNO_2} [Rh(9S3)_2]^{2+} \tag{10b}$$

$$[Rh(9S3)_2]^{2+} + e^- \xrightarrow{MeNO_2} [Rh(9S3)_2]^{+} \tag{10c}$$

$$[RhCl(COD)]_2 + 9S3 \xrightarrow{\text{toluene}} [Rh(9S3)(COD)]^{+} \tag{10d}$$

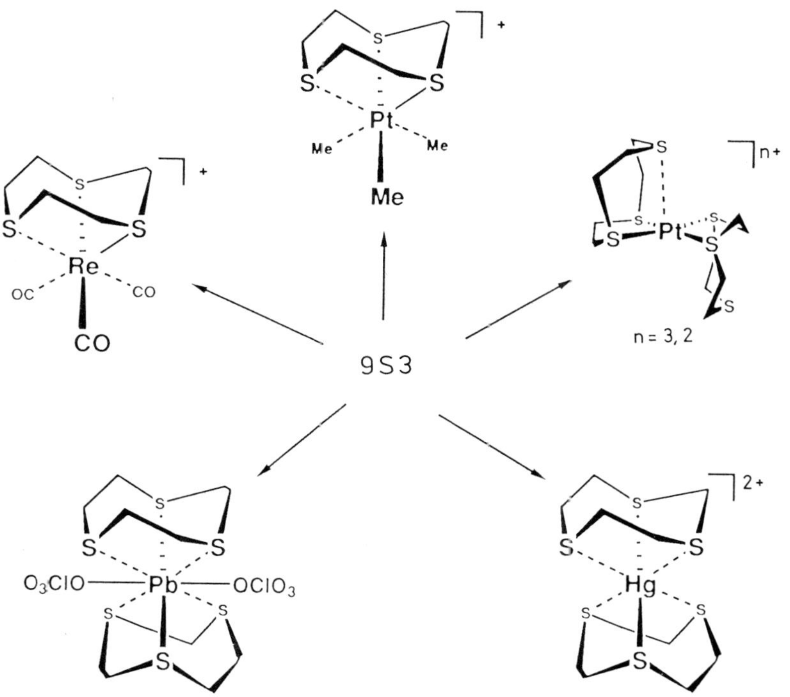

Fig. 7. Third row complexes of 9S3

Electrochemistry demonstrates the remarkable existence of a reasonably long-lived monomeric Rh(II) species. Cyclic voltammetry shows that $[Rh(9S3)_2]^{3+}$ undergoes reduction in two quasi-reversible one-electron processes rather than one two-electron process (Table 2) [128–130] (cf. the one-electron redox processes of the congeneric $[Co(9S3)_2]^{n+}$ (n = 3, 2, 1) couples) [128, 131]. Controlled potential coulometry confirms the one-electron nature of both redox processes, and thus buttresses the existence of Rh(II) as a discrete monomeric species of appreciable lifetime (hours).

While this intriguing Rh(II) complex has thus far eluded structural characterization, EPR studies suggest an axially elongated structure. At 77 K electrochemically generated solutions of the straw-colored $[Rh(9S3)_2]^{2+}$ cation yield rhombic spectra (Table 5) with small ^{103}Rh hyperfine splitting on g_1. As in the isoelectronic Co(II) complex, the g value pattern (g_1, $g_2 > 2$; $g_3 \approx 2$) is consistent with axial elongation of this Jahn-Teller active ion to give a d_{z^2} ground state. Consonant with this assignment, the crystal structure of the isoelectronic $[Pd(9S3)_2]^{3+}$ cation reveals an axially elongated MS_6 coordination sphere [132].

By similar reasoning, the structurally uncharacterized Rh(I) complex probably resembles one of the isoelectronic d^8 complexes $[M(9S3)_2]^{2+}$ (M = Pd [106], Pt [107]), which display quasi-octahedral and square pyramidal geometry, respectively.

Reaction of $[RhCl(COD)]_2$ with 9S3 gives $[Rh(9S3)(COD)]^+$ (Eq. 10d), in which a tridentate 9S3 caps a square pyramidal $\{Rh(COD)\}^+$ unit (Table 1) [130] that structurally resembles $[\{Rh(COD)\}_2(20S6)]^{2+}$ [133]. In $[Rh(benzo_3\text{-}9S3)(NO_3)_3]$ [134] Rh(III) coordinates to a tridentate macrocycle in the conical "crown" conformation, and to three monodentate NO_3^- groups to yield a complex with crystallographically-imposed threefold symmetry. The unique Rh–S distance falls well short (0.11 Å) of that found in $[Rh(9S3)_2]^{3+}$ [128, 129].

Use of Rh(II) carboxylates as starting materials upon reaction with 9S3 yields apparently polymeric adducts with 3:2 stoichiometry (i.e., $[\{Rh_2(O_2CR)_4\}_3(9S3)_2]_n$) in which each S atom of the macrocycle is thought to bind to a different Rh ion [135].

3.1.2.4 Palladium

While 9S3 promotes trigonal coordination, Pd(II) with few exceptions adopts square planar coordination. Thus the interaction of these two species sets their stereochemical preferences in opposition. Reaction of Pd(II) acetate (or $K_2[PdCl_4]$) with 9S3 in methanol gives $[Pd(9S3)_2]^{2+}$ (which as the PF_6^- salt crystallizes in blue hydrated and green anhydrous forms) [106, 136]. Note that this reaction constitutes a rare example of halide displacement by a crown thioether. In both forms 9S3 coordinates in a tridentate fashion to yield a quasi-octahedral coordination sphere. Bond lengths to the axial S atoms ($Pd\text{–}S_{axial} > 2.95$ Å) substantially exceed those in the equatorial plane (by 0.63 Å), and thereby establish that these donors only "semi-coordinate" (Fig. 6; Table 1). In addition, the apical S atoms lie appreciably ($\approx 6°$) off the normal to the PdS_{eq} plane. It is not clear why the two 9S3 molecules do not each coordinate in a bidentate fashion (as one 9S3 does in $[Pt(9S3)_2]^{2+}$ [107]) to yield simple square planar coordination. Oxidation of $[Pd(9S3)_2]^{2+}$ either chemically (with, e.g. 70% $HClO_4$) or electrochemically gives the red $[Pd(9S3)_2]^{3+}$ cation [136]. EPR spectroscopy indicates the metal as the site of oxidation, as demonstrated by both the anisotropy of the g values and the presence of ^{105}Pd hyperfine coupling (Table 5). The curious oxidation by perchloric acid underscores the potential hazard associated with this "inert" anion.

$$Pd(OAc)_2 + \text{excess } 9S3 \xrightarrow{\text{MeOH}} [Pd(9S3)_2]^{2+} \tag{11a}$$

$$[Pd(9S3)_2]^{2+} \longrightarrow [Pd(9S3)_2]^{3+} + e^- \tag{11b}$$

$$Pd(OAc)_2 \xrightarrow{9S3} \xrightarrow{Br^-} cis\text{-}[PdBr_2(9S3)] \tag{11c}$$

$$K_2[PdCl_4] + 9S3 \xrightarrow{\text{MeOH}} cis\text{-}[PdCl_2(9S3)] \tag{11d}$$

Structurally $[Pd(9S3)_2]^{3+}$ consists of a centrosymmetric tetragonally elongated six-coordinate complex, as expected for this Jahn-Teller active (low-spin d^7) ion (Fig. 6) [132]. Axial Pd–S distances exceed equatorial ones by 0.18 Å (Table 1). Oxidation of $[Pd(9S3)_2]^{2+}$ shortens the axial Pd–S distance by over 0.4 Å, but equatorial ones by only 0.04 Å. In its electrochemical behavior the

$[Pd(9S3)_2]^{3+/2+}$ couple (Table 2) [136] resembles the isoelectronic $[Rh(9S3)_2]^{2+/+}$ complexes. No evidence has yet been found for Pd(IV), which would complete the series analogous to the isoelectronic Rh(III), (II) and (I) complexes.

Addition of NaBr to a solution containing $Pd(OAc)_2$ and 9S3 in methanol gives the orange-brown $[PdBr_2(9S3)]$ complex (Eq. 11c), in which Pd(II) interacts with two halide ions and a tridentate 9S3 to yield square pyramidal coordination (Fig. 6; Table 1) [106]. As in the bis(9S3) complex, the coordination sphere contains a semi-coordinated S atom in the apical position [106]. The chloride analogue [137] has a similar structure.

3.1.2.5 Cadmium

Interaction of Cd(II) perchlorate with 9S3 in MeCN gives $[Cd(9S3)_2]^{2+}$ [142]. No structural data have yet been published, although it has been used as a diamagnetic host lattice for EPR studies on the analogous Cu(II) complex [113]. Use of CdX_2 salts (X = Cl, NO_3) in a similar procedure gives complexes of $[Cd(9S3)X_2]$ stoichiometry, which have been characterized by elemental analysis and vibrational spectroscopy [142].

3.1.2.6 Silver

The well-known affinity of Ag(I) for thioether coordination results in rich coordination chemistry for this ion with 9S3. Reaction of Ag(triflate) with two equivalents of 9S3 in methanol yields $[Ag(9S3)_2]^+$ (Eq. 12a). Despite the propensity of Ag(I) for linear and tetrahedral coordination, structural work [114–116] reveals a six-coordinate Ag(I) complex (Fig. 6; Table 1). The spherical symmetry of this d^{10} ion and the attendant malleability of its coordination sphere facilitates attainment of six-coordination. The limited bite of 9S3, however, necessitates substantial trigonal elongation of the octahedron (e.g. chelating S–Ag–S angles average 80°).

$$Ag^+ + 2\,9S3 \xrightarrow{\text{MeOH}} [Ag(9S3)_2]^+ \tag{12a}$$

$$[Ag(9S3)_2]^+ \xrightarrow{\text{MeOH}} [Ag(9S3)_2]^{2+} + e^- \tag{12b}$$

$$Ag^+ + 9S3 \xrightarrow{\text{MeOH}} [Ag_3(9S3)_3]^{3+} \tag{12c}$$

$$AgCl + 9S3 \xrightarrow{\text{MeOH}} [AgCl(9S3)] \tag{12d}$$

Metal–ligand bond lengths in $[Ag(9S3)_2]^+$ show the danger of reasoning too closely from ionic radii. Silver-sulfur distances (average 2.72 Å; Table 1) fall well short of those expected from the sum of ionic radii (2.99 Å) [104]. This difference reflects appreciable covalency in the Ag–S interaction as well as (to a lesser extent) conformational constraints imposed by 9S3. Clearly arguments from ionic radii must be viewed with caution, particularly for interactions between thioethers and transition metals in the second or third rows, or those in lower oxidation states.

Imposition of a hexakis(thioether) environment results in surprising redox behavior. Electrochemical studies reveal a quasi-reversible one-electron oxidation at + 1.30 V vs SHE [114–116]. Electrochemical or chemical (Ce(IV)) oxidation of $[Ag(9S3)_2]^+$ affords an unstable paramagnetic blue species of unknown structure. Besides increasing the electron density at the metal ion (because of the high coordination number), coordination to 9S3 may promote this oxidation by binding tightly. Oxidation may decrement the stability of the complex, but not so much that the oxidized form cannot exist in solution.

Reaction of Ag(I) perchlorate with less than two equivalents of 9S3 affords a complex that crystallizes as the cyclic trimer $[Ag_3(9S3)_3]^{3+}$ (Eq. 12c) (Fig. 6; Table 1). This compound features crystallographically-imposed threefold symmetry between the three {Ag(9S3)} units. In each unit 9S3 acts as a tridentate ligand to one Ag^+ ion; with one S atom also bridging to another, to complete idealized tetrahedral [3 + 1] coordination about each Ag(I) ion [115].

Despite its affinity for silver(I), 9S3 fails to displace halide ions from the metal. Reaction of AgCl with 9S3 in MeOH affords the colorless, sparingly soluble complex [AgCl(9S3)]; conductivity studies indicate that the complex does not dissociate in CH_3NO_2 [116]. Here 9S3 caps an AgCl unit to yield a four-coordinate complex that strongly resembles the closely related complex [CuI(9S3)] [115]. The resulting AgS_3Cl coordination sphere deviates from idealized tetrahedral geometry (c.f. Cu(9S3)I) [115] largely through a substantial trigonal stretching ($\angle$S–Ag–Cl 129°). The reduction in coordination number from $[Ag(9S3)_2]^+$ to [Ag(9S3)Cl] decreases Ag–S distances (by 0.12 Å) and chelating S–Ag–S angles (by 4.5°). In contrast to this discrete complex, reaction of AgBr with 18S6 affords an oligomer with a network structure (vide infra).

3.1.3 9S3 – Third-row Metals

3.1.3.1 Rhenium

In $[Re(9S3)(CO)_3]^+$, prepared by reaction of $[Re(CO)_5Br]$ with 9S3, three facially bound CO groups and a tridentate 9S3 offer octahedral coordination to the metal ion (Fig. 7; Table 1) [138]. The kinetic inertness of this complex has frustrated attempts to replace the remaining carbonyl ligands with 9S3.

3.1.3.2 Platinum

Reaction of $PtCl_2$ or K_2PtCl_4 with 9S3 affords the green $[Pt(9S3)_2]^{2+}$ cation (Eq. 13a) [107]. Two crystallographically distinct cations occupy the asymmetric unit. Both contain a square pyramidal coordination sphere that comprises one tridentate and one bidentate 9S3; the two cations differ primarily in their Pt–S_{apical} distances. Distances to the semi-coordinated S atom from the tridentate ligand differ appreciably: 2.88 and 2.92 Å (Fig. 7; Table 1). In both cases, also, the sixth

S atom (from the bidentate 9S3) fails to bind (Pt · · · S 4.04 Å) because this macrocycle adopts an unusual conformation in which the remaining donor points *away* from the metal ion. As in the Pd(II) analogue, the Pt–S_{ax} vector deviates from normality with the PtS_{eq} plane by approximately 5°. This presumably reflects the limited bite of 9S3. Structurally both $[M(9S3)_2]^{2+}$ (M = Pd, Pt) cations reflect a compromise: axial coordination somewhat stabilizes the complex electronically, but it necessitates a small conformational change in the ligand. Oxidation of $[Pt(9S3)_2]^{2+}$ yields the red mononuclear species $[Pt(9S3)_2]^{3+}$, where the modest reversibility of this process presumably reflects substantial structural rearrangement. The syntheses of *cis*-$[PtCl_2(9S3)]$ and *cis*-$[Pt(PPh_3)_2(9S3)]^{2+}$ have also been described [136]. These compounds presumably have structures analogous to that of $[PdBr_2(9S3)]$ [106].

$$K_2PtCl_4 + 9S3 \longrightarrow [Pt(9S3)_2]^{2+} \tag{13a}$$

$$[Pt(9S3)_2]^{2+} \longrightarrow [Pt(9S3)_2]^{3+} + e^- \tag{13b}$$

$$[PtClMe_3]_4 + 9S3 \xrightarrow{CHCl_3} [PtMe_3(9S3)]^+ \tag{13c}$$

The utility of 9S3 as a capping ligand in organometallic chemistry has recently been exploited in the synthesis of $[Pt(9S3)Me_3]^+$. This colorless diamagnetic complex was prepared by reaction of $[PtClMe_3]_4$ with 9S3 (Eq. 13c), and it contains an idealized octahedral coordination sphere (Fig. 7; Table 1) [139]. This synthesis represents a rare example of halide displacement by a thioether ligand. In this case such displacement probably occurs because the three strongly σ-donating methyl groups mitigate the poor charge neutralization offered by thioethers.

In $[PtCl_2((MeO)6\text{-tribenzo-}9S3)]$ a macrocycle in the crown conformation envelops a square pyramidal PtS_3Cl_2 core (Table 1) [140]. The complex resembles a square planar $PtCl_2S_2$ coordination sphere with addition of the third S atom in an apical position. The Pt–S distance to this apical atom (2.86 Å) greatly exceeds that of the two equatorial S atoms (2.24 Å). This 0.62 Å elongation reflects the tenuous nature of the Pt–S_{apical} interaction. Because of steric constraints imposed by the ligand the apical S atom lies appreciably (≈ 10°) off the pseudo-fourfold axis of the molecule. Here the rigidity introduced by fusion of three phenylene groups further exacerbates the difficulty of coordination of the third S atom.

3.1.3.3 Gold

In aqueous acetone 9S3 reacts with [AuCl(thiodiglycol)] to give $\{[AuCl(9S3)]\}_x$ [141]. This oligomeric compound features square planar coordination at the Au atoms. Each Au binds to one S atom from a 9S3 molecule, which must twist into a strained conformation ([225] in the nomenclature of Dale) [76]. *Trans* to this S atom lies the Cl^-. Bonds to two *trans* Au atoms completes the $AuSClAu_2$ coordination sphere about any given Au ion, and results in a nearly linear infinite polymer. NMR studies in solution indicate interaction of the Au ion with three S atoms that are equivalent on the NMR time scale (at least down to 240 K).

3.1.3.4 Mercury

9S3 avidly binds Hg(II) to yield $[Hg(9S3)_2]^{2+}$ (Eq. 14), which, like the isoelectronic Ag(I) analogue, deviates from octahedral geometry principally through trigonal elongation: chelating S–Hg–S angles average 82°, while non-chelating ones average 98° (Fig. 7; Table 1) [142]. This complex particularly highlights the danger of uncritical application of ionic radii: Hg–S_{ave} (2.70 Å) here falls substantially below the sum of ionic radii (2.86 Å) [104].

$$Hg(CF_3SO_3)_2 + 2\,9S3 \xrightarrow{MeOH} [Hg(9S3)_2]^{2+} \tag{14}$$

Table 4. Synthetic Studies of Crown Thioethers

Ligand[a]	Route	Conditions	Yield	Year	Ref.
9S3[b]	$BrCH_2CH_2Br + Na_2S$	EtOH	—	1886	[40]
9S3[c]	$BrCH_2CH_2Br + KSH$	EtOH	—	1920	[41]
9S3	$Na_2(SCH_2CH_2)_2S + ClCH_2CH_2Cl$	EtOH; < 5°	0.04%	1977	[11]
9S3	$(NBzMe_3)_2(SCH_2CH_2)_2S + Cl(CH_2)_2Cl$	MeOH; 65°	4.4	1983	[62]
9S3	$[Mo(CO)_3(L)] + BrCH_2CH_2Br$ ($L = S(CH_2CH_2S)_2$	MeCN; 25°	60	1984	[63, 64]
9S3	$Cs_2S(CH_2CH_2S)_2 + ClCH_2CH_2Cl$	DMF; 100°	50	1987	[59]
benzo-9S3	$[Ru(CO)(L)] + S(CH_2CH_2Br)_2$	DMF; 140°	45[d]	1988	[126]
$(benzo)_3$-9S3	1,2-Br_2-4,5-$(MeO)_2C_6H_3$ + 1,2-$(SH)_2$-4,5-$(MeO)_2C_6H_3$	Cu_2O; DMA 165°	70	1979	[66]
12S3[e]	$Br(CH_2)_3Br + HSCH_2CH_2SH$	EtOH	—	1886	[40]
12S3	$Na_2(S(CH_2)_3)_2S + Br(CH_2)_3Br$	EtOH;,78°	3	1970	[51]
12S4	$Na_2(S(CH_2)_2S(CH_2)_2 + Br(CH_2)_2Br$	EtOH; 78°	4	1970	[51]
12S4	$S(CH_2CH_2Cl)_2 + Na_2(SCH_2CH_2)_2S)$	n-BuOH; < 25°	6.3	1974	[55]
12S4	$Cs_2(S(CH_2)_2S(CH_2))_2 + BrCH_2CHBr$	DMF; 50°	88	1981	[57]
$(benzo)_4$-12S4	$[2,4,5\text{-}Br(MeO)_2C_6H_2]_2S$ $[2,4,5\text{-}HS(MeO)_2C_6H_2]S$	Cu_2O; DMA 165°	20	1985	[67]
13S4	$Na_2(S(CH_2)_2S(CH_2)_2 + Br(CH_2)_2Br$	EtOH; 78°	16	1970	[51]
13S4	$Cs_2(S(CH_2)_2S(CH_2)_2 + Br(CH_2)_3Br$	DMF; 50°	72	1981	[57]
14S4	$Na_2(SCH_2CH_2S) + Br(CH_2)_3Br$	EtOH; 25°	1	1933	[46]
14S4	$Na_2(S(CH_2)_2S(CH_2)_2 + Br(CH_2)_3Br$	EtOH; 78°	7.5	1969	[49]
14S4	$Na_2(S(CH_2)_2S(CH_2)_2 + Br(CH_2)_3Br$	EtOH; 78° high dil.	55	1974	[50]
14S4	$Na_2(SCH_2CH_2CH_2S) + CH_2(CH_2S(CH_2)_2Cl)_2$	n-BuOH; 25°	22.1	1974	[55]
14S4	$Cs_2(S(CH_2)_2S(CH_2))_2CH_2 + Br(CH_2)_3Br$	DMF; 50°	76	1981	[57]
benzo-14S4	$Cs_2(S(CH_2)_2S(CH_2))_2$ + 1,2-$(BrCH_2)_2C_6H_4$	DMF; 50°	69	1981	[57]
benzo-15S4	$Na_2(S(CH_2)_2S(CH_2))_2CH_2$ + 1,2-$(BrCH_2)_2C_6H_4$	EtOH; 78°	38	1969	[49]
benzo-15S4	$Cs_2(S(CH_2)_2S(CH_2))_2CH_2$ + 1,2-$(BrCH_2)_2C_6H_4$	DMF; 50°	85	1981	[57]
15S5	$Na_2(SCH_2CH_2)_2S + (CH_2S(CH_2)_2Cl)_2$	n-BuOH; 60°	11	1974	[55]

Table 4. (continued)

Ligand[a]	Route	Conditions	Yield	Year	Ref.
16S4	$Br(CH_2)_3Br + Na_2(S(CH_2)_3S)$	EtOH; 25°	1.0	1934	[47]
16S4	$Na_2(S(CH_2)_3S) + Br(CH_2)_3Br$	1:1 EtOH/ *n*-BuOH; 15°	6.2	1974	[55]
18S4	$Br(CH_2)_4Br + Na_2(S(CH_2)_3S)$	EtOH; 25°	1.8	1934	[47]
18S6	$BrCH_2CH_2Br + Na_2(SCH_2CH_2S)$	EtOH; 25°	1.7	1934	[47]
18S6	$BrCH_2CH_2Br + Na_2(SCH_2CH_2)_2S$	EtOH; 70°	31	1969	[52, 53]
18S6	$S(CH_2CH_2Cl)_2 + Na_2(SCH_2CH_2)_2S$	*n*-BuOH; < 25°	32.8	1974	[55]
18S6	$Na_2(SCH_2CH_2)_2S + BrCH_2CH_2Br$	*n*-BuOH; < 25°	8.1	1974	[55]
18S6	$Na_2(SCH_2CH_2S) + S(CH_2CH_2Cl)_2$	*n*-BuOH; < 25°	21.7	1974	[55]
(benzo)$_2$-18S6	$[Fe(L)_2(CO)_2] + S(CH_2CH_2Br)_2$ ($L = 1,2\text{-}(SH)_2C_6H_4$)	THF; 80°	68[f]	1986	[68, 69]
20S4	$Na_2(S(CH_2)_4S) + Br(CH_2)_4Br$	1:1 EtOH/ *n*-BuOH; 15°	3.9	1974	[55]
21S6	$Na_2(S(CH_2)_2SCH_2)_2CH_2 + ((CH_2)S(CH_2)_3Cl)_2$	*n*-BuOH; 25°	9.7	1974	[55]
22S4	$Br(CH_2)_6Br + Na_2(S(CH_2)_3S)$	EtOH; 25°	1.1	1934	[47]
24S4	$Na_2(S(CH_2)_5S) + Br(CH_2)_5Br$	1:1 EtOH/ *n*-BuOH; 15°	5.3	1974	[55]
24S6	$Na_2(S(CH_2)_3S) + Br(CH_2)_3Br$	1:1 EtOH/ *n*-BuOH; 15°	7.3	1974	[55]
28S4	$Na_2(S(CH_2)_6S) + Br(CH_2)_6Br$	1:1 EtOH/ *n*-BuOH; 15°	3.9	1974	[55]
36S6	$Na_2(S(CH_2)_5S) + Br(CH_2)_5Br$	1:1 EtOH/ *n*-BuOH; 15°	6.8	1974	[55]
42S6	$Na_2(S(CH_2)_6S) + Br(CH_2)_6Br$	1:1 EtOH/ *n*-BuOH; 15°	3.2	1974	[55]

[a] For structures of ligands see Figure 1; Systematic names: benzo$_4$-12S4 (2,3,7,8,12,13,17,18-octamethoxy-10,15,20-tetrathiatetrabenzo[a, d, g, j]cyclododecine); benzo$_2$-18S6 (2,3,11,12-dibenzo-1,4,7,10,13,16-hexathiacyclo-octadeca-2,11-diene)
[b] Later shown to be incorrect. See references 44 and 45
[c] Later shown to be incorrect. See references 44, 45, 47 and 11
[d] As metal complex
[e] Not verified
[f] After two alkylations of complex with $S(CH_2CH_2Br)_2$.

3.1.3.5 Lead

Unlike its mercury(II) analogue, the Pb(II) complex of 9S3 does not assume octahedral coordination. Instead, reaction of $Pb(ClO_4)_2$ with 2 equivalents of ligand in CH_3CN affords $[Pb(9S3)_2(OClO_3)_2]$ (Eq. 15) [119]. X-ray diffraction reveals a distorted square antiprismatic geometry, with coordination of six S atoms and two O atoms from two tridentate ligands and two monodentate perchlorate groups (Table 1). In the eight-coordinate PbS_6O_2 coordination sphere the particularly long Pb–S distances (Table 1) reflect the weakness of the Pb $\cdots$ S interaction.

$$Pb(ClO_4)_2 + 2\,9S3 \xrightarrow{MeCN} [Pb(9S3)_2(OClO_3)_2] \qquad (15)$$

3.1.4 12S3 – First-Row Metals

3.1.4.1 Nickel

In 1970 Rosen and Busch [51] reported the synthesis of $[Ni(12S3)_2]^{2+}$ from the reaction of $Ni(BF_4)_2 \cdot 6H_2O$ with 12S3 in $MeNO_2/Ac_2O$) (Eq. 16). In the blue high-spin $[Ni(12S3)_2]^{2+}$ cation (Table 3; Table 5) [49, 51] Ni(II) ion coordinates to six thioether groups in an essentially regular octahedral fashion (Fig. 8) [23]. As

Table 5. Crown Thioether Complexes: Magnetic Data

Complex	μ^1	g(A)	Comments	Ref.
		9S3		
$[Cr(9S3)Cl_3]$	4.13		293 K	[94]
$[Cr(9S3)(triflate)_3]$	3.95		293 K	[94]
$[Fe(9S3)_2]^{3+}$	2.46		2.09 μ_B 110 K	[74]
$[Co(9S3)_2]^{2+}$	1.82	$g_\parallel$ 2.136 g 2.07 $A_\parallel$ 0.698 $A^\perp$ 0.241	 $\times 10^{-4} cm^{-1}$	[99–100, 102]
$[Ni(9S3)_2]^{2+}$	3.05			[23, 62, 96]
$[Pd(9S3)_2]^{3+}$	—	$g_\parallel$ 2.008 $g_\perp$ 2.04 $A_\parallel$ 0.5 G, $A_\perp$ 20 G		[62, 102]
$[Pt(9S3)_2]^{3+}$			$g_\parallel$ 1.98 $g_\perp$ 2.044	[107]
		18S6		
$[Ni(18S6)]^{2+}$	2.77		298 K	[23]
$[Co(18S6)]^{2+}$	1.81	g_x 2.044 g_y 2.121 g_z 2.135	298 K	[98, 99]
$[MoCl_3(18S6)]$	3.44			[155]
$[(MoCl_3)_2(18S6)(THF)_3]$	3.46			[155]
$[(MoOCl_3)_2(18S6)]$	1.70	1.954	300 K; g_{iso}	[154]
$[(MoCl_4)_2(18S6)]$	2.31		300 K	[154]
		12S3		
$[Ni(12S3)_2]^{2+}$	3.19			[23]
		14S4		
$[Ni(14S4)Cl_2]^{2+}$	3.04			[50]
$[Ni(14S4)Br_2]^{2+}$	3.18			[50]
$[Ni(14S4)I_2]^{2+}$	3.10			[50]
$[Ni(14S4)(NCS)_2]^{2+}$	3.11			[50]
$[(MoOCl_3)_2(14S4)(THF)_2]$	1.66			[154]
$[MoOCl_3(14S4)]$; 15	1.65	1.954	g_{iso}	[154]
$[(MoCl_4)_2(14S4)]$	2.32		154	
$[MoCl_4(14S4)]$	2.21		154	
		Miscellaneous		
$[Ni_2(NCS)_4(28S8)]$	3.08			[192]
		16S4		
$[MoBr_2(Me_8\text{-}16S4)]$	3.1		$CHCl_3$	[187]
$[MoCl_2(Me_8\text{-}16S4)]$	3.1		$CHCl_3$	[187]
fac-$[MoCl_3(Me_8\text{-}16S4)]$	3.9		$CHCl_3$	[187]

[1] At room temperature unless otherwise noted

a consequence of the larger ring size Ni–S_{ave} increases by 0.05 A from that in the corresponding 9S3 complex (Table 1).

$$Ni(HOAc)_6 + 12S3 \longrightarrow [Ni(12S3)_2]^{2+} \qquad (16)$$

Conformational effects manifest themselves in the solution chemistry of this complex. Substantial rearrangement – with its attendant cost in enthalpy – must occur for 12S3 to function as a tridentate ligand. Hence conformational factors clearly disfavor coordination. Contact with even traces of water immediately hydrolyzes the complex (to yield the hexaaquo ion and free ligand) [23]. By contrast the 9S3 analogue – where ligand conformational preferences favor coordination – withstands recrystallization from boiling water [23].

Optical studies of Ni(II) complexes $[Ni(L)_2]^{2+}$ (L = 9S3, 12S3; Table 3) show that the increase in average Ni–S bond length decreases the ligand field splitting by approximately 10% [23].

3.1.4.2 Copper

Addition of 12S3 to Cu(II) in aqueous MeOH forms a complex (or mixture of complexes) with an exceptionally high reduction potential (Table 2) [143]. In view of the water sensitivity of other first-row $[M(12S3)_2]^{2+}$ cations, this as yet uncharacterized complex probably does not contain 12S3 bound in a tridentate fashion.

$$Cu(II) + 12S3 \xrightarrow{\text{aq MeOH}} [Cu(12S3)(OH_2)_x]^{2+}? \qquad (17a)$$

$$CuCl_2 + 12S3 \xrightarrow{\text{THF}} [CuCl_2(12S3)_2] \qquad (17b)$$

$$[Cu(MeCN)_4]^+ + 12S3 \longrightarrow [Cu_2(12S3)_3]^{2+} \qquad (17c)$$

Treatment of 12S3 with $CuCl_2$ in THF yields $[CuCl_2(12S3)_2]$ as a red-brown crystalline precipitate [34]. It contains a centrosymmetric $CuCl_2S_2$ unit that bridges monodentate 12S3 rings, each of which essentially retains the conformation of the free ligand (Fig. 8). Surprisingly, the $CuCl_2$ moiety binds a sulfur atom in a side rather than the one in the more protruding corner position.

With Cu(I), 12S3 affords $[Cu_2(12S3)_3]^{2+}$, where the cation features a central 12S3 (in the free ligand conformation) donating one S atom to each of two $\{Cu(12S3)\}^+$ units (Fig. 8) [117]. Within each $\{Cu(12S3)\}^+$ unit 12S3 coordinates in a tridentate fashion to complete an approximately tetrahedral CuS_4 coordination sphere.

3.1.5 12S3 – Second-Row Metals

3.1.5.1 Ruthenium

Synthesis of $[Ru(12S3)_2]^{2+}$ proceeds analogously to that of the 9S3 analogue (Eq. 18a) to yield an octahedral complex structurally similar to $[Ni(12S3)_2]^{2+}$ (Fig. 8;

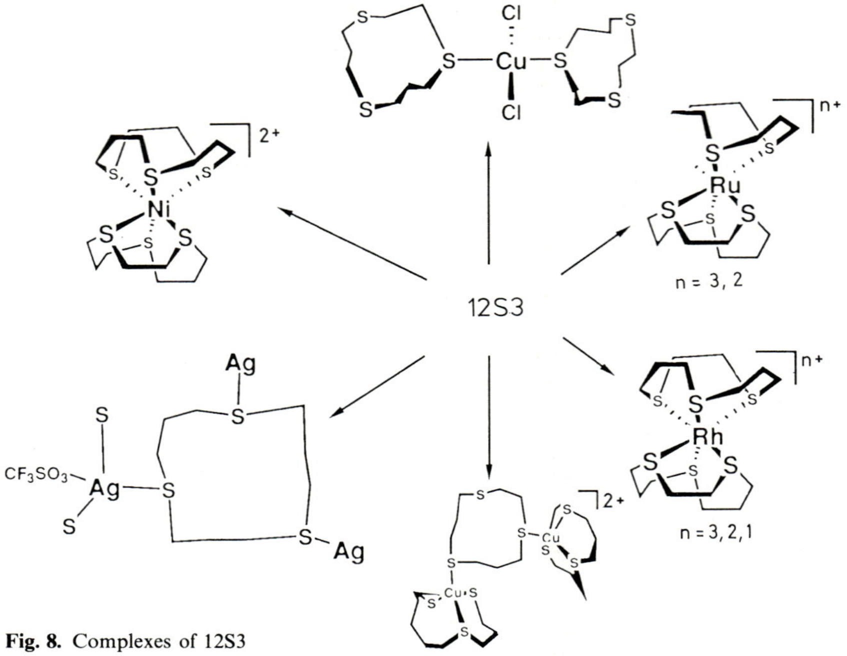

Fig. 8. Complexes of 12S3

Table 1) [121]. Unlike $[Ni(12S3)_2]^{2+}$, however, the Ru(II) complex resists attack by water; indeed, it can be recrystallized from it. As in the case of the 9S3 complex, this difference probably reflects thermodynamic factors as well as the obvious kinetic ones. Ruthenium-sulfur bond lengths exceed those in $[Ru(9S3)_2]^{2+}$ by approximately 0.03 Å, comparable with the difference found in the corresponding Ni(II) complexes (Table 1) [121].

$$[Ru(Me_2SO)_6]^{2+} + 2\,12S3 \xrightarrow{MeOH} [Ru(12S3)_2]^{2+} \tag{18a}$$

$$[Ru(12S3)_2]^{2+} \longrightarrow [Ru(12S3)_2]^{3+} + e^- \tag{18b}$$

Optical spectroscopic studies on the d^6 species $[Ru(12S3)_2]^{2+}$ (Table 3) yield a value of Δ, the ligand field splitting, 29570 cm^{-1}, little different from that of the 9S3 analogue (30760 cm^{-1}). Despite the close similarity in electronic spectra, the two complexes differ appreciably in their electrochemical behavior. Cyclic voltammetry shows a quasi-reversible one-electron wave for the $[Ru(12S3)_2]^{3+/2+}$ couple [121], at a potential 230 mV less oxidizing than in $[Ru(9S3)_2]^{2+}$ (Eq. 18b; Table 2).

3.1.5.2 Rhodium

As discussed earlier, synthesis of homoleptic Rh(III) complexes from the commercially available "$RhCl_3 \cdot x\,H_2O$" requires prior removal of the halide ions by treatment with Ag(triflate). Reaction of the resulting rhodium triflate solution with 12S3 affords $[Rh(12S3)_2]^{3+}$ (Eq. 19a) [130]. Structural characterization confirms the presence of an octahedral coordination sphere (Fig. 8; Table 1) in which Rh–S distances only slightly exceed (by 0.01 Å) those in the 9S3 analogue (Table 1). Disorder in the chelate groups precludes detailed discussion of the ligand conformation.

$$\text{“}RhCl_3 \cdot xH_2O\text{”} \xrightarrow[MeOH]{Ag+} \xrightarrow{12S3} [Rh(12S3)_2]^{3+} \tag{19a}$$

$$[Rh(12S3)_2]^{3+} + e^- \longrightarrow [Rh(12S3)_2]^{2+} \tag{19b}$$

$$[Rh(12S3)_2]^{2+} + e^- \longrightarrow [Rh(12S3)_2]^{+} \tag{19c}$$

The most interesting aspect of $[Rh(12S3)_2]^{3+}$ concerns its electrochemistry. Cyclic voltammetry of $[Rh(L)_2]^{3+}$ (L = 12S3) reveals reversible behavior for the 3+/2+ wave (Eq. 19b) but not for the 2+/+ process (Eq. 19c) (cf. the L = 9S3 complex [128]) (Table 2). Compared to $[Rh(9S3)_2]^{n+}$ both the 3+/2+ and particularly the 2+/1+ waves of $[Rh(12S3)_2]^{n+}$ system shift to more oxidizing potentials. Consequently, for the reaction K_{comp} (the comproportionation constant) for $[Rh(12S3)_2]^{2+}$ falls short of that of $[Rh(9S3)_2]^{2+}$ by three orders of magnitude.

$$[RhL_2]^{3+} + [RhL_2]^{+} \longrightarrow 2[RhL_2]^{2+} \qquad K_{comp}$$

3.2 Hexadentate Crown Thioethers

Black and McLean pointed out that a hexadentate ligand can yield either of two isomeric octahedral complexes of *meso* or *racemic* stereochemistry (Fig. 9) [52, 53]. These differ primarily in that the former consists entirely of facially coordinating tridentate loops; the latter comprises two meridional loops (i.e., those in which three adjacent donor atoms, as well as the metal ion, lie in a plane). In all encapsulated complexes reported to date, 18S6 coordinates in the centrosymmetric meso form. This behavior contrasts with the racemic coordination found for corresponding hexaaza macro-cycles [144–147].

This difference apparently arises at least in part from the differing propensity of the two ligands to adopt *gauche* placement at the C-heteroatom bonds [80, 86]. In

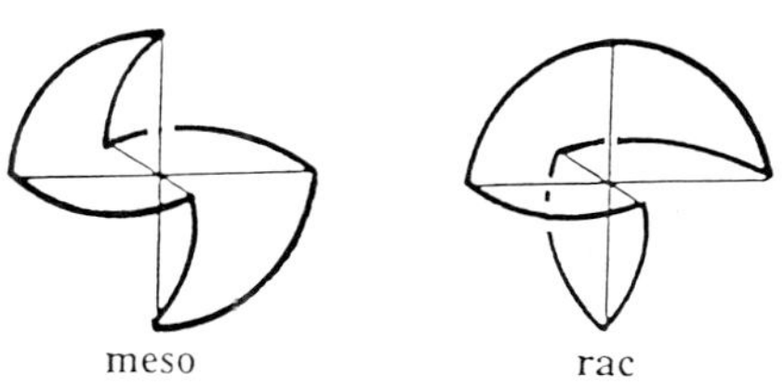

Fig. 9. *meso* and *rac* isomers of an $[M(18S6)]^{n+}$ complex

the *meso* isomer all 12 of the C–E bonds assume *gauche* placement, whereas in the isomeric *racemic* structure, only eight of the 12 do so. Since aza macrocycles preferentially assume *anti* placement at the C–N atom bonds, the stereochemical difference found for complexes of 18S6 and 18N6 parallels the conformational preferences of the component carbon–heteroatom bonds [145, 147].

3.2.1 18S6 – First-row Metals

3.2.1.1 Nickel

In the picrate salt of $[Ni(18S6)]^{2+}$ the ligand envelops a high-spin Ni(II) ion to yield an octahedral NiS_6 coordination sphere (Fig. 10; Table 1) [23, 148]. This complex was first prepared by Black and McLean [52, 53] by reaction of nickel(II) picrate with 18S6 in acetone (Eq. 21a); it also results from interaction of nickel(II) trifluoroacetate with the ligand in nitromethane containing trifluoracetic anhydride (Eq. 21b). As a whole the complex possesses *meso* stereochemistry, consonant with centrosymmetric structure. All six $MSCH_2CH_2S$ five-membered chelate rings

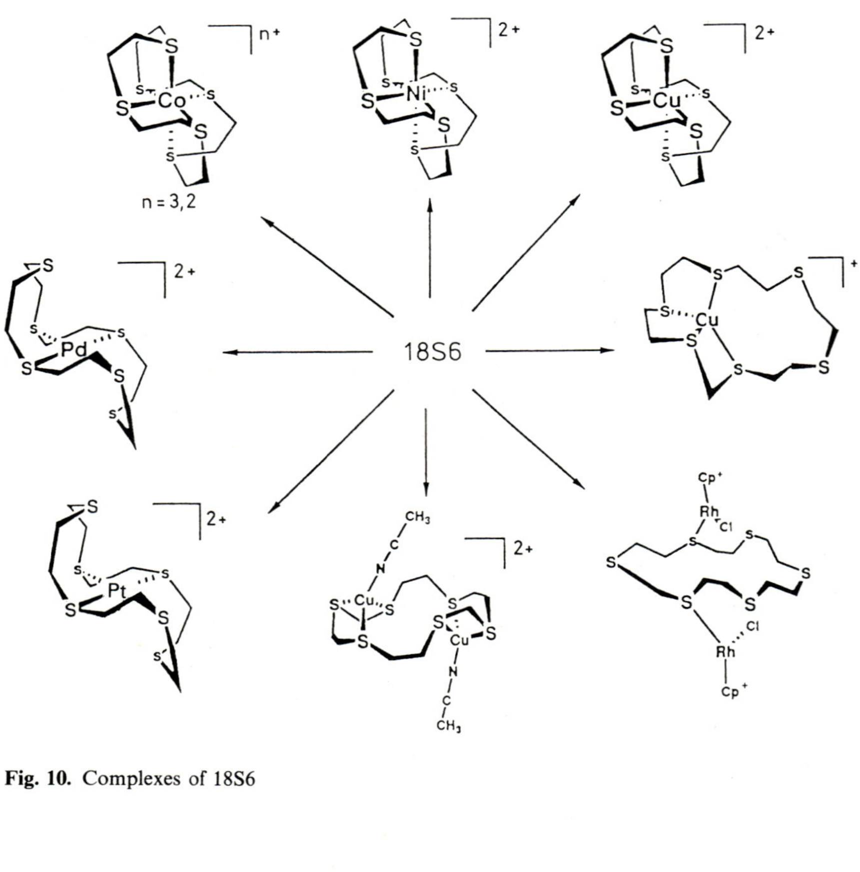

Fig. 10. Complexes of 18S6

adopt "lel" orientation, in which the CH_2CH_2 linkages lie parallel to the pseudo threefold axis of the molecule (as opposed to the "ob" or transverse orientation).

$$Ni(picrate)_2 + 18S6 \longrightarrow [Ni(18S6)](picrate)_2 \quad (21a)$$

$$Ni(CF_3COO)_2 + 18S6 \xrightarrow{MeNO_2/(CF_3CO)_2O} [Ni(18S6)]^{2+} \quad (21b)$$

As in the analogous bis(9S3) complex, coordination to the ethyl-linked macrocycle entails appreciable compression of the M–L bond lengths (the "macrocyclic constriction effect" [16–18]) ($Ni\text{–}S_{ave} = 2.39$ Å; cf. the sum of covalent radii (2.44 Å)) [2]. Electronic spectroscopy reflects the compression in the molecular structure: the ligand field splitting in $[Ni(18S6)]^{2+}$ exceeds those in propyl-linked or open-chain analogues by approximately 10% [23].

3.2.1.2 Cobalt

A reaction path analogous to that used for Ni(II) affords the $[Co(18S6)]^{2+}$ cation (Eq. 22a) (Fig. 10; Table 1), where 18S6 wraps around the metal ion in an all facial fashion to yield the *meso* isomer. Several lines of evidences indicate adoption of the low-spin (^2E_g) state. These include observation of a pronounced Jahn-Teller axial elongation (0.21 Å) as well as Co–S distances consistent with the ionic radius of the low-spin ion [104]. Magnetic susceptibility measurements confirm the low-spin formulation (Table 5) [98, 99]. EPR g values ($g_{\parallel}$ approx. 2; $g_{\perp} > 2$; Table 5) establish the existence of a d_{z^2} ground state (consistent with axial elongation of a low-spin six-coordinate d^7 system).

$$Co(picrate)_2 + 18S6 \xrightarrow{MeCN/acetone} [Co(18S6)]^{2+} \quad (22a)$$

$$[Co(18S6)]^{2+} \longrightarrow [Co(18S6)]^{3+} + e^- \quad (22b)$$

In view of the compression seen for the Ni(II) complex, the low-spin state of $[Co(18S6)]^{2+}$ (and $[Co(9S3)_2]^{2+}$) might be attributed to macrocyclic constriction of the metal ion. Subsequent investigation showed that $[Co(ttn)_2)]^{2+}$ (where ttn = 2,5,8-trithianonane) is also low-spin [99]. Thus the low-spin state in $[Co(18S6)]^{2+}$ derives from the electronic consequences of thioether coordination, and not from macrocyclic constriction of the metal ion.

Cyclic voltammetric measurements reveal a reversible one-electron oxidation process (Eq. 22b; Table 2) [99]. Unlike the 9S3 analogue [98, 99, 102], $[Co(18S6)]^{2+}$ exhibits no reversible reduction to the Co(I) oxidation state.

3.2.1.3 Copper

Like its isostructural Co(II) analogue, the octahedral $[Cu(18S6)]^{2+}$ cation exhibits a pronounced Jahn-Teller distortion ($\Delta d(Cu\text{–}S) > 0.2$ Å; Fig. 10; Table 1) [93].

This compound results from the reaction of Cu(II) salts (e.g. picrate [93], BF_4^-) with the ligand in $MeNO_2$ (Eq. 23a).

$$Cu(II) + 18S6 \xrightarrow{MeNO_2} [Cu(18S6)]^{2+} \quad (23a)$$

$$[Cu(MeCN)_4]^+ + 18S6 \xrightarrow{MeCN} [Cu(18S6)]^+ \quad (23b)$$

$$[Cu(18S6)]^{2+} + e^- \xrightarrow{MeNO_2} [Cu(18S6)]^+ \quad (23c)$$

$$2[Cu(MeCN)_4]^+ + 18S6 \xrightarrow{CH_2Cl_2/MeCN} [Cu_2(MeCN)_2(18S6)]^{2+} \quad (23d)$$

$$CuCl + 18S6 \longrightarrow [CuCl(18S6)]_x \quad (23e)$$

Reduction of $[Cu(18S6)]^{2+}$ (or reaction of 18S6 with $[Cu(MeCN)_4](BF_4)$) yields $[Cu(18S6)]^+$ (Fig. 10; Eq. 23b, c), in which Cu(I) coordinates tetrahedrally with four thioether groups from one 18S6 molecule (Table 1) [93]. The other two S atoms remain free. In light of the severely distorted coordination sphere ($\angle$S–Cu–S = 138°), this complex can also be described as derived from a diagonally coordinated Cu(I) ion by addition of two thioether groups [93]. Similar results have been reported by Rorabacher et al. for the Cu(I) complexes of a variety of crown thioethers [149]. An electrochemically reversible process interrelates the Cu(II) and Cu(I) complexes (Table 2; Eq. 23b). As in the numerous $[CuL]^{2+/+}$ couples of tetradentate thioether ligands [149], the high reduction potential reflects the ability of thioethers to stabilize low oxidation states.

Lower ligand : metal stoichiometry affords $[(Cu(MeCN)_2(18S6)]^{2+}$ (Eq. 23d) [150], where 18S6 encompasses two {CuMeCN} units (Cu ··· Cu 4.25 Å). Each Cu(I) ion coordinates to three adjacent S atoms from 18S6, as well as an N atom from MeCN. The resulting centrosymmetric coordination sphere deviates from iealized tetrahedral geometry, with two chelating S–Cu–S angles near 90°. Evidently by virtue of its ethylene linkages 18S6 cannot easily span the coordination sites of a tetrahedron. Rorabacher and coworkers have previously made this same point for 14S4 [151, 152].

Conformationally, coordination to the {Cu(MeCN)} unit requires considerable rearrangement from free 18S6 [80, 86]. Coordination to two {Cu(MeCN)} groups requires half of the C–S bonds to change to *anti* placement, and all of the C–C bonds to adopt *gauche* placement.

Bidentate and monodentate coordination coexist in $[CuCl(18S6)]_x$ [153] (and also the isostructural $[AgBr(18S6)]_x$). In both cases a given 18S6 molecule coordinates to two MX units. To one it chelates in a bidentate fashion, and to the other as a monodentate ligand. In the resulting oligomer 18S6 bridges M–X units to yield a staircase-like network structure (Fig. 10; Table 1) [116]. In neither case does 18S6 succeed in displacing the halide: instead three S atoms combine with it to generate a distorted tetrahedral MS_3X coordination sphere.

3.2.2 18S6 – Second- and Third-row Metals

3.2.2.1 Molybdenum

Evidence for coordination between 18S6 and Mo in oxidation states 6+ to 3+ (Eqs. 24a–e) does not, unfortunately, include structural data [154, 155]. Instead the compounds were characterized by elemental analysis and magnetic measurements, as well as infrared and mass spectroscopy. Addition of 18S6 suspended in Et_2O to a solution of $[MoO_2Cl_2]$ in the same solvent gives $[(MoO_2Cl_2)_2(18S6)]$ as a yellow powder (Eq. 24a). Reaction of $[MoOCl_3(THF)_2]$ or $[MoCl_4]$ with 18S6 in Et_2O yields $[(MoOCl_3)_2(18S6)]$ as green crystals (Eq. 24b), while use of $[MoCl_4(PrCN)_2]$ gives brown crystals of $[(MoCl_4)_2(18S6)]$ (Eq. 24c).

Molybdenum(III) starting materials give similar products [155]. Treatment of $[MoCl_3(THF)_3]$ (or $[MoCl_3(PrCN)_3]$) in CH_2Cl_2 with excess 18S6 gives $[(MoCl_3)(18S6)]$ (Eq. 24d). Infrared spectroscopy of this compound suggests meridional stereochemistry for the $MoCl_3S_3$ coordination sphere. One 18S6 was suggested to bridge two $\{MoCl_3\}$ units by acting as a monodentate ligand to one, and as a bidentate ligand to the other (as in $[AgCl(18S6)]_x$). In THF the same reaction with $[MoCl_3(solv)_3]$ gives $[(MoCl_3)_2(18S6)(THF)]$ (Eq. 24e), for which a similar bridging mode was postulated. Both compounds exhibit magnetic moments/Mo near 3.5 μ_B (Table 5); hence Mo · · · Mo exchange interaction is weak or nonexistent.

$$[MoO_2Cl_2] + 2\,18S6 \xrightarrow{Et_2O} [(MoO_2Cl_2)_2(18S6)] \tag{24a}$$

$$[MoOCl_3(THF)_2] + 2\,18S6 \xrightarrow{Et_2O} [(MoOCl_3)_2(18S6)] \tag{24b}$$

$$[MoCl_4(PrCN)_2 + 18S6 \xrightarrow{CH_2Cl_2} [(MoCl_4)_2(18S6)] \tag{24c}$$

$$[MoCl_3(solv)_3 + \text{excess } 18S6 \xrightarrow{CH_2Cl_2} [MoCl_3(18S6)] \tag{24d}$$

(solv = THF, PrCN)

$$[MoCl_3(THF)_3 + \text{excess } 18S6 \longrightarrow [(MoCl_3)_2)(18S6)(THF)] \tag{24e}$$

Each of these compounds apparently result from simple adduct formation between 18S6 and the strongly Lewis acidic Mo centers. Structurally they probably resemble discrete complexes such as $[(HgCl_2)_2(14S4)]$ [35], and $[(NbCl_5)_2(14S4)]$ [31–33] or oligomers such as $[HgI_2(14S4)]^{2+}$ [156], where an exodentate ligand coordinates to the Lewis acidic site through one or two donor atoms. In view of the low solubility of these complexes oligomeric structures seem particularly likely.

3.2.2.2 Rhodium

In $[(\eta^5\text{-}Me_5CpRhCl)_2 \cdot 18S6]^{2+}$ 18S6 donates two S atoms to each of two $(\eta^5\text{-}Me_5CpRhCl)^+$ units to give a binuclear cation (Table 1) [157]. This orange compound results from reaction of $[Rh(\eta^4\text{-}C_5Cp_5)Cl_2]_2$ with an equivalent of 18S6 in MeOH.

3.2.2.3 Palladium

As in the corresponding 9S3 complexes, the octahedral coordination offered by 18S6 conflicts with the preference of Pd(II) for square planar coordination. Reaction of $PdCl_2$ with 18S6 affords the yellow-brown centrosymmetric $[Pd(18S6)]^{2+}$ cation (Fig. 10; Table 1) (Eq. 25) [136]. Note the unusual displacement of halide from the metal coordination sphere. This complex consists of a square planar PdS_4 coordination sphere augmented by axial semi-coordination (d(Pd $\cdots$ S) 3.27 Å) with the two remaining S atoms (cf. Pd–S_{eq} 2.31 Å). Owing to the constraints conferred by the short ethylene linkages, the apical S atoms do not lie directly above and below the Pd ion. Instead they lie approximately 15° off the axis that would yield tetragonally elongated octahedral coordination.

$$PdCl_2 + 18S6 \xrightarrow{CH_2Cl_2/MeCN} [Pd(18S6)^{2+}] \tag{25}$$

3.2.2.4 Platinum

With 18S6 platinum(II) yields $[Pt(18S6)]^{2+}$ [108], which is isostructural and otherwise essentially identical with its Pd(II) analogue (Fig. 10). The two differ primarily in M–S_{ax} distances, which in the Pt complex exceed those in the Pd analogue by about 0.1 Å (Table 1).

Bond lengths to the equatorial S atoms exceed that to the apical thioether (3.38 Å) by over 1 Å. As in the corresponding Pd(II) complex, the axial S atoms lie off the tetragonal axis by approximately 15°. This angular distortion presumably reflects the conflict between ligand conformational preferences and M–L bonding in determining geometry. Ordinarily the former represents a much smaller energetic term than the latter (as in, e.g. $[Ni(18S6)]^{2+}$) [148]; for Pd(II) and Pt(II), however, where axial ligands bind at most weakly, conformational factors influence the coordination geometry. This is particularly apparent in comparison with the 9S3 analogues, where the conformation of 9S3 imposes much shorter axial M–S distances.

3.2.2.5 Lanthanides

In acetonitrile 18S6 fails to coordinate to lanthanide(III) perchlorates, but it does so in CH_2Cl_2 to yield complexes of stoichiometry $Ln[ClO_4]_3 \cdot 18S6 \cdot H_2O$ (Ln = Sm, Eu, Yb) [158]. Although no structural data are available, electronic spectroscopic studies of these complexes establish interaction between Ln(III) and 18S6.

3.3 Other Hexadentate Ligands

3.3.1 24S6

Reaction of 24S6 with $[Ni(EtOH)_6]^{2+}$ yields the turquoise complex $[Ni(24S6)]^{2+}$ (Table 1) [23, 159]. Traces of moisture or donor solvents quickly decompose the complex, which reflects the weakness of the long Ni–S bonds. Adoption of meridional stereochemistry here contrasts with the facial coordination in $[M(18S6)]^{2+}$ (M = Co, Ni, Cu).

Optical spectroscopy (Table 3) shows that 24S6 exerts a ligand field strength approximately 10% smaller than that of 9S3 or 18S6. This difference arises from the absence of a significant macrocyclic constriction effect for the larger ring ligand, which gives Ni–S distances comparable with those observed for acyclic thioethers [23, 159].

3.3.2 20S6

Treatment of $[Rh(1,5\text{-cyclooctadiene})Cl]_2$ with $AgPF_6$ in acetone, followed by addition of 20S6 (0.5 equiv per Rh), gives $[\{Rh(COD)\}_2(20S6)]$ $(PF_6)_2$ as red crystals (Table 1) [133]. The centrosymmetric cation comprises two $\{Rh(\eta^4\text{-COD})\}^+$ units that each bind to three adjacent S atoms of an inside-out 20S6 unit. Note the parallel to $[\{HgCl_2\}_2(14S4)]$ [35], where coordination of two metal ions to one macrocycle forces extension of the ligand to minimize M $\cdots$ M repulsion.

Each five-coordinate Rh(I) ion binds to the three S atoms from an ethyl-linked $SCH_2CH_2SCH_2CH_2S$ run (rather than one that includes a propyl-linkage). A strong *trans* influence lengthens the Rh–C bonds *trans* to one thioether group (by 0.11 Å; approximately 22 times the esd) while shortening the relevant Rh–S distance (by 0.15 Å). Neither hydrogenation of the COD groups nor treatment with such good ligands as PPh_3 displace either ligand from this surprisingly stable complex.

3.4 Tetradentate Ligands

3.4.1 14S4

Early work on 14S4 [49, 51, 160] arose its similarity to cyclam (14N4). While 14S4 finds its natural constituency among square-planar metal ions, it also forms octahedral $[M(14S4)X_2]^{n+}$ complexes of both *cis* and *trans* stereochemistry.

A macrocycle such as 14S4 can girdle a square-planar metal ion (or the equatorial plane of an octahedral one) in two ways [161]. These are best described by reference to the six-membered $SCH_2CH_2CH_2SM$ chelate rings, which bear an obvious relation to the chair form of cyclohexane. In *syn* coordination the central methylene groups of the two six-membered $(SCH_2CH_2CH_2SM)$ rings lie on the same side of the MS_4 plane (Fig. 11). This leads to *R*, *S*, *R*, *S* stereochemistry at the S atoms. In the centrosymmetric *anti* isomer they lie on opposite sides of the equatorial plane (*R*, *S*, *S*, *R* stereochemistry).

Alternatively, *syn* and *anti* conformers can be designated by reference to the relative positions of the lone pairs on sulfur. This approach suffers from reliance upon classification of molecules by a property that is not directly observed. Nevertheless, with this caveat in mind *syn* coordination corresponds to an "all up" disposition of S lone pairs; *anti* implies an "up, up, down, down" arrangement (Fig. 11).

Which coordination mode a complex adopts depends critically upon the size of the metal ion [162]. *Anti* stereochemistry, with its concomitant inversion symmetry, requires the metal ion to lie in the plane of the four donor atoms. Only the smaller metal ions can fit in this plane at normal M–S distances. *Syn* coordination better accommodates larger metal ions, which can (and invariably do) lie out of the S_4 plane. It therefore predominates when the ionic radius of the metal exceeds the optimum for the ligand cavity [162]. For example, $[M(14S4)]^{2+}$ complexes of first-row metals uniformly adopt *anti* coordination (as do $[M(14N4)]^{2+}$ complexes [162]). On the other hand, *syn* stereochemistry occurs in, e.g. $[Hg(14S4)(OH_2)]^{2+}$ [35], where it arises from the attempt to circumscribe the large Hg(II) ion.

Six-coordinate complexes of 14S4 can have either *cis* or *trans* geometry. Those with idealized D_{4h} symmetry result from addition of two axial ligands to an *anti* $[M(14S4)]^{n+}$ complex. Steric repulsion impedes the corresponding process in a *syn* complex, since the macrocyclic ligand restrict access to the sixth coordination site. Accordingly, *syn* complexes accept at most one additional ligand (to achieve square pyramidal coordination, as in $[Hg(14S4)(OH_2)]^{2+}$) [35].

Alternatively, addition of two ligands to an $[M(14S4)]^{n+}$ unit can lead to *cis* stereochemistry, where 14S4 folds to bind to three equatorial and one apical site. Only two examples, *cis*-$[RuCl_2(14S4)]$ [163] and *cis*-$[Hg(14S4)(picrate)_2]$ [164], are known thus far.

Both coordination mode and ring size affect the properties of the resulting complexes. For example, in Cu(II) complexes of cyclic tetrathioethers the stability constants decrease as the ring size increases from 14 to 18 [165]. Kinetic measurements reveal that the difference arises almost entirely in the dissociation rate constants.

3.4.2 First-row Metals

3.4.2.1 Cobalt

In 1974 Travis and Busch reported the preparation of the red-brown $[Co(14S4)]^{2+}$ cation from interaction of 14S4 with $[Co(MeCN)_6]^{2+}$ in nitromethane. The Co(II) complex probably resembles $[Co(dth)_2(OClO_3)_2]$ (dth = 2,5-dithiahexane) [92], in which two ClO_4^- ions bind to the axial positions of a square-planar CoS_4 unit. In light of the unusual properties of other Co-crown thioether complexes the electronic structures and electrochemical behavior of the $[Co(14S4)]^{n+}$ complexes

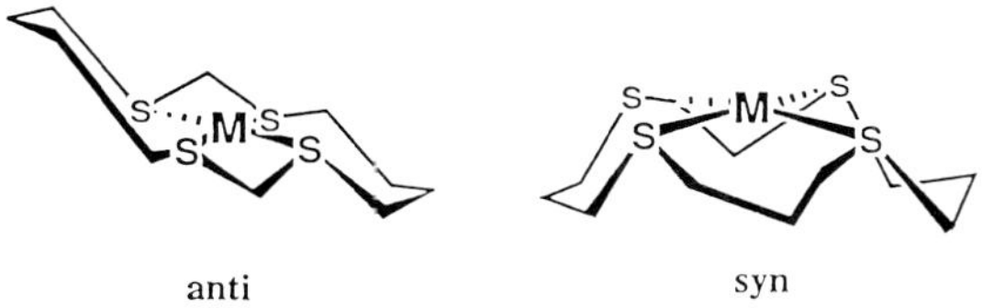

Fig. 11. *Syn* and *anti* stereoisomers of an $[M(14S4)]^{n+}$ complex

warrant further examination.

$$[Co(MeCN)_6]^{2+} + 14S4 \xrightarrow{CH_3NO_2} [Co(14S4)]^{2+} \tag{26a}$$

$$[Co(14S4)]^{2+} + 2X^- \xrightarrow{O_2} cis\text{-}[Co(14S4)X_2]^+ \tag{26b}$$

$(X^- = Br^-, Cl^-, NO_2^-, NCS^-, oxalate^{2-})$

$$[Co(14S4)]^{2+} + 2I^- \xrightarrow{O_2} cis\text{-}[Co(14S4)I_2]^+ \tag{26c}$$

$(X^- = I^-)$

Air oxidation of $[Co(14S4)]^{2+}$ in the presence of LiCl affords $[Co(14S4)Cl_2]^+$ [50], where *cis* stereochemistry is established by the facile substitution of Cl^- by bidentate ligands (e.g. oxalate). *Cis*-$[Co(14S4)X_2]^+$ ($X^- = Br^-, Cl^-, NO_2^-, NCS^-$, oxalate^{2-}) and *trans*-$[Co(14S4)X_2]^+$ ($X^- = I^-$) complexes [50] can be made similarly. The *cis* complexes also result from substitution of X for Cl^- in *cis*-$[Co(14S4)Cl_2]^+$. Cobalt complexes of benzo-15S4 behave similarly.

3.4.2.2 Nickel

Many donor solvents displace 14S4 from Ni(II). Accordingly Rosen and Busch developed new synthetic methods that have since been widely used for preparation of thioether complexes [49, 51]. Reaction of $[Ni(HOAc)_6]^{2+}$ with 14S4 in $MeNO_2$ yields the red low-spin $[Ni(14S4)]^{2+}$ cation [3]. X-ray diffraction reveals a square-planar NiS_4 coordination sphere (Table 1) with *anti* stereochemistry [166]. ^{13}C NMR studies show that in CD_3NO_2 solution, however, this complex exists as an approximately 50:50 equilibrium mixture of the *anti* and *syn* forms [167], where the paramagnetic *syn* form coordinates a water molecule in the apical position [35].

$$[Ni(HOAc)_6]^{2+} + 14S4 \xrightarrow{CH_3NO_2} [Ni(14S4)]^{2+} \tag{27}$$

Unlike the salts of non-coordinating anions (BF_4^- or ClO_4^-), coordinating anions such as Cl^-, Br^-, I^-, and NCS^- give paramagnetic six-coordinate complexes with S = 3/2 ground states (as shown by magnetic measurements; Table 5). Conductivity measurements indicate that the NCS^-, Cl^-, and Br^- complexes remain intact as nonelectrolytes in $MeNO_2$ solution, but that the I^- complex does not. Instead it participates in the coordinative equilibria

$$[Ni(14S4)I_2] = [Ni(14S4)I]^+ + I^- \qquad K_1$$

$$[Ni(14S4)I]^+ = [Ni(14S4)]^{2+} + I^- \qquad K_2$$

where $K_1 = 6.3 \times 10^{-2} M^{-1}$ and $K_2 = 2.5 \times 10^{-4} M^{-1}$ [48, 50].

In solution $[Ni(14S4)]^{2+}$ displays relatively low stability. Addition of a wide variety of ligands – including any solvent of significant donor ability – immediately displaces the crown thioether. Unlike 14N4, 14S4 shows minimal macrocyclic effect towards Ni(II) [13]. That is, the stability of $[Ni(L)]^{2+}$ (L = 14S4) barely exceeds (180-fold) that of a complex with a comparable acyclic tetrathioether ligand (cf. $[Ni(14N4)]^{2+}$, for which the macrocyclic effect exceeds 10^6) [13]. Margerum and Smith [13] emphasized the role of differential ligand solvation (i.e., the loss of solvation of the free ligand on coordination) as the source of this difference in behavior. Rorabacher and coworkers pointed out that while solvation effects clearly play an important role for macrocyclic amines, they minimally influence complexes of macrocyclic thioethers [37].

Schrauzer and coworkers found that reaction of norbornadiene with $[Ni(diphenyldithiolene)_2]$, followed by treatment with α,α′-dibromo-*o*-xylene results in formation of macrocyclic tetrathioether (template condensation). The resulting paramagnetic complex of this unsaturated 14S4 analogue solvolyzes in MeOH to give the free ligand [168, 169].

3.4.2.3 Copper

Rorabacher and coworkers have extensively studied the electronic and molecular structures of Cu complexes with 14S4 and other tetrathioether ligands as models for the Cu-methionine interaction in the blue copper protieins [149, 170].

In $[Cu(14S4)(OClO_3)_2]$ four sulfur atoms from 14S4 and two axially bound monodentate perchlorate anions complete an octahedral coordination sphere (Fig. 12; Table 1) [89, 171]. The resulting structure resembles that of $[Co(2,5\text{-dithiahexane})_2(OClO_3)_2]$, in which a square planar cobalt(II) ion binds two perchlorate anions [92]. Metal–sulfur distances on average substantially exceed those in $[Ni(14S4)]^{2+}$ (2.30 vs 2.18 Å) (Table 1). As in the Ni(II) analogue, $[Cu(14S4)(OClO_3)_2]$ adopts the centrosymmetric *anti* (*R*, *S*, *S*, *R*) conformation. Since Cu(II) is smaller than Ni(II) [104], $[Cu(14S4)]^{2+}$ should be less prone to adoption of *syn* stereochemistry.

Electrochemical reduction of $[Cu(14S4)(OClO_3)_2]$ [151, 152] gives the corresponding Cu(I) complex, which crystallizes as an oligomer. This structure apparently results from the incompatibility of 14S4 with tetrahedral ion (at normal M–S distances). Each Cu(I) ion achieves distorted tetrahedral coordination through three S atoms from one ligand, and a fourth from an adjacent 14S4 (Fig. 12; Table 1). Bond angles at copper indicate that the unique S atom coordinates with minimal strain (angles approximately 109°); on the other hand, those between S atoms from the same ligand (one nearly 130°, another near 90°), reflect the unsuitability of 14S4 for tetrahedral coordination [172].

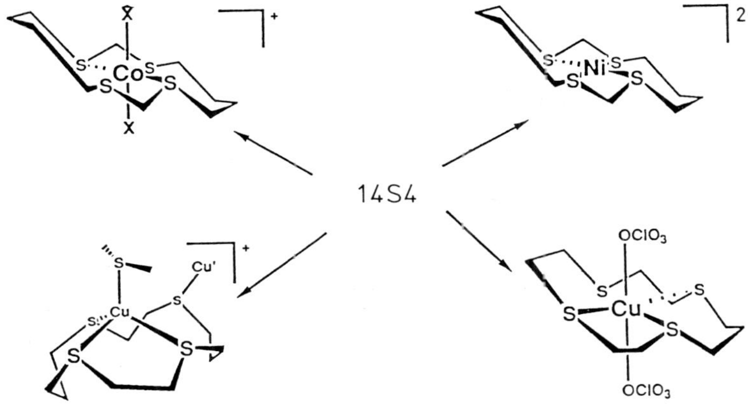

Fig. 12. Coordination complexes of 14S4 with first row metal ions

Measurements on complexation kinetics in a series of tetradentate crown thioether complexes of Cu(II) indicate that second-bond formation probably represents the rate-limiting step [173]. Formation rate constants vary in accord with the change in difficulty of wrapping the ligand around the metal ion. Both the ring-size and the macrocyclic effect manifest themselves in the final steps of complexation; accordingly, their magnitudes tend to mirror those of the dissociation rate constants. Introduction of hydroxyl groups in the propyl chains slightly slows the rate of complexation, apparently in part through steric effects [174].

Strain-energy calculations on $[Cu(nS4)(OH_2)_2]^{2+}$ complexes (n = 12–16) [175] parallel the experimental results: an increase in ligand strain energy decreases both the stability constant and the dissociation rate constant (in a linear manner, the origin of which is unclear). Comparison of the various complexes indicates that the "strain-free" free energy of formation of a Cu(II)-tetrathioether complex is $-8.65(0.66)\ \text{kcal mol}^{-1}$, and that the strain energy is surprisingly small: $2.9\text{–}5.5\ \text{kcal mol}^{-1}$.

Electrochemical measurements of the Cu(II/I) potentials with the nS4 ligands (n = 12–16) indicate that the Cu(II) and Cu(I) species each exist in two different conformational states [170]. Conformational rearrangement may either precede or succeed electron transfer. Rorabacher and coworkers interpreted their results in light of a "square" mechanistic scheme that neatly reconciles the sweep rate dependence of the cyclic voltammograms with the requisite change in coordination geometry at Cu. Kinetic studies on the electron transfer [149, 170, 176–177] support this scheme; application of the Marcus cross relationship to reduction of Cu(II) and oxidation of Cu(I) yields widely discrepant values, presumably because of the different conformational states involved.

Spectroscopic work on $[Cu(14S4)]^{2+}$ (Table 3) revealed visible bands strongly reminiscent of those in the blue copper proteins, and thereby stimulated investigation of how thioethers affect the electronic structure of this ion [178]. Strong

LMCT bands characterize the visible spectrum of this and other thioether complexes of Cu(II). Resonance Raman spctroscopy on $[Cu(14S4)]^{2+}$ and copper(II) complexes of other crown thioethers extended the parallel with the blue copper proteins, and assigned the Cu–S stretch to vibrations around 250 cm^{-1} [179].

3.4.3 14S4 Second- and Third-Row Metals

3.4.3.1 Molybdenum

Reaction of 14S4 with molybdenum halides and oxohalides affords a series of crown thioether complexes of Mo(IV), (V), or (VI), depending on the molybdenum-containing starting material. Thus reaction of $MoCl_4$ with 1 or 2 equivalents of 14S4 in Et_2O or CH_2Cl_2 gives the brown complexes $[(MoCl_4)_n(14S4)]$ with n = 1 or 2, respectively. A similar procedure affords the green adducts $[(MoOCl_3)(14S4)]$ and $[(MoOCl_3)_2(14S4)(THF)_2]$ from $MoOCl_3$, as well as the yellow complex $[MoO_2Cl_2(14S4)]$ from MoO_2Cl_2 [154]. In this last complex, infrared studies indicate retention of the *cis*-dioxo structure of the starting material.

$$[MoO_2Cl_2] + 14S4 \xrightarrow{Et_2O \text{ or } CH_2Cl_2} [MoO_2Cl_2(14S4)] \quad (29a)$$

$$[MoOCl_3(THF)_2] + 14S4 \xrightarrow{Et_2O \text{ or } CH_2Cl_2} [(MoOCl_3)_2(14S4)(THF)] \quad (29b)$$

$$[MoCl_4] + 14S4 \xrightarrow{Et_2O-CH_2Cl_2} [MoOCl_3(14S4)] \quad (29c)$$

$$[MoCl_4(PrCN)_2] + 14S4 \xrightarrow{CH_2Cl_2} [(MoCl_4)_n(14S4)]\ (n = 1, 2) \quad (29d)$$

Regardless of the stoichiometry used no more than one MoO_2Cl_2 unit will coordinate to a single 14S4. Presumably this is attributable to steric repulsion between the *cis*-dioxo groups, since 2:1 adducts readily form with $MoOCl_3$ and $MoCl_4$ [154]. Furthermore, $[(MoO_2Cl_2)_2(L)]$ forms for L = 18S6. Indeed, in this case the converse situation obtains: *only* the 2:1 adduct forms, for reasons that are not clear.

In the absence of crystallographic studies the structures of these compounds remain a matter of speculation. Their low solubility suggests that these complexes may have an oligomeric structure. As the products from the reaction of a crown thioether with a highly Lewis acidic metal complex, these compounds are reminiscent of, e.g., the adducts of 14S4 with $NbCl_5$.

3.4.3.2 Ruthenium

Reaction of 14S4 with $K_2[RuCl_5(OH_2)]$ in 2-methoxyethanol yields orange crystals of $[Ru(14S4)Cl_2]$, where reduction accompanies coordination (Eq. 30a) [180]. X-ray diffraction reveals a *cis* stereochemistry, in correction of an earlier assignment (based upon infrared spectroscopy) to *trans* stereochemistry (Fig. 13; Table 1) [163, 181]. Sulfur atoms trans to other S atoms exhibit considerably longer Ru–S distances (0.1 Å) than those *trans* to Cl. The ligand adopts a folded conformation with both the $RuSCH_2CH_2CH_2S$ rings in chair conformations. Treatment of the Ru(II) complex with $HClO_4$ in dilute HCl oxidizes it to the corresponding Ru(III) complex, $[RuCl_2(14S4)]^+$ (Eq. 30b), which presumably also has *cis* stereochemistry.

$$K_2[RuCl_5(OH_2)] + 14S4 \xrightarrow{\text{2-methoxyethanol}} cis\text{-}[RuCl_2(14S4)] \quad (30a)$$

$$cis\text{-}[RuCl_2(14S4)] + HClO_4 \longrightarrow cis\text{-}[RuCl_2(14S4)]^+ \quad (30b)$$

Reactivity studies emphasize the important role of stereochemistry in *cis*- and *trans*-$[RuCl_2(14S4)]$ [231].

3.4.3.3 Rhodium

Preparation of the congeneric Rh(III) compounds proceeds from reaction of rhodium trichloride or tribromide (dissolved in the minimum amount of hot water) with an ethanolic solution containing excess ligand (Eq. 31a). Thus reaction of rhodium trihalides with 14S4 in refluxing ethanol yields the yellow complex $[Rh(14S4)X_2]^+$ (Fig. 13) [6, 50]. As in the isoelectronic $[Ru(14S4)Cl_2]$ case, infrared spectroscopy suggests *cis* stereochemistry for this complex. In view of the difficulty of such assignments [20], however, this conclusion should be considered tentative.

$$\text{“}RhX_3 \cdot nH_2O\text{”} + 14S4\ (xs) \xrightarrow{\text{EtOH}} [RhX_2(14S4)]^+\ (X = Cl, Br) \quad (31a)$$

$$[RhX_2(14S4)]^+ + Y^- \longrightarrow [RhY_2(14S4)]^+\ (Y = I, NO_2) \quad (31b)$$

$$[RhCl_2(14S4)]^+ + NaBH_4 \longrightarrow [Rh(14S4)]^+ \quad (31c)$$

$$[Rh(COD)_2Cl]_2 + 2\,14S4 \xrightarrow{\text{toluene}} [Rh(14S4)]^+ \quad (31d)$$

$$[Rh(14S4)]^+ + CH_2Cl_2 \longrightarrow trans\text{-}[Rh(14S4)(CH_2Cl)(Cl)]^+ \quad (31e)$$

Subsequent ligand exchange of $[Rh(14S4)X_2]^+$ (X = Cl, Br) with LiI or $LiNO_2$ gives $[Rh(14S4)I_2]^+$ and $[Rh(14S4)(NO_2)_2]^+$, respectively (Fig. 13; Eq. 31b). Travis and Busch [50] emphasized the necessity of using refluxing solvent to avoid formation of an intractable precipitate. Elemental analysis indicates the composition $[Rh(14S4)X_2]X$, while infrared studies suggest the presence of Rh–Cl–Rh bridges.

Reduction of $[Rh(14S4)Cl_2]^+$ with sodium borohydride yields the yellow square-planar Rh(I) species $[Rh(14S4)]^+$, of unknown stereochemistry (Eq. 31c; Fig. 13; Table 1) [6]. A compound of the same stoichiometry, but different color (red-brown) results from the reaction of 14S4 with $[Rh(COD)_2Cl]_2$ in toluene at room temperature (Eq. 31d) [182]. In this latter complex Rh(I) lies 0.13 Å out of the S_4 plane, consistent with adoption of *syn* stereochemistry. Interaction of the Rh(I)

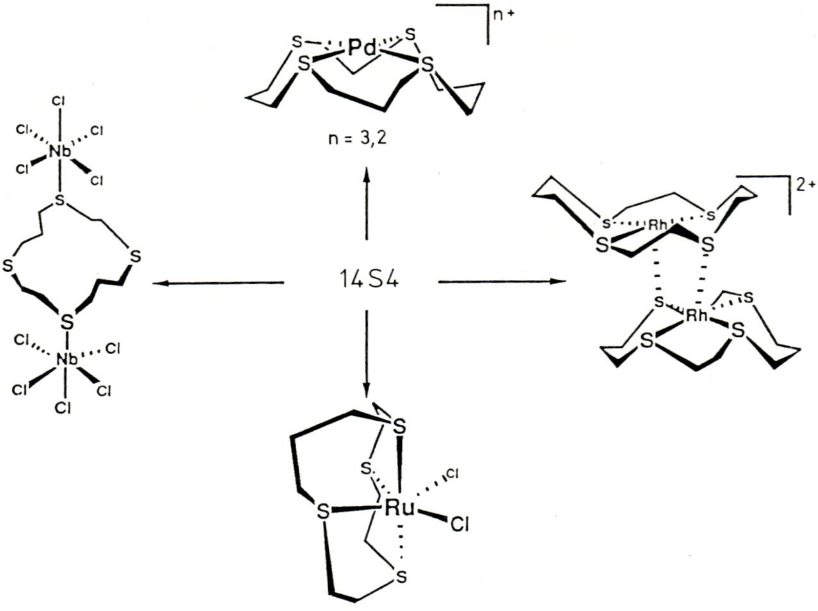

Fig. 13. Coordination complexes of 14S4 with second row metal ions

ion with a neighboring RhS_4 unit further stabilizes *syn* stereochemistry. Despite the stereochemistry, however, this complex has shorter M–S distances than those in $[Cu(14S4)]^{2+}$, which exists as the more constrained *anti* isomer. This difference presumably reflects the strength of the Rh–S bonds. As in $[Ni(14S4)]^{2+}$, the *syn* and *anti* isomers of $[Rh(Me_4 14S4)]^+$ both *syn* and *anti* stereoisomers have been observed [182].

$[Rh(14S4)]^+$ reacts with BF_3, SO_2, NO^+, O_2, TCNE, and H^+ [6], and it also oxidatively adds such species MeI, benzyl bromide, and acetyl chloride [6], as well as CH_2Cl_2 [182]. These products of oxidative addition presumably adopt *trans* stereochemistry [6, 182].

Yoshida and coworkers [182] have examined the coordination chemistry of Rh(I) with 14S4 and with its tetramethyl analogue, Me_4-14S4 (6,6,13,13-tetramethyl-tetrathiacyclotetradecane). Reaction of $[Rh(COD)Cl]_2$ with either ligand affords the complexes [RhL] as their chloride salts. In the dimeric $[Rh(14S4)]_2^{2+}$ cation the two centrosymmetrically-related Rh lie above the S_4 plane (by 0.13 Å); both Rh · · · Rh (3.313(1) Å) and Rh · · · S (3.70–3.82 Å) intermolecular interactions stabilize the dimeric structure [182]. The $Me_4$14S4 complex differs from the 14S4 parent in that it does not form an intermediate dimeric Rh(II) species on oxidation. Interestingly, the red-brown *syn* conformer of $[Rh(Me_4 14S4)]^+$ readily converts to the yellow *anti* form in MeCN (but not in DMSO). No analogous *anti* conformer of the parent compound $[Rh(14S4)]^+$ has yet been observed.

Structural investigation of $[Rh(14S4)]_2Cl_2$ confirms deductions from ^1H NMR data that this complex (as well as its $Me_4$14S4 analogue) assumes *syn* stereochemistry [182]. For the $[Rh(Me_414S4)]_2^{2+}$, however, metathesis with $NaBPh_4$ results in isomerization from the *syn* to the *anti* form (as shown by ^1H NMR). Oxidative addition of CH_2Cl_2 to [Rh(L)]Cl (L = 14S4 or $Me_4$14S4) or $[Rh(L)]BPh_4$ (L = $Me_4$14S4) gives the corresponding *trans*-$[RhCl(CH_2Cl)L]$ complex. The greater reactivity of Rh-thioether complexes toward CH_2Cl_2 compared with Rh-phosphine and -isonitrile complexes was attributed to the lesser π-acidity of thioethers. Conformational factors influence this reactivity: $[Rh(Me_414S4)]Cl$ (*syn*), in which the Rh ion protrudes from the S_4 plane, attacks CH_2Cl_2 seven-fold faster than $[Rh(Me_4\text{-}14S4)]BPh_4$ (*anti*) [182] (although the counterion could play a role).

3.4.3.4 Palladium

Treatment of $K_2[PdCl_4]$ with 14S4 in $MeCN/CH_2Cl_2$ affords $[Pd(14S4)]^{2+}$ as yellow crystals after recrystallization from water and metathesis to the hexafluorophosphate salt (Eqs. 32a, b). X-ray diffraction confirms the expected square-planar geometry of a cation with *syn* stereochemistry (Fig. 13; Table 1) [129, 136].

$$K_2[PdCl_4] + 14S4 \xrightarrow{MeCN/CH_2Cl_2} [PdCl_2(14S4)] \tag{32a}$$

$$[PdCl_2(14S4)] \xrightarrow{H_2O} [Pd(14S4)]^{2+} \tag{32b}$$

3.4.3.5 Mercury

Reaction of 14S4 with Hg(II) salts gives different products depending on the nature of the counterion. Indeed, the sweeping differences in coordination mode observed with different counterions makes clear why X-ray diffraction studies have proven indispensable for characterization of crown thioether complexes. In aqueous methanol $Hg(ClO_4)_2$ with 14S4 yields $[Hg(14S4)(OH_2)]^{2+}$, in which 14S4 supplies a basal plane of four thioethers to the square pyramidal coordination sphere (Eq. 33a). In this complex Hg(II) lies 0.48 Å above the S_4 plane, with an apical water molecule completing the square pyramidal coordination sphere (Table 1; Fig. 14) [35]. The large size of Hg(II) necessitates *syn* stereochemistry.

$$Hg(ClO_4)_2 + 14S4 \xrightarrow{\text{aq MeOH}} [Hg(14S4)(OH_2)]^{2+} \tag{33a}$$

$$HgCl_2 + 14S4 \xrightarrow[\text{reflux}]{CH_3NO_2} [(HgCl_2)_2(14S4)] \tag{33b}$$

$$HgI_2 + 14S4 \longrightarrow [HgI_2(14S4)] \tag{33c}$$

Use of $HgCl_2$ instead of the perchlorate salt affords $[(HgCl_2)_2(14S4)]$ [35]. Here two $HgCl_2$ moieties coordinate to an exodentate 14S4, which chelates each

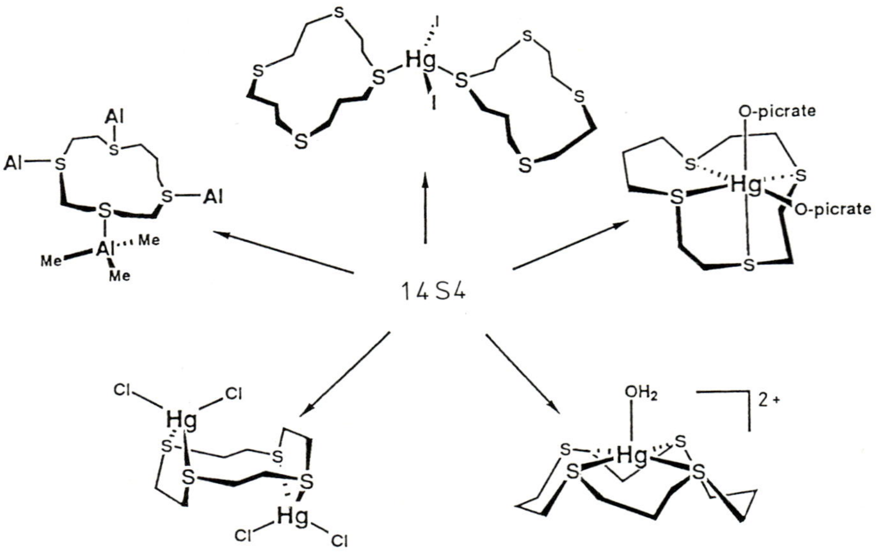

Fig. 14. Third-row and miscellaneous complexes of 14S4

Hg^{2+} ion in a bidentate fashion through an SCH_2CH_2S unit. In the resulting centrosymmetric 2:1 complex the Hg ions approach idealized tetrahedral microsymmetry (Fig. 14; Table 1). Conformationally 14S4 in this complex differs from the free ligand primarily in the orientation of the C_2 linkages, which adopt *gauche* placement (cf. *anti* in free 14S4) to permit coordination to the Hg ion.

Unlike either the chloro or bromo analogues, $[HgI_2(14S4)]$ [156] consists of an oligomer with HgI_2 units bridging exodentate 14S4 ligands through diametrically opposite S atoms. (The other two S atoms remain free.) Tetrahedral HgI_2S_2 units straddle adjacent exodentate ligands, each of which binds in a monodentate fashion (Table 1). This contrasts with the discrete $[(HgCl_2)_2(14S4)]$ units found for the chloro analogue. Presumably the larger halides generate unacceptable X · · · H repulsions that obviate adoption of the chloro structure.

Mercury–sulfur distances in the iodo complex (2.75 Å) considerably exceed those in the chloro analogue (2.53 Å); evidently the stronger Hg–X interaction for X = I takes place at the expense of the Hg–S bonding. Angles at Hg further support this inference: ∠I–Hg–I opens from the tetrahedral value to 136° (with concomitant closing of ∠S–Hg–S to 84°). Since all of the ligands are monodentate, bond angles in this complex lack obvious constraints. Thus the large ∠I–Hg–I angle (which does not arise from I · · · I interaction: d(I · · · I) = 4.94 Å, cf. sum of van der Waals radii, 4.30 Å) suggests that coordination to the two thioether groups only perturbs somewhat the strong linear I–Hg–I bonding of the HgI_2 precursor (cf. $[Cu(18S6)]^+$).

In yet another structural motif, use of $Hg(picrate)_2$ instead of either a halide or perchlorate anion yields a fourth structure. In *cis*-$[Hg(14S4)(picrate)_2]$ [164] mercury accepts four S atoms from the macrocycle along with the two picrate

Table 6. Crown Thioethers: Stability Constant Determinations

Metal	Ligand	Conditions	log K	Ref.
Co(II)	9S3	MeCN	log K_1 6.96 log K_2 7.00	[100]
Cu(II)	12S4	0.1 M $HClO_4$; 25°	3.48	[37, 170]
Cu(II)	13S4	0.1 M $HClO_4$; 25°	3.41	[37, 170]
Cu(II)	14S4	0.1 M $HClO_4$; 25°	4.34	[37, 170]
Cu(II)	15S4	0.1 M $HClO_4$; 25°	3.17	[37, 170]
Cu(II)	16S4	0.1 M $HClO_4$; 25°	≈ 2.2	[37, 170]
Cu(II)	15S5	0.1 M $HClO_4$; 25°	4.07	[37, 170]
Hg(II)	13S4	80% MeOH/0.1 M $HClO_4$	≈ 10.3	[b]
Hg(II)	14S4	80% MeOH/0.1 M $HClO_4$	9.55(4)	[189][b]
Hg(II)	15S4	80% MeOH/0.1 M $HClO_4$	9.33(4)	[189][b]
Hg(II)	16S4	80% MeOH/0.1 M $HClO_4$	10.48(3)	[189][b]
Hg(II)	18S4	80% MeOH/0.1 M $HClO_4$	8.88(2)	[189][b]
Hg(II)	20S4	80% MeOH/0.1 M $HClO_4$	7.88(3)	[189][b]
Hg(II)	21S4	80% MeOH/0.1 M $HClO_4$	≈ 8.4	[189][b]
Hg(II)	u-20S6	80% MeOH/0.1 M $HClO_4$	≈ 13.6	[b, c]
Hg(II)	21S6	80% MeOH/0.1 M $HClO_4$	12.26(5)	[230]

[b] Sokol, L. S. W. L.; Rorabacher, D. B.; personal communication.
[c] The ligand designated as "u-20S6" (where u stands for "unsymmetrical") has two propyl bridges separated by a single ethyl bridge, with three other ethyl bridges completing the macrocycle.

phenolate O atoms. Owing to the weak stereochemical preferences of this d^{10} ion, the HgS_4O_2 coordination sphere deviates substantially from octahedral symmetry: for example, one *cis* ∠S–Hg–S is 77°, while a *trans* ∠S–Hg–S is 162° (Fig. 14; Table 1). The two picrate groups lie parallel to each other, with ∠O–Hg–O 77°. It is not clear why this complex prefers *cis* octahedral stereochemistry over square pyramidal (with an axial picrate) with the ligand in *syn* conformation.

3.4.4 14S4 – Miscellaneous Complexes

The lack of a substantial macrocyclic effect for 14S4 (and its weak binding affinities generally) implies an antipathy to chelation. Part of the reluctance of 14S4 to coordinate in a planar fashion probably results from the unfavorable conformation necessary to do so. Compared to the free ligand, 14S4 in its planar complexes must change conformation in every one of its 14 bonds. The enthalpic terms associated with this change doubtless decrement the free energy of complexation. Nevertheless non-chelated 14S4 complexes have so far been established only where the metal offers fewer than four coordination sites.

Reaction of 14S4 with strongly Lewis acidic compounds leads to adducts in which an exodentate 14S4 acts as a monodentate ligand. Addition of $NbCl_5$ to 14S4 in benzene affords the centrosymmetric adduct $[(NbCl_5)_2(14S4)]$ [31–33]. Here an exodentate 14S4 bridges two $NbCl_5$ units by acting as a monodentate ligand to each. In this case adduct formation does not appreciably perturb the conformation of the macrocycle from that observed in the free state. The $NbCl_5S$ coordination sphere closely approaches octahedral microsymmetry. Curiously, the four equatorial Cl atoms bend *toward* the S atom; ∠S–Nb–Cl angles range from 80 to 85°, despite four close contacts with the macrocyclic ring (d(C · · · Cl) 3.31–3.70 Å). This "dishing" of the $NbCl_4$ plane occurs in conjunction with a short Nb–Cl_{ax} bond (2.25 Å; Nb–Cl_{eq} average 2.32 Å). This short distance may arise from the poor σ-donor ability of the trans thioether group, which in turn may necessitate closer approach of the axial chloride ion to compensate for the high formal charge on the metal ion.

$$2(AlMe_3)_2 + 14S4 \xrightarrow{\text{toluene}} [(AlMe_3)_4(14S4)] \tag{34a}$$

$$2\,NbCl_5 + 14S4 \xrightarrow{\text{benzene}} [(NbCl_5)_2(14S4)] \tag{34b}$$

Similarly, in $[(AlMe_3)_4(14S4)]$ (prepared by reaction of $(AlMe_3)_2$ and 14S4 in benzene) one $AlMe_3$ unit adds to each of the four S exodentate atoms of the ligand (Fig. 14; Table 1) [183]. The structures adopted by these compounds and others such as $[(HgCl_2)_2(14S4)]$ probably represent a good model for the intermediate formed in the first step of complexation by 14S4.

Partial chelation of 14S4 also occurs in $[PtMe_3(14S4)]^+$, which results from reaction of 14S4 with $[PtMe_3]_4$ in $CHCl_3$ [139]. NMR measurements (both ^1H and ^{13}C-$\{^1H\}$) show that intramolecular rotation of the ligand exchanges the free and bound S atoms with $\Delta G^‡ = 56.79(2)$ kJ mol^{-1} ($T_{coalescence} = 333$ K).

3.4.5 *16S4*

3.4.5.1 Copper

Solutions of $[Cu(16S4)]^{2+}$ in aqueous MeOH in the presence of ClO_4^- afford $[Cu(16S4)(OClO_3)_2]$, in which a centrosymmetric CuS_4 unit coordinates to two axial monodentate ClO_4^- anions [89]. Comparison with the analogous 14S4 complex, which also has *anti* stereochemistry, shows that the larger ring size dilates the CuS_4 coordination sphere by 0.06 Å (Table 1) [89, 152, 171]. Two of the Cu–S considerably exceed the other two in length (0.05 Å).

Stability constant determinations show that the axial Cu–$OClO_3^-$ interactions are important in solution as well [36]. Perchlorate and other "non-coordinating" anions such as BF_4^- and $CF_3SO_3^-$ increase the formation constants of Cu(II)-thioether complexes in aqueous solution [36, 37]. Gorewit and Musker had previously noted that non-coordinating counterions give square planar Cu(II) complexes, while potentially coordinating ones such as chloride apparently yield binuclear complexes or simple adducts [184].

Electrochemical studies show that 16S4 yields the highest Cu(II/I) potential of any tetradentate crown thioether nS4 (n = 12–16) [143, 170]. The high redox potential of Cu complexes of large ring crown thioethers results both from the ability of the propyl linkages to span the coordination sites of tetrahedral Cu(I) and especially from the decreased ligand field in the Cu(II) complex.

3.4.5.2 Molybdenum

Treatment of 16S4 with $Mo_2(OAc)_4$ in the presence of CF_3SO_3H cleaves the Mo–Mo quadruple bond to yield a variety of Mo(II) and Mo(IV)-containing products [185]. DeSimone and co-workers isolated three of them and characterized them by structural and other means.

$$[Mo_2(CF_3SO_3)_2(OH_2)_4]^{2+} + 16S4 \xrightarrow{EtOH} sym\text{-}[Mo^{II}(SH)(16S4)]_2^{2+} \quad (35a)$$

$$+ [Mo^{IV}(OEt)(16S4)\text{–}O\text{–}Mo^{IV}(O)(16S4)]^{3+} \quad (35b)$$

$$+ [Mo^{IV}(O)(SH)(16S4)]^{+} \quad (35c)$$

$$\tfrac{1}{2}\{[MoBr_2(CO)_4]\}_2 + Me_8\text{-}16S4 \xrightarrow{toluene} [MoBr_2(Me_8\text{-}16S4)] \quad (35d)$$

$$[MoBr_2(Me_8\text{-}16S4)] + CO + Na \xrightarrow{THF} [Mo(CO)_2(Me_8\text{-}16S4)] \quad (35e)$$

$$\tfrac{1}{2}\{[MoCl_2(CO)_4]\}_2 + Me_8\text{-}16S4 \xrightarrow{toluene} fac\text{-}[MoCl_3(Me_8\text{-}16S4)] \quad (35f)$$

$$fac\text{-}[MoCl_3(Me_8\text{-}16S4)] + Zn \xrightarrow{CH_2Cl_2} [MoCl(Me_8\text{-}16S4)] \quad (35g)$$

Several curious features characterize the orange diamagnetic *sym*-$[Mo_2(SH)_2(16S4)_2]^{2+}$ cation (Eq. 35a). First, in this centrosymmetric complex each Mo(II) binds to a hydrosulfide (^-SH) group, which apparently results from cleavage of the ligand during the reaction. The absence of halide-containing materials in the preparation militates against the more intuitively more appealing possibility that the terminal group is in fact Cl^- instead of ^-SH. (X-ray diffraction cannot easily distinguish between S and Cl because of their similar scattering power for X-rays.) Nevertheless, in the absence of elemental analytical data the possibility cannot be discounted entirely.

In addition to the terminal hydrosulfide group, *sym*-$[Mo_2(SH)_2(16S4)_2]^{2+}$ is also unusual in possessing bridging thioether groups. Few other examples have been reported. In each {Mo(16S4)} unit all four sulfur atoms coordinate to the chelated Mo(II); in addition, one of them also coordinates to the Mo(II) ion in the other {Mo(16S4)} unit. Thus each dimeric cation features a rectangular Mo · · · S · · · Mo · · · S core. Surprisingly, bridging Mo–thioether distances are much shorter (2.320 (1) and 2.380 (1) Å) than non-bridging ones (average values 2.49 Å). The ligands coordinate in *syn* fashion, in which diametrically opposite six-membered Mo–S–CH_2–CH_2–CH_2–S rings lie on the same side of the MS_4 plane.

The second product, the blue diamagnetic $[(OEt)(16S4)Mo^{IV}\text{–}O\text{–}Mo^{IV}(O)(16S4)]^{3+}$ cation (Eq. 35b), contains two octahedral Mo(IV) ions bridged by an oxo group (∠ Mo–O–Mo 177°) [185, 186]. The two Mo ions lie out of their respective

S_4 planes (by 0.10 Å and 0.30 Å for the Mo(OEt) and Mo(O) fragments, respectively) so as to increase the distance between them.

The third product, $[Mo^{IV}(O)(SH)(16S4)]^+$ (Eq. 35c), contains a red-brown diamagnetic mononuclear cation that apparently results from decomposition of *sym*-$[Mo_2(SH)_2(16S4)_2]^{2+}$. Like its progenitor, it too contains a coordinated hydrosulfide (^-SH) group. As discussed above, this assignment must be viewed with caution. Interaction with the *trans* oxo group (*trans* influence) slightly lengthens the Mo–SH bond relative to that in the dimeric analogue.

In all three complexes the 16S4 units adopt *syn* stereochemistry. In the two dimeric complexes excessive steric crowding probably precludes adoption of the *anti* form. No such consideration constrains the ligand conformation in the monomeric $[Mo^{IV}(O)(SH)(16S4)]^+$ ion, however.

Of the three complexes only the first, *sym*-$[Mo_2(SH)_2(16S4)_2]^{2+}$, shows even moderately reversible cyclic voltammetric behavior (Table II); the other two complexes show totally irreversible redox processes.

Yoshida and coworkers [187] found that reaction of $[MoX_2(CO)_4]_2$ (X = Br) with Me_8-16S4 in refluxing toluene gives $[MoBr_2(Me_8\text{-}16S4)]$ as paramagnetic orange crystals (Eq. 36a). A similar reaction carried out with X = Cl does not give the chloro analogue, but rather *fac*-$[MoCl_3(Me_8\text{-}16S4)]$ (also paramagnetic and orange) (Eq. 36b). The chloro analogue *trans*-$[MoCl_2(Me_8\text{-}16S4)]$ can be isolated as paramagnetic yellow crystals following reduction of *fac*-$[MoCl_3(Me_8\text{-}16S4)]$ with Zn in CH_2Cl_2 (Eq. 36c). Reduction (chemically or electrochemically) of *trans*-$[MoBr_2(16S4)]$ under an atmosphere of CO gives *trans*-$[Mo(CO)_2(16S4)]$, as the *syn* stereoisomer (Eq. 36d). All of these compounds have high redox potentials, consistent with the ubiquitous stabilization of lower oxidation states observed for thioether complexes.

$$[MoBr_2(CO)_4]_2 + Me_8\text{-}16S4 \xrightarrow{\text{toluene}} trans\text{-}[MoBr_2(Me_8\text{-}16S4)] \quad (36a)$$

$$[MoCl_2(CO)_4]_2 + Me_8\text{-}16S4 \xrightarrow{\text{toluene}} fac\text{-}[MoCl_3(Me_8\text{-}16S4)] \quad (36b)$$

$$fac\text{-}[MoCl_3(Me_8\text{-}16S4)] + Zn \xrightarrow{CH_2Cl_2} trans\text{-}[MoCl_2(Me_8\text{-}16S4)] \quad (36c)$$

$$trans\text{-}[MoBr_2(Me_8\text{-}16S4)] + 2e^- \xrightarrow{CO} trans\text{-}[Mo(CO)_2(Me_8\text{-}16S4)] \quad (36d)$$

$$trans\text{-}[MoBr_2(Me_8\text{-}16S4)] + Na/Hg \xrightarrow{N_2} trans\text{-}[Mo(N_2)_2(Me_8\text{-}16S4)] \quad (36e)$$

$$trans\text{-}[Mo(N_2)_2(Me_8\text{-}16S4)] + MeBr \xrightarrow{\text{benzene}} trans\text{-}[Mo(N_2Me_2)(Me_8\text{-}16S4)]^+ \quad (36f)$$

In an exciting recent development Yoshida and coworkers have also found that a Mo(0)–crown thioether complex binds N_2 [188]. Reduction of *trans*-$[MoBr_2(Me_8\text{-}16S4)]$ with 40% Na amalgam under N_2 gives *trans*-$[Mo(N_2)_2$-

$(Me_8$-16S4)] as orange red crystals (Eq. 36e). This nearly octahedral complex of *syn* stereochemistry features the four S atoms in equatorial positions and two axially end-on bound N_2 molecules (Table 1). Molybdenum–sulfur distances are unexpectedly short (compared to other Mo(0)–thioether complexes), presumably because of the limited size of the macrocyclic ring. Perhaps as an additional consequence of this cation-cavity size mismatch the Mo atom moves out of the S_4 toward the C atoms (i.e., toward the more congested side of the molecule) by 0.1 Å.

Both experimental and theoeretical evidence indicates lesser π-acidity (or greater π-donor ability) of the thioether ligand compared to phosphines. Cyclic voltammetric measurements on the irreversible Mo(I/0) couple are consistent with greater electron richness for Mo in this complex than in analogous phosphine complexes. Further evidence supporting this contention comes from the lower ν_{NN} frequencies compared to phosphine analogues. MO calculations indicate that Me_8-16S4 acts as a p_π donor toward the metal center. Last, *trans*-$[Mo(N_2)_2$ $(Me_8$-16S4)] reacts readily with MeBr to afford the dimethylhydrazido complex *trans*-$[Mo(N_2Me_2)(Me_8$-16S4)]$^+$ as brown crystals (Eq. 36f). This facile alkylation clearly indicates relative electron richness of the Mo center – and consequently of the coordinated N_2 molecules – in the parent compound.

3.4.3.5 Mercury

In $[Hg(16S4)(ClO_4)_2]$, unlike $[Hg(14S4)(OH_2)]^{2+}$ [35], Hg(II) lies in the S_4 plane (Table 1). Four S atoms from the macrocycle, as well as one monodentate and one bidentate ClO_4^- ion, combine to complete seven-coordination about Hg (cf. the five-coordinate 14S4 analogue) [189]. The ligand assumes *syn* conformation to generate a puckered S_4 coordination sphere.

16S4 shows greater affinity for Hg(II) (and $HgMe^+$) than 14S4, but not as great as that of open chain tetradentate ligands [14]. The superiority of 16S4 over 14S4 as a ligand for Hg(II) contrasts with the situation for first-row transition metal ions. This contrast arises from the ring size and flexibility of 16S4, which enables it to chelate the large Hg(II) ion more easily. Acyclic ligands enjoy even more conformational freedom than 16S4, and consequently accommodate the large Hg(II) ion still more effectively.

3.4.6 Other Tetradentate Crown Thioethers

In the series $[Cu(nS4)]^{2+}$ (n = 12–16) plus $[Cu(2,5\text{-dithiahexane})_2]^{2+}$ Cu–S distances increase systematically. This trend reflects the conformational mobility associated with successive replacement of by propyl linkages [89].

In $[CuCl((MeO)_8benzo_4$-12S4)] [190] (Table 1) four sulfur atoms from the ligand form the base of a square pyramidal coordination sphere that is capped by an apical Cl ion. Two mutually *cis* S atoms bind more strongly than do the other two, as reflected in the Cu–S distances (which differ by 0.2 Å). "Doming" of the CuS_4Cl coordination sphere places the Cu ion 0.121 Å above the S_4 basal plane.

Two of the four phenylene units lie approximately in the idealized S_4 plane, while the other two lie below it, on the side opposite the Cu ion.

In their early work Rosen and Busch [49] prepared the diamagnetic square-planar cation $[Ni(benzo\text{-}15S4)]^{2+}$ as its tetrafluoroborate salt. They also isolated Ni(II) complexes of 12S4 and 13S4 [51], for which conductivity studies and elemental analysis suggest the presence of dimeric octahedral species $[Ni_2(nS4)_3]^{4+}$. The contrast with $[Ni(14S4)]^{2+}$ derives apparently from the inability of 12S4 and 13S4 to encircle square-planar Ni(II).

Travis and Busch [50] prepared Co(III) complexes of benzo-15S4 by reaction of Co(II) perchlorate with the ligand in $MeNO_2$. Aerial oxidation of the product in the presence of LiX ($X^- = Cl^-, Br^-$) affords (green for $X^- = Cl^-$; green-brown for $X^- = Br^-$) *trans*-$[CoX_2(benzo\text{-}15S4)]^+$ as its perchlorate salt. Note that the *trans* stereochemistry of these complexes (which was assigned from their infrared and electronic spectra) differs from that of the analogous 14S4 complexes, both of which are *cis* (for $X^- = Cl^-, Br^-$). Reaction of rhodium(III) chloride with benzo-15S4 in refluxing EtOH gives *cis*-$[RhCl_2(benzo\text{-}15S4)]Cl$ as a bright yellow crystals.

3.5 Pentadentate Crown Thioethers – 15S5

Relatively little work on this ligand has been reported. In future investigation of reactivity this ligand may well take on considerable importance, since the sixth coordination site is available for interaction with substrate molecules.

To assess the structural factors bearing on the Cu(II/I) self-exchange rate [170] of the 15S5 complexes Rorabacher and coworkers examined the crystal structures of the two halves of the couple [191]. $[Cu(15S5)]^{2+}$ contains a square pyramidal CuS_5 coordination sphere in which the Cu atom lies 0.41 Å above the mean plane of the four equatorial S atoms. The apical S atom coordinates not along the idealized C_4 axis, but rather 12.8° away. This distortion apparently results from the inability of 15S5 to span the five coordination positions without strain. Strain results from three factors: 1) the displacement of the Cu atom from the equatorial plane, 2) the shortness of the C_2 linkages between S atoms, and 3) the greater length of the apical Cu–S bond. The classes of complexes $[M(15S5)]^{n+}$ and $[M(18S6)]^{n+}$ differ significantly in that the former necessarily contain a meridional loop (i.e., one in which three adjacent S atoms lie in a plane containing the metal ion), whereas the latter need not. Meridional coordination of one S_3 loop further tightens the ligating band constricting the metal ion, and thereby exacerbates the difficulty of spanning the five coordination positions. As a further consequence of this "tightness", the apical Cu–S distance exceeds the equatorial ones by only 0.08 Å (cf. 0.27 Å in $[Cu(18S6)]^{2+}$) [93].

Strain-energy calculations [175] show that $[Cu(15S5)]^{2+}$ deviates from the simple pattern of the S_4 ligands. This result parallels earlier experimental results, which reveal anomalous results for both its stability constant and dissociation rate constants with Cu(II) [36, 149]. As suggested earlier [37], this difference probably arises from a more favorable entropic term for complexation, as well as the greater enthalpic contribution from formation of an additional Cu–S bond.

In $[Cu(15S5)]^+$ [191] (Table 1) one S atom fails to coordinate (Cu–S 3.5 Å), which leads to a distorted tetrahedral CuS_4 coordination sphere. Owing to the constraints imposed by the ring, S–Cu–S angles deviate considerably from tetrahedral values. In general, S–Cu–S angles involving S atoms adjacent in the ring (i.e., S–C–C–S–Cu) fall short of 109°, while that in which another S atom intervenes exceed the tetrahedral value. The greatest deviation (136°) occurs for the S–Cu–S angle involving the S atoms flanking the uncoordinated thioether group.

Resonance Raman measurements indicate that the S atoms in $[Cu(15S5)]^{2+}$ migrate rapidly between apical and equatorial positions [192]. Presumably the strain evident in the ring pushes this complex well up the potential curve toward the transition state for rearrangement, and thereby facilitates this stereochemical scrambling.

3.6 *Miscellaneous Ligands*

Travis and Busch [192] isolated 28S8 as a by-product from the synthesis of 14S4. This ligand coordinates to Ni(II) to yield the red low-spin $[Ni_2(28S8)]^{2+}$ cation. Addition of KNCS affords $[Ni_2(NCS)_4(28S8)]$ as a blue-green neutral complex that contains high-spin octahedral Ni(II) (Table 3; Table 5). Both complexes react with water with expulsion of the ligand. Reaction with $[MCl_4]^{2-}$ (M = Pd, Pt) gives $[M_4Cl_8(28S8)]$, where 28S8 functions as a bidentate ligand toward each metal ion [193].

Reaction of α,α′-dibromo-*o*-xylene with ethane- or propanedithiol affords low yields of dibenzo-16S4 and dibenzo-18S4 [194]. Complexes of these ligands with $PdBr_2$ and $PtBr_2$ were prepared and characterized by elemental analysis.

4 Conclusions

Taken together, the work summarized here indicates that thioethers exhibit a marked preference for the lower, "softer" oxidation states. Put another way, they strongly stabilize lower oxidation and spin states of metal ions. They do so by accepting electron density from the metal back into σ* orbitals on the thioether that are of π symmetry with respect to the metal. This delocalization manifests itself not only in the redox properties of thioether complexes, but also in their magnetic and EPR behavior.

In addition to the intrinsic properties of thioethers as ligands, crown thioethers often display peculiar effects caused by incorporation of the donor atoms into a macrocyclic structure. These effects can be subdivided into two groups: 1) those (such as macrocyclic constriction/dilation) that are common to all macrocycles, and 2) those that derive from imposition of a coordination sphere dominated by thioethers (irrespective of whether or not they are incorporated in a macrocycle).

The first group includes the compression of metal coordination spheres observed in, e.g. $[Ni(9S3)_2]^{2+}$ and $[Ni(18S6)]^{2+}$. The second is in many respects more interesting. Use of crown thioethers can afford complexes in which thioethers dominate the coordination sphere. This may induce unusual chemistry not directly because of the crown, but because of the number of thioether groups it imposes on the metal ion. Nevertheless, because such coordination spheres are often unattainable without use of crown thioethers, it is appropriate to attribute the unusual behavior to use of these ligands.

5 Applications and Future Directions

Much of the motivation for investigation of thioether coordination chemistry comes from the potential parallel of these ligands to phosphines. Both phosphines and thioethers possess large polarizable second-row elements as donor groups; accordingly similarities in their chemistry are to be expected. This is an exciting prospect in view of the extensive and industrially important coordination chemistry of phosphines. Despite early interest in the use of thioether-based systems for catalytic hydrogenation of olefins [5, 6], relatively little recent work has been devoted to this application. In part this may be because of the reputation of sulfur-containing compounds as "poisons" for catalysts. This impression springs from use of bulk noble metal catalysts; it obviously lacks relevance to a sulfur-containing catalyst. The use of thioether-based catalytic systems clearly deserves more attention; crown thioethers represent an especially attractive means of imposing and controlling thioether coordination. In this context, recent work by Kellogg and coworkers [194, 195] on the cross-coupling of Grignard reagents catalyzed by Ni(II) complexes of chiral crown thioethers deserves special mention.

In addition to stabilizing lower oxidation states, crown thioethers can also be used to manipulate the coordination geometry of a metal ion. The elegant work of Rorabacher, Ochrymowycz, and coworkers demonstrates the use of closely related crown thioethers to study how coordinative plasticity affects the thermodynamics and kinetics of electron transfer [149, 170]. The same approach could be used with equal profit on fundamental studies on the interrelation of ligand conformation and binding affinity. The importance of such studies transcends crown thioether chemistry, which merely provides ideal systems in which to work out the requisite concepts.

Design and synthesis of ligands for use in analysis (e.g. in sensors), sequestration of toxic metal ions (in both industrial and clinical contexts), specific delivery of metal ions in biological systems (e.g. for therapeutic or radioimaging purposes), hydrometallurgy, and models for metalloenzymes demands fundamental insight into the factors (e.g. ligand conformation, solvation) governing binding affinity. Apart from any direct applications of crown thioethers in any of these areas, investigation of why some of these ligands bind so much better than others will provide information vital to rational design of ligands.

Crown thioethers present obvious opportunities as the basis for the extraction of metal ions. Their preference for soft "b" metals makes them complementary to many of the chelating agents (e.g. EDTA) currently in use. Sevdic and coworkers have reported use of 14S4, 18S6, and 28S8 as extractants for Ag(I) and Hg(II) [196–201]. Similarly, Sekido and coworkers [202–212], and others [213–217], have reported the use of thiacrowns for extraction of silver, copper, and mercury. On the other hand, 14S4 and 16S4 bind $MeHg^+$ weakly, and without a macrocyclic effect [14]. Moyer and coworkers have recently summarized progress in this field [218].

The affinity of crown thioethers for heavy metals – and their antipathy to the biologically important ions, such as Na^+, Ca^{2+}, and Mg^{2+} – suits crown thioethers particularly well for decorporation of toxic metal ions. Similarly, crown thioethers may find potential use in the hydrometallurgical winning of precious metals such as silver, gold, and platinum, or as the basis for ion-selective electrodes [219]. Other potential applications [220] may arise from use of crown thioethers as structural building blocks or capping members (e.g. {M(9S3)}) for synthesis of metal cluster compounds.

Other applications involve the use of crown thioethers as binding groups in ligands that incorporate other types of reactivity. Polythiametallocenophanes, for example, couple the redox activity of the Cp_2M (M = Fe [221, 222], Ru [223–227]) group with the soft metal binding ability of the oligo thioether loop. Similarly, incorporation of an azobenzene linkage into a thiacrown yields a ligand that binds soft metal ions with an affinity that can be modulated by illumination [228]. A third possibility is the potential utility of crown thioethers supported on a polymeric backbone [229, 230]. Such second-generation ligands may prove useful in the specific sequestration of Hg(II) from effluents, for example. Another application centers on use of such chelating resins in the winning of precious metals such as gold, silver, and the platinum metals. Yet another future application might lie in the use of macrocyclic thioethers for chelation and biological delivery (through, e.g. conjugation to monoclonal antibodies) of second- and third-row transition metal radionuclides for either diagnostic or therapeutic purposes. The recent upswing in activity in crown thioethers has laid the foundation for such future developments.

Several properties of crown thioethers commend them for these applications. Their straightforward preparation by general routes, their chemical robustness and inherent achirality (cf. macrocyclic phosphines) all facilitate practical applications of these ligands. In addition, crown thioethers largely ameliorate the relatively weak coordinative properties of acyclic thioethers. Last, crown thioethers neither solvate strongly nor hydrogen bond (cf. amines), and thus greatly simplify calorimetric studies of binding affinity.

Noteworthy by their virtual absence in this review are the lanthanides [158] and actinides, whose crown thioether chemistry is as yet untouched. The propensity of these ligands to stabilize low oxidation states if carried over to the lanthanides engenders the possibility of synthesizing, e.g. complexes of divalent lanthanides (e.g. Sm(II), Eu(II), Yb(II)). While the lanthanides and actinides as class a elements (in their usual oxidation states) exhibit little affinity for soft class b ligands, imposition of a soft coordination sphere may induce greater class b behavior in them. Similarly, rich coordination chemistry for the lower-valent actinides (e.g. U(III)) can be expected. This area represents an exciting frontier of research in crown thioether chemistry.

Acknowledgements. We are grateful to the Petroleum Research Fund, administered by the American Chemical Society, for support. We would also like to thank Drs. Dave Rorabacher, Karl Wieghardt, Heinz-Josef Küppers, Dieter Sellmann and Bruce Moyer for communication to us of unpublished work. One of us (SRC) would also like to thank the Japan Industrial Technology Association for a summer research fellowship, and especially Dr. Yohmei Okuno and the National Chemistry Laboratory for Industry, Tsukuba, Japan, for their hospitality during which this review was prepared. We are also grateful to Dr. Mike Mingos for his encouragement. This work was supported in part by the Petroleum Research Fund (administered by the American Chemical Society) and by the U.S. Department of Energy through DE-AC03-76SF00472.

6 References and Notes

1. Pedersen CJ (1978) In: Izatt RM, Christensen J (eds) Synthetic multidentate macrocyclic compounds, Academic, New York, p 1
2. Murray SG, Hartley FR (1981) Chem. Rev. 81: 365
3. Bradshaw JS, Hui JYK (1974) J. Heterocyclic Chem. 11: 649
4. Christensen JJ, Eatough DJ, Izatt RM (1974) Chem. Rev. 74: 351
5. Chatt J, Leigh GJ, Storace AP: J. Chem. Soc. A 1971: 1380
6. Lemke W, Travis K, Takvoryan N, Busch DH (1976) Adv. Chem. Ser. 150: 358
7. Beinert H (1980) Coord. Chem. Rev. 33: 55
8. Colman PM, Freeman HC, Guss JM, Murata M, Norris VA, Ramshaw JAM, Venkatappa MP, (1978) Nature 272: 319
9. Adman ET, Stenkamp RE, Sieker LC, Jensen LH (1978) J. Mol. Biol. 123: 35
10. Norris GE, Anderson BF, Baker EN (1983) J. Mol. Biol. 165: 501
11. Gerber D, Chongsawangvirod P, Leung AK, Ochrymowycz LA (1977) J. Org. Chem. 42: 2644
12. Cabbiness DK, Margerum DW (1969) J. Amer. Chem. Soc. 91: 6540
13. Margerum DW, Smith GF: J. Chem. Soc., Chem. Commun., 1975: 807
14. Bach RD, Vardhan HB (1986) J. Org. Chem. 51: 1609
15. Livingstone SE (1965) Quart. Rev., Chem. Soc. 19: 386
16. Hung Y, Martin LY, Jackels SC, Tait AM, Busch DH (1977) J. Am. Chem. Soc. 99: 4029
17. Martin LY, Sperati CR, Busch DH (1977) J. Am. Chem. Soc. 99: 2967
18. Martin LY, DeHayes LJ, Zompa LJ, Busch DH (1974) J. Am. Chem. Soc. 96: 4046
19. Abel EW, Bennett MA, Wilkinson G: J. Chem. Soc. 1959: 2323
20. Cotton FA, Zingales F (1962) Inorg. Chem. 1: 145
21. Ainscough EW, Birch EJ, Brodie AM (1976) Inorg. Chim. Acta 20: 187
22. Reinen D, Ozarowski A, Jakob B, Pebler J, Stratemeier H, Wieghardt K, Tolksdorf I (1987) Inorg. Chem. 26: 4010
23. Cooper SR, Rawle SC, Hartman JR, Hintsa EJ, Admans GA (1988) Inorg. Chem. 27: 1209
24. Poon C-K, Kwong S-S, Che C-M, Kan Y-P: J. Chem. Soc., Dalton Trans. 1982: 1457
25. Weiner MA, Lattman M (1978) Inorg. Chem. 17: 1084
26. Baker EN, Larsen NG: J. Chem. Soc., Dalton Trans. 1976: 1769
27. Root MJ, Sullivan BP, Meyer TJ, Deutsch E (1985) Inorg. Chem. 24: 2731
28. Xiao S-X, Trogler WC, Ellis DE, Berkovich-Yellin Z (1983) J. Am. Chem. Soc. 105: 7033
29. Marynick DS (1984) J. Am. Chem. Soc. 106: 4064
30. Orpen AG, Connelly NG: J. Chem. Soc., Chem. Commun., 1985: 1310
31. DeSimone RE, Glick MD (1975) J. Am. Chem. Soc. 97: 942
32. DeSimone RE, Glick MD (1976) J. Coord. Chem. 5: 181

33. DeSimone RE, Tighe TM (1976) J. Inorg. Nucl. Chem. 38: 1623
34. Rawle SC, Admans G, Cooper SR: J. Chem. Soc., Dalton Trans. 1988: 93
35. Alcock NW, Herron N, Moore P: J. Chem. Soc., Dalton Trans. 1978: 394; J. Chem. Soc., Chem. Commun. (1976): 886
36. Young IR, Ochrymowycz LA, Rorabacher DB (1986) Inorg. Chem. 25: 2576
37. Sokol LSWL, Ochrymowycz LA, Rorabacher DB (1981) Inorg. Chem. 20: 3189
38. Nikles DE, Anderson AB, Urbach FL (1983) In: Karlin KD, Zubieta J (eds), Copper coordination chemistry: Biochemical and inorganic perspectives, Adenine Press, Guilderland, New York, p 203
39. Volkov VB, Yatsimirskii KB (1979) Teor. Eksp. Khim. 15: 711; CA 92: 163406a
40. Mansfeld W (1886) Ber. 19: 696
41. Ray PC (1920) J. Am. Chem. Soc. 117: 1090
42. Ray PC (1922) J. Chem. Soc. 121: 1279
43. Ray PC (1923) J. Chem. Soc. 123: 2174
44. Bennett GM (1922) J. Chem. Soc. 121: 2139
45. Bennett GM, Berry WA (1925) J. Chem. Soc. 127: 910
46. Tucker NB, Reid EE (1933) J. Am. Chem. Soc. 55: 775
47. Meadow, JR, Reid EE (1934) J. Am. Chem. Soc. 56: 2177
48. Rosen W, Busch DH: J. Chem. Soc., Chem. Commun. 1969: 148
49. Rosen W, Busch DH (1969) J. Am. Chem. Soc. 91: 4694
50. Travis K, Busch DH (1974) Inorg. Chem. 13: 2591
51. Rosen W, Busch DH (1970) Inorg. Chem. 9: 262
52. Black D St C, McLean IA (1969) Aust. J. Chem. 22: 3961
53. Black D St C, McLean IA (1971) Aust. J. Chem. 24: 1401
54. Ochrymowycz LA (personal communication); Cooper SR (unpublished work)
55. Ochrymowycz LA, Mak C-P, Michna JD (1974) J. Org. Chem. 39: 2079
56. Buter J, Kellogg RM: J. Chem. Soc., Chem. Commun. 1980: 466
57. Buter J, Kellogg RM (1981) J. Org. Chem. 46: 4481; (1987) Org. Synth. 65: 150
58. Wolf RE Jr, Hartman JR, Ochrymowycz LA, Cooper SR (1989) Inorg. Syn. 25: 125
59. Blower PJ, Cooper SR (1987) Inorg. Chem. 26: 2009
60. Dijkstra G, Kruizinga WH, Kellogg RM (1987) J. Org. Chem. 52: 4230
61. Cooper SR (1988) Acc. Chem. Res. 21: 141
62. Setzer WN, Ogle CA, Wilson GS, Glass RS (1983) Inorg. Chem. 22: 266
63. Sellmann D, Zapf LS (1984) Angew. Chem. Int. Ed. Engl. 23: 807
64. Sellmann D, Zapf L (1985) J. Organomet. Chem. 289: 57
65. Fujihara H, Imaoka K, Furukawa N, Oae S: J. Chem. Soc., Perkin Trans. I 1986: 465
66. Weiss T, Klar G (1979) Z. Naturforsch. B Anorg. Chem. 34B: 448
67. Von Deuten K, Hinrichs W, Weiss T, Klar G: J. Chem. Res., Synop. 1985: 52
68. Sellmann D, Frank P (1986) Angew. Chem., Int. Ed. Engl. 25: 1107
69. Sellmann D, Frank P, Knoch FJ (1988) J. Organomet. Chem. 339: 345
70. Chen B, Li Y, Yang J, Peng Q (1984) Huaxue Xuebao 42: 701; CA 101: 183147z
71. Li Y, Wang D, Wu L, Luo S, Yang J (1984) Huaxue Xuebao 42: 313; CA 101: 72076f
72. Aldrich Chemical Co. now sells 9S3, 14S4, 18S6 and other crown thioethers
73. Flint CD, Goodgame M: J. Chem. Soc., A 1968: 2178
74. Küppers H-J, Wieghardt K, Nuber B, Weiss J, Bill E, Trautwein AX (1987) Inorg. Chem. 26: 3762
75. Olmstead MM, Musker WK, Kessler RM (1981) Inorg. Chem. 20: 151
76. Dale J (1980) Isr. J. Chem. 20: 3 and references therein
77. Dale J (1973) Acta Chem. Scand. 27: 1115
78. DeSimone RE, Albright MJ, Kennedy WJ, Ochrymowycz LA (1974) Org. Magn. Res. 6: 583
79. DeSimone RE, Glick MD (1976) J. Am. Chem. Soc. 98: 762
80. Wolf RE Jr, Hartman JR, Storey JME, Foxman BM, Cooper SR (1987) J. Am. Chem. Soc. 109: 4328
81. Glass RS, Wilson GS, Setzer WN (1980) J. Am. Chem. Soc. 102: 5068
82. Ashby MT, Lichtenberger DL (1985) Inorg. Chem. 24: 636
83. Von Deuten K, Klar G (1981) Z. Naturforsch., B: Anorg. Chem., Org. Chem. 36B: 1526
84. Von Deuten K, Kopf J, Klar G (1979) Cryst. Struct. Commun. 8: 569
85. Robinson GH, Sangokoya SA (1988) J. Am. Chem. Soc. 110: 1494
86. Hartman JR, Wolf RE Jr, Foxman BM, Cooper SR (1983) J. Am. Chem. Soc. 105: 131
87. Wolf RE Jr, Kabanos TE, Rawle SC, Cooper SR (manuscript in preparation): Wolf RE Jr, Cooper SR (American Chemical Society, Seattle, March 1983, abstract INOR 173.)
88. For recent summaries of the work of specific groups see: Schröder M (1988) Pure Appl. Chem. 60: 517 and reference 61

89. Pett VB, Diaddario LL Jr, Dockal ER, Corfield PW, Ceccarelli C, Glick MD, Ochrymowycz LA, Rorabacher DB (1983) Inorg. Chem. 22: 3661
90. Even 9S3, however, which is conformationally well-suited for chelation, fails to displace strongly bound-anions (e.g. in [AgCl(9S3)]). The phenomenon is particularly common in conformationally ill-suited ligands (e.g. 12S3), which cannot displace even poorly binding anions such as triflate
91. Sawyer DT, Roberts JL Jr (1974) Experimental electrochemistry for chemists, Wiley, New York, p 170
92. Cotton FA, Weaver DL (1965) J. Amer. Chem. Soc. 87: 4189
93. Hartman JR, Cooper SR (1986) J. Amer. Chem. Soc. 108: 1202
94. Küppers H-J (1987) Dissertation, University of Bochum
95. Elias H, Schmidt G, Küppers H-J, Saher M, Wieghardt K, Nuber B, Weiss J (submitted for publication)
96. Wieghardt K, Küppers H-J, Weiss J (1985) Inorg. Chem. 24: 3067
97. Carlin RL, Weissberger E (1964) Inorg. Chem. 3: 611
98. Hartman JR, Hintsa, EJ, Cooper SR: J. Chem. Soc., Chem. Commun. 1984: 386
99. Hartman JR, Hintsa EJ, Cooper SR (1986) J. Amer. Chem. Soc. 108: 1208
100. Wilson GS, Swanson DD, Glass RS (1986) Inorg. Chem. 25: 3827
101. Cotton FA, Wilkinson G, Advanced inorganic chemistry, 4th edn, Wiley, New York, p 772
102. Küppers H-J, Neves A, Pomp C, Ventur D, Wieghardt K, Nuber B, Weiss J (1986) Inorg. Chem. 25: 2400
103. Wieghardt K, Schmidt W, Herrmann W, Küppers HJ (1983) Inorg. Chem. 22: 2953
104. Shannon RD (1976) Acta Crystallogr. A32: 751
105. Küppers H-J, Wieghardt K, Steenken S, Nuber B, Weiss J: Z. Anorg. Allg. Chem. (in press)
106. Wieghardt K, Küppers H-J, Raabe E, Krüger C (1986) Angew. Chem., Int. Ed. 25: 1101
107. Blake AJ, Gould RO, Holder AJ, Hyde TI, Lavery AJ, Odulate MO, Schröder M: J. Chem. Soc., Chem. Commun. 1987: 118
108. Blake AJ, Gould RO, Lavery AJ, Schröder M (1986) Angew. Chem., Int. Ed. Engl. 25: 274
109. Udupa MR, Krebs B (1981) Inorg. Chim. Acta 52: 215
110. Hill NL, Hope H (1974) Inorg. Chem. 13: 2079
111. Zompa LJ, Margulies TN (1980) Inorg. Chim. Acta 45: L263 and references therein
112. Chaudhuri P, Wieghardt K (1987) Prog. Inorg. Chem. 35: 329
113. Glass RS, Reedijk J (unpublished work); Reedijk J (personal communication)
114. Clarkson JA, Yagbasan R, Blower PJ, Rawle SC, Cooper SR: J. Chem. Soc., Chem. Commun. 1987: 950
115. Küppers H-J, Wieghardt K, Tsay Y-H, Krüger C, Nuber B, Weiss J (1987) Angew. Chem., Int. Ed. 26: 575
116. Blower PJ, Clarkson J, Rawle SC, Hartman JR, Wolf RE Jr, Yagbasan R, Bott SG, Cooper SR (1989) Inorg. Chem. 28: 4040
117. Clarkson JA, Yagbasan R, Blower PJ, Cooper SR (1989) J. Chem. Soc., Chem. Commun. 1244
118. Von Deuten K, Kopf J, Klar G (1979) Cryst. Struct. Commun. 8: 721
119. Küppers H-J, Wieghardt K, Nuber B, Weiss J (personal communication)
120. Rawle SC, Cooper SR: J. Chem. Soc., Chem. Commun. 1987: 308
121. Rawle SC, Sewell TJ, Cooper SR (1987) Inorg. Chem. 26: 3769
122. Bell MN, Blake AJ, Schröder M, Küppers H-J, Wieghardt K (1987) Angew. Chem. Int. Ed. Engl. 26: 250
123. Rawle SC, Cooper SR (unpublished work)
124. Wieghardt K, Herrmann W, Köppen M, Jibril I, Huttner G (1984) Z. Naturforsch. B 39: 1335
125. Rawle SC (1988) Ph.D. thesis, University of Oxford
126. Sellmann D, Knoch F, Wronna C (1988) Angew. Chem., Int. Ed. Engl. 27: 691
127. Bell EV, Bennett GM, Hock AL: J. Chem. Soc. 1927: 1803
128. Rawle SC, Yagbasan R, Prout K, Cooper SR (1987) J. Amer. Chem. Soc. 109: 6181
129. Bell MN, Blake AJ, Gould RO, Holder AJ, Hyde TI, Lavery AJ, Reid G, Schröder M (1987) J. Inclusion Phenomena 5: 169
130. Rawle SC, Marchant CA, Yagbasan R, Bott SG, Cooper SR (submitted for publication)
131. Blake AJ, Gould AJ, Hyde TI, Schröder M: J. Chem. Soc., Dalton Trans. 1988: 1861
132. Blake AJ, Holder AJ, Hyde TI, Schröder M: J. Chem. Soc., Chem. Commun. 1987: 987
133. Riley DP, Oliver JD (1983) Inorg. Chem. 22: 3361
134. Kopf J, Von Deuten K, Klar G (1979) Cryst. Struct. Commun. 8: 1011
135. Holder AJ, Schröder M, Stephenson TA (1987) Polyhedron 6: 461
136. Blake AJ, Holder AJ, Hyde TI, Roberts YV, Lavery AJ, Schröder M (1987) J. Organomet. Chem. 323: 261

137. Blake AJ, Holder AJ, Roberts YV, Schröder M (1988) *Acta Crystallogr., Sect. C; Cryst. Struct. Commun.* C44: 360
138. Pomp C, Drüeke S, Küppers H-J, Wieghardt K, Krüger C, Nuber B, Weiss J (1988) Z. Naturforsch. 43B: 299
139. Abel EW, Beer PD, Moss I, Orrell KG, Sik V, Bates PA, Hursthouse MB: J. Chem. Soc., Chem. Commun. 1987: 978; J. Organomet. Chem. (1988) 341: 559
140. Von Deuten K, Klar G (1981) Cryst. Struct. Comm. 10: 757
141. Parker D, Roy PS, Ferguson G, Hunt MM: XXVI International Conference on Coordination Chemistry, Porto, Portugal, August 1988, abstract B4
142. Blower PJ, White JCP, Cooper SR (submitted for publication)
143. Dockal ER, Jones TE, Sokol WF, Engerer RJ, Rorabacher DB, Ochrymowycz LA (1976) J. Am. Chem. Soc. 98: 4322
144. Yoshikawa Y: Chem. Lett. 1978: 109
145. Hay RW, Jeragh B, Lincoln SF, Searle GH (1978) Inorg. Nucl. Chem. Lett. 14: 435
146. Sato S, Saito Y (1975) Acta Cryst. B31: 2456
147. Hay RW, Jeragh B, Lincoln SF, Searle GH (1978) Inorg. Nucl. Chem. Lett. 14: 435
148. Hintsa EJ, Hartman JR, Cooper SR (1983) J. Amer. Chem. Soc. 105: 3738
149. Rorabacher DB, Martin MJ, Koenigbauer MJ, Malik M, Schroeder RR, Endicott JF, Ochyrmowycz LA (1983) In: Karlin KD, Zubieta J (eds) Copper coordination chemistry: Biochemical and inorganic perspectives, Adenine: Guilderland, NY, p 167
150. Gould RO, Lavery AJ, Schröder M: J. Chem. Soc., Chem. Commun. 1985: 1492
151. Dockal ER, Diaddario LL, Glick MD, Rorabacher DB (1977) J. Am. Chem. Soc. 99: 4530
152. Diaddario LL Jr, Dockal ER, Glick MD, Ochrymowycz LA, Rorabacher DB (1985) Inorg. Chem. 24: 356
153. Wolf RE Jr, Cooper SR (unpublished work)
154. Sevdic D, Fekete L (1982) Inorg. Chim. Acta 57: 111
155. Sevdic D, Fekete L (1985) Polyhedron 4: 1371
156. Galesic N, Herceg M, Sevdic D (1986) Acta Cryst., C (Cryst. Struct. Comm.) 42: 565
157. Bell MN, Blake AJ, Schröder M, Stephenson TA: J. Chem. Soc., Chem. Commun. 1986: 471
158. Ciampolini M, Mealli C, Nardi N: J. Chem. Soc., Dalton Trans. 1980: 376
159. Rawle SC, Hartman JR, Watkin DJ, Cooper SR: J. Chem. Soc., Chem. Commun. 1986: 1083
160. Lindoy LF, Busch DH (1971) Preparative inorganic reactions, vol VI, Jolly WL (ed) Wiley, New York, p 1
161. Bosnich B, Poon CK, Tobe ML (1965) Inorg. Chem. 4: 1102
162. Alcock NW, Curson EH, Herron N, Moore P: J. Chem. Soc., Dalton Trans. 1979: 1987
163. Lai T-F, Poon C-K: J. Chem. Soc., Dalton Trans. 1982: 1465
164. Herceg M, Matkovic B, Sevdic D, Matkovic-Cologvic D, Nagl A (1984) Croat. Chem. Acta 57: 609
165. Jones TE, Zimmer LL, Diaddario LL, Rorabacher DB, Ochrymowycz LA (1975) J. Am. Chem. Soc. 97: 7163
166. Davis PH, White LK, Belford RL (1975) Inorg. Chem. 14: 1753
167. Herron N, Howarth OW, Moore P (1976) Inorg. Chim. Acta 20: L43
168. Schrauzer GN, Ho RKY, Murillo RP (1970) J. Amer. Chem. Soc. 92: 3508
169. Schrauzer GN (1972) In: Sulfur research trends Adv. Chem. Ser. 110 p 73, Miller DJ, Wiewiorowski TK (eds), American Chemical Society
170. Rorabacher DB, Bernado MM, Vande Linde AMQ, Leggett GH, Westerby BC, Martin MJ, Ochrymowycz LA (1988) Pure Appl. Chem. 60: 501
171. Glich MD, Gavel DP, Diaddario LL, Rorabacher DB (1976) Inorg. Chem. 15: 1190
172. Recent evidence indicates that monoaza-trithia-cyclotetradecane can accommodate tetrahedral coordination about Cu. Bernado MM (1987) Ph.D. dissertation, Wayne State University, Rorabacher DB (personal communication)
173. Diaddario LL, Zimmer LL, Jones TE, Sokol LSWL, Cruz RB, Yee EL, Ochrymowycz LA, Rorabacher DB (1979) J. Am. Chem. Soc. 101: 3511
174. Pett VB, Leggett GH, Cooper TH, Reed PR, Situmeang D, Ochrymowycz LA, Rorabacher DB (1988) Inorg. Chem. 27: 2164
175. Brubaker GR, Johnson DW (1984) Inorg. Chem. 23: 1591
176. Augustin MA, Yandell JK, Addison AW, Karlin KD (1981) Inorg. Chim. Acta 55: L35
177. Martin MJ, Endicott JF, Ochrymowycz LA, Rorabacher DB (1987) Inorg. Chem. 26: 3012
178. Jones TE, Rorabacher DB, Ochrymowycz LA (1975) J. Am. Chem. Soc. 97: 7485
179. Ferris NS, Woodruff WH, Rorabacher DB, Jones TE, Ochrymowycz LA (1978) J. Am. Chem. Soc. 100: 5939
180. Poon C-K, Che C-M: J. Chem. Soc., Dalton Trans. 1980: 756

181. Poon C-K, Che C-M: J. Chem. Soc., Dalton Trans. 1981: 495
182. Yoshida T, Ueda T, Adachi T, Yamamoto K, Higuchi T: J. Chem. Soc., Chem. Commun. 1985: 1137
183. Robinson GH, Zhang H, Atwood JL (1987) Organometallics 6: 887
184. Gorewit BV, Musker WK (1976) J. Coord. Chem. 5: 67
185. Cragel J Jr, Pett VB, Glick MD, DeSimone RE (1978) Inorg. Chem. 17: 2885
186. DeSimone RE, Cragel J Jr, Ilsley WH, Glick MD (1979) J. Coord. Chem. 9: 167
187. Yoshida T, Adachi T, Ueda T, Watanabe M, Kaminaka M, Higuchi T (1987) Angew. Chem., Int. Ed. Engl. 26: 1171
188. Yoshida T, Adachi T, Kaminaka M, Ueda T (1988) J. Am. Chem. Soc. 110: 4872
189. Jones TE, Sokol LSWL, Rorabacher DB, Glick MD: J. Chem. Soc., Chem. Commun. 1979: 140
190. Von Deuten K, Klar G (1981) Cryst. Struct. Comm. 10: 765
191. Corfield PWR, Ceccarelli C, Glick MD, Moy IW-Y, Ochrymowycz LA, Rorabacher DB (1985) J. Am. Chem. Soc. 107: 2399
192. Travis K, Busch DH: J. Chem. Soc., Chem. Commun. 1970: 1041
193. Allen DW, Braunton PN, Millar IT, Tebby JC: J. Chem. Soc., C. 1971: 3454
194. Vriesema BK, Lemaire M, Buter J, Kellogg RM (1986) J. Org. Chem. 51: 5169
195. Kellogg RM (1984) Angew. Chem., Int. Ed. Engl. 23: 782
196. Sevdic D, Jovanovac L, Meider-Gorican H: Mikrochim. Acta 1975: 235
197. Sevdic D, Meider H (1977) J. Inorg. Nucl. Chem. 39: 1403
198. Sevdic D, Meider H (1977) J. Inorg. Nucl. Chem. 39: 1409
199. Sevdic D, Fekete L, Meider H (1980) J. Inorg. Nucl. Chem. 42: 885
200. Sevdic D, Meider H (1981) J. Inorg. Nucl. Chem. 43: 153
201. Sevdic D (1974) Proc. Int. Solvent Extr. Conf. 3: 2733, Jeffreys GV (ed), Soc. Chem. Ind. London; CA 83: 153296q
202. Saito K, Masuda Y, Sekido (1984) Bull. Chem. Soc. Jpn. 57: 189
203. Saito K, Masuda Y, Sekido E (1983) Anal. Chim. Acta 151: 447
204. Sekido E, Saito K, Naganuma Y, Kumazaki H (1985) Anal. Sci. 1: 363
205. Muroi M, Hamaguchi A, Sekido E (1986) Anal. Sci. 2: 351
206. Sekido E, Suzuki K, Hamada K (1987) Anal. Sci. 3: 505
207. Sekido E, Chayama K, Muroi M (1985) Talanta 32: 797
208. Sekido E (1984) Kagaku (Kyoto) 39: 854; CA: 102(20) 178008w
209. Chayama K, Sekido E (1987) Anal. Sci. 3: 535
210. Sekido E, Kawahara H, Tsuji K (1988) Bull. Chem. Soc. Jpn. 61: 1587
211. Sekido E: Bunseki 1987: 161; CA 106: 226387
212. Sekido E, Chayama K: Nippon Kagaku Kaishi 1986: 907; CA 106: 213912
213. Ohki A, Takagi M, Veno K (1984) Anal. Chim. Acta 159: 245
214. Tsukube H, Takagi K, Higashiyama T, Iwachido T, Hayama N (1985) Tet. Lett. 26: 881
215. Oue M, Kimura K, Shono T (1987) Anal. Chim. Acta 194: 293
216. Afzaletdinova NG, Chujakova GR, Murinov YI, Lerman BM, Komissarova NG, Tolstikov GA (1987) Zh. Neorg. Khim. 32: 2757; CA 108: 101689k
217. Zolotov YA, Poddubnykh LP, Dmitrienko SG, Kuz'min NM, Formanovskii AA (1986) Zh. Anal. Khim. 41: 1046; CA 105: 164013u
218. Moyer BA, Westerfield CL, McDowell WJ, Case GN (1988) Sep. Sci. Technol. 23: 000
219. Xi Z, Li J, Yu L, Zhang D, Yang J, Luo S, Wu B, Cun L (1986) Huaxue Xuebao 44: 951; CA 105: 237485m.
220. Goh SH, Lee SY, Seah CL (1988) Thermochem. Acta 126: 149
221. Sato M, Tanaka S, Akabori S, Habata Y (1986) Bull. Chem. Soc. Jpn. 59: 1515
222. Sato M, Akabori S, Katada M, Motoyama I, Sano H: Chem. Lett. 1987: 1847
223. Akabori S, Habata Y, Munegami H, Sato M (1987) Chem. Lett. 4: 1991
224. Akabori S, Munegami H, Habata Y, Sato S, Kawazoe K, Tamura C, Sato M (1985) Bull. Chem. Soc. Jpn. 58: 2185
225. Akabori S, Sato S, Tokuda T, Habata Y, Kawazoe K, Tamura C, Sato M (1986) Bull. Chem. Soc. Jpn. 59: 3189
226. Akabori S, Habata Y, Sato M (1984) Bull. Chem. Soc. Jpn. 57: 68
227. Akabori S, Munegumi H, Sato S, Sato M: J. Organomet. Chem. 272: C54
228. Ammon HL, Bhattacharjee SK, Shinkai S, Honda Y (1984) J. Am. Chem. Soc. 106: 262
229. Oue M, Ishigaki A, Kimura K, Matsui Y, Shono Y (1985) J. Polym. Sci., Polym. Chem. Ed. 23: 2033
230. Tomoi M, Abe O, Takasu N, Kakiuchi H (1983) Makromol. Chem. 184: 2431
231. Veda T, Yamanaka H, Adachi T, Yoshida T (1988) Chem. Lett. 525

Hybridization Schemes for Co-ordination and Organometallic Compounds

D. Michael P. Mingos and Lin Zhenyang

Inorganic Chemistry Laboratory, University of Oxford, South Parks Road, Oxford OX1 3QR, United Kingdom

The important hybridization schemes for co-ordination and organometallic compounds have been derived using a methodology based on a spherical harmonic expansion of the hybridized orbitals. For spherical co-ordination compounds ML_n it is possible to define a set of n hybrids which have their maxima in the metal-ligand directions and 9-n *d* orbitals or hybrids which have nodes along the metal-ligand bonds. The latter are important for π-bonding to the ligands.

Structure and Bonding 72

1 Introduction

The first quantum mechanical theory of the electron pair bond was developed by Heitler and London for the hydrogen molecule [1]. Their *valence bond* method retained a valuable connection with the classical localized electron pair bonding description of the chemical bond proposed by Lewis. This *valence bond* approach was developed and popularized by Pauling in the 1930s [2]. It proved to be particularly effective for describing the stereochemistries of simple organic and inorganic molecules, and in Pauling's hands these geometric conclusions were brilliantly extended even to complex and biologically important molecules [3]. The *valence bond* approach has lost some ground relative to the molecular orbital method since the 1960s [4]. The magnetic and electronic properties of transition metal complexes were not always satisfactorily accounted for using *valence bond theory* and this led to a popularization of crystal and ligand field theories [5]. Similarly the application of *valence bond* theory to aromatic organic and highly delocalized organometallic molecules such as ferrocene required either subtle additional knowledge concerning details about the Hamiltonian integrals or the consideration of a very large number of canonical forms. Interestingly, a recent spin-coupled valence-bond treatment of the benzene molecule has led to the conclusion that the Kekulé description of the molecule may be more accurate than a description in terms of the delocalized molecular orbitals [6].

Although much of the current thinking in co-ordination and organometallic chemistry is couched in molecular orbital terms and geometric conclusions are derived from Walsh type analyses [7], it is apparent that many of the conclusions could also have been derived within a localized bonding framework. For example, the *isolobal* relationships [8] between transition metal and main group fragments can be derived using octahedral hybrids for $M(CO)_n$ fragments and tetrahedral hybrids for EH_n fragments (M = a transition metal and E = a main group atom). In addition the conformational preferences for $M(CO)_n(\eta$-polyene) ($n = 2, 3, 4$) with 18 and 16 valence electrons can be rationalized using a localized bonding description [9]. The aim of this paper is to simplify the derivation of hybridization schemes for transition metal complexes and thereby provide a link between the valence bond and molecular orbital approaches. The fact that the qualitative conclusions derived from both theories depend on the properties of spherical harmonic functions to describe the angular properties of the wavefunctions ensures a close correspondence between the two methodologies [10].

The hybridization scheme in *valence bond* theory is a very useful concept for chemists since it permits a localized view of the bonding. The most general method for generating hybridized orbitals is based on defining a bond wavefunction (a linear combination of atomic orbitals) in a specific bond direction (usually the z-axis direction). Then the second and subsequent hybrids are obtained by a rotation transformation. Orthogonality conditions are then used to evaluate the hybrid coefficients. These bond wavefunctions are defined as equivalent because they differ from one another only by a rotation. Generally, the first bond wavefunction is defined as [11]:

$$as + bp_z + cd_{z^2}$$

along the bonding direction. This is the only *s–p–d* combination with cylindrical symmetry about the bond, and is referred to as a cylindrical bond function.

In 1932 Hultgren [12] demonstrated that no more than six equivalent cylindrical bond functions could be defined which are mutually orthogonal for *s*, *p* and *d* valence orbitals. Therefore for 7, 8 and 9 co-ordinate complexes which occur commonly in transition metal chemistry the methodology described above is not applicable. Pauling has described how the directional character of the generalized hybrid orbitals can be utilized to discuss the bond angles in these high co-ordination number compounds [13–17]. Since the metal-ligand bonds in co-ordination compounds are not always symmetry equivalent, e.g. in trigonal bipyramidal and pentagonal bipyramidal complexes, the equatorial and axial bonds cannot be transformed by symmetry operations of the D_{nh} (n = 3 or 5) point groups, this provides an additional difficulty for qualitative valence bond theory. In 1940 Kimball [18] defined the range of alternative hybridization possibilities for co-ordination compounds using group theoretical principles, but his methodology did not provide a chemical basis for defining the optimum hybridization possibilities. In 1960, Murrell [19] developed a general method for constructing the best hybrids based on the principle of maximum overlap. However this method has not been widely applied. In the 1970s, Pauling developed a method for maximizing the bond strengths of hybridized orbitals which do not necessarily have cylindrical symmetry [13–17]. He applied the results to high coordination number complexes and metal–metal bonded compounds [13–17]. The history of important developments in the theory of hybrid bond orbitals and its application to valence-bond theory have been reviewed by Herman [10] and most recently by Yang [20]. The methodologies developed previously were either based on the maximizing bond strengths [17, 21] or overlap with the ligands [19].

In this paper, an alternative general methodology based on the spherical harmonic expansion of hybridized orbitals is developed to generate hybridized orbitals for most of the situations commonly encountered in co-ordination and main group chemistry. The method developed in this paper is based on the transformation properties of hybrids and the atomic orbitals from which they are derived. The angular information is slightly different from those obtained by Pauling and his co-workers, because a different criterion for evaluating the best hybrids is used.

2 Equivalent Hybridized Orbitals in ML_n Compounds

In the majority of coordination compounds, ML_n, the ligands, L, are distributed evenly on the surface of a spherical shell. Consequently the hybridized orbitals forming the M–L bonds in the bonding directions are equivalent to each other. The hy1, hy2, . . . , hyn are assigned to represent the n hybrids in the spherical ML_n compounds. The symmetry-adapted linear combinations of these hyi hybrids can be expressed in terms of the following spherical harmonic expansion:

$$\psi_{lm}(hy) = N' \sum_i C_{lm}(\theta_i, \phi_i) hyi \quad (1)$$

$$l = 0, 1, 2, \ldots$$

$$m = 0, 1c, 1s, 2c, 2s, \ldots$$

where θ_i and ϕ_i represent the bonding directions in the spherical polar coordinates and N′ is a normalizing constant, and C_{lm} is the modified spherical harmonic wavefunction, $C_{lm} = [4\pi/(2l + 1)]^{1/2} Y_{lm}$. The modified functions are given in Table 1. The linear combinations $\psi_{lm}(hy)$ can be labelled analogously to atomic wavefunctions, i.e. $l = 0$, $m = 0$, *s*; $l = 1$, $m = 0$, p_z; etc. In a perfectly spherical situation, S(hy), P(hy), D(hy), . . . , etc., belong to different irreducible representations. Comparing them with the *s*, *p*, *d*, . . . atomic orbitals on the central atom, there is a one-to-one mapping between $\psi_{lm}(hy)$ from Eq. (1) and ϕ_{lm} of the central atom. Since the hyi is a linear combination of atomic orbitals on the central atom, the spherical harmonic expansion in Eq. (1) has the following important implication: the atomic orbitals (*s*, *p*, *d*, . . .) on the central atom can be expanded as a linear combination of hybridized orbitals in terms of spherical harmonic functions. This becomes highly significant when we want to generate the hybridized orbitals. For example, for a tetrahedral molecule, CH_4(**1**), the following equations

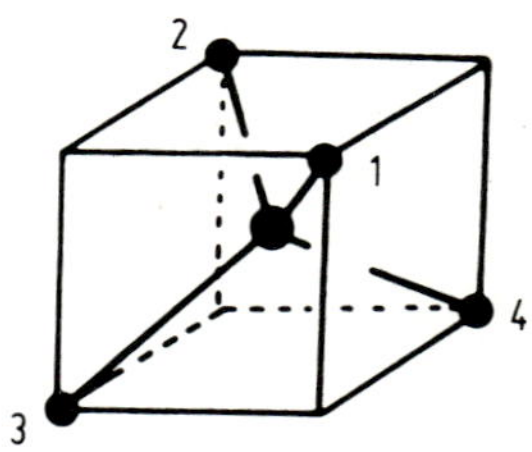

(**1**)

Table 1. Polar forms of the modified spherical harmonic functions

C_{lm}	Polar form	C_{lm}	Polar form
C_{00}	1	C_{30}	$1/2(5\cos^3\theta - 3\cos\theta)$
C_{10}	$\cos\theta$	C^c_{31}	$(3/8)^{1/2}\sin\theta(5\cos^2\theta - 1)\cos\phi$
C^c_{11}	$\sin\theta\cos\phi$	C^s_{31}	$(3/8)^{1/2}\sin\theta(5\cos^2\theta - 1)\sin\phi$
C^s_{11}	$\sin\theta\sin\phi$	C^c_{32}	$(15/4)^{1/2}\cos\theta\sin^2\theta\cos 2\phi$
C_{20}	$1/2(3\cos^2\theta - 1)$	C^s_{32}	$(15/4)^{1/2}\cos\theta\sin^2\theta\sin 2\phi$
C^c_{21}	$(3)^{1/2}\cos\theta\sin\theta\cos\phi$	C^c_{33}	$(5/8)^{1/2}\sin^3\theta\cos 3\phi$
C^s_{21}	$(3)^{1/2}\cos\theta\sin\theta\sin\phi$	C^s_{33}	$(5/8)^{1/2}\sin^3\theta\sin 3\phi$
C^c_{22}	$(3/4)^{1/2}\sin^2\theta\cos 2\phi$		
C^s_{22}	$(3/4)^{1/2}\sin^2\theta\sin 2\phi$		

can be obtained from Eq. (1) and the C_{lm} spherical harmonics of Table 1. The bond directions are defined by the polar coordinates, θ_i, ϕ_i: (54.73°, 45°), 54.73°, 225°), (125.27°, − 45°) and (125.27°, − 225°), i.e. the z axis is defined down the C_2 axis.

$$\psi_{00}(\text{hy}) = s: \quad 1/2(\text{hy1} + \text{hy2} + \text{hy3} + \text{hy4})$$

$$\psi_{10}(\text{hy}) = p_z: \quad 1/2(\text{hy1} + \text{hy2} - \text{hy3} - \text{hy4})$$

$$\psi_{11c}(\text{hy}) = p_x: \quad 1/2(\text{hy1} - \text{hy2} + \text{hy3} - \text{hy4})$$

$$\psi_{11s}(\text{hy}) = p_y: \quad 1/2(\text{hy1} - \text{hy2} - \text{hy3} + \text{hy4})$$

$$\begin{pmatrix} s \\ p_z \\ p_x \\ p_y \end{pmatrix} = \begin{pmatrix} 0.5 & 0.5 & 0.5 & 0.5 \\ 0.5 & 0.5 & -0.5 & -0.5 \\ 0.5 & -0.5 & 0.5 & -0.5 \\ 0.5 & -0.5 & -0.5 & 0.5 \end{pmatrix} \begin{pmatrix} \text{hy1} \\ \text{hy2} \\ \text{hy3} \\ \text{hy4} \end{pmatrix} \rightarrow \mathbf{A} \begin{pmatrix} \text{hy1} \\ \text{hy2} \\ \text{hy3} \\ \text{hy4} \end{pmatrix}$$

Since the linear combinations generated from Eq. (1) are orthogonal to each other because of the equivalent property, **A** is an Unitary matrix. Thus,

$$\begin{pmatrix} \text{hy1} \\ \text{hy2} \\ \text{hy3} \\ \text{hy4} \end{pmatrix} = (\mathbf{A})^* \begin{pmatrix} s \\ p_z \\ p_x \\ p_y \end{pmatrix}$$

where $(\mathbf{A})^*$ is the transpose of the matrix **A**. Therefore,

$$\begin{aligned} \text{hy1} &= 0.5(s + p_z + p_x + p_y) \\ \text{hy2} &= 0.5(s + p_z - p_x - p_y) \\ \text{hy3} &= 0.5(s - p_z + p_x - p_y) \\ \text{hy4} &= 0.5(s - p_z - p_x + p_y) \end{aligned} \tag{2}$$

When the 3-fold axis is defined as z axis, the following linear expansions can be obtained from Eq. (1):

$$\psi_{00}(\text{hy}) = s: \quad 0.5(\text{hy1} + \text{hy2} + \text{hy3} + \text{hy4})$$

$$\psi_{10}(\text{hy}) = p_z: \quad 0.866(\text{hy1} - 0.333(\text{hy2} + \text{hy3} + \text{hy4}))$$

$$\psi_{11c}(\text{hy}) = p_x: \quad (1/6)^{1/2}(2\text{hy2} - \text{hy3} - \text{hy4})$$

$$\psi_{11s}(\text{hy}) = p_y: \quad 0.707(\text{hy3} - \text{hy4})$$

Therefore,

$$\begin{aligned} \text{hy1} &= 0.5s + 0.866p_z \\ \text{hy2} &= 0.5s - 3^{1/2}/6p_z + 2/6^{1/2}p_x \\ \text{hy3} &= 0.5s - 3^{1/2}/6p_z - 1/6^{1/2}p_x + 0.707p_y \\ \text{hy4} &= 0.5s - 3^{1/2}/6p_z - 1/6^{1/2}p_x - 0.707p_y \end{aligned} \tag{2'}$$

The hybrids in (2) and (2)′ are related by a simple rotation transformation. In each case the hybrid has $1/4\,s$ character and $3/4\,p$ character.

The 6 equivalent hybrids in an octahedral molecule (**2**) provide another example. From Eq. (1), we have

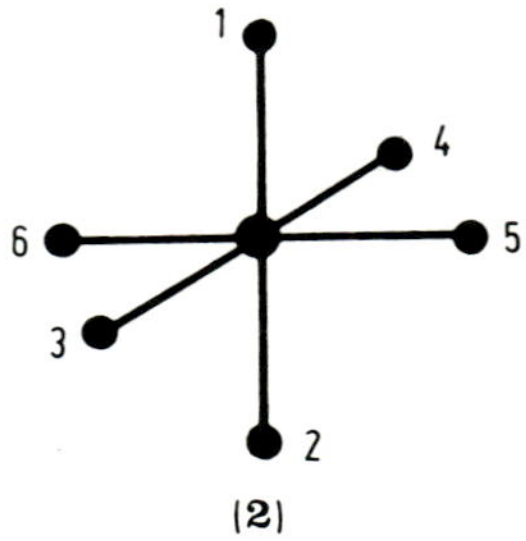

(**2**)

$$\begin{matrix} s: \\ p_z: \\ p_x: \\ p_y: \\ d_{z^2}: \\ d_{x^2-y^2}: \end{matrix} \begin{pmatrix} 0.408 & 0.408 & 0.408 & 0.408 & 0.408 & 0.408 \\ 0.707 & -0.707 & 0.000 & 0.000 & 0.000 & 0.000 \\ 0.000 & 0.000 & 0.707 & -0.707 & 0.000 & 0.000 \\ 0.000 & 0.000 & 0.000 & 0.000 & 0.707 & -0.707 \\ 0.577 & -0.288 & -0.288 & -0.288 & -0.288 & 0.577 \\ 0.000 & 0.000 & 0.500 & 0.500 & -0.500 & -0.500 \end{pmatrix} \begin{pmatrix} hy1 \\ hy2 \\ hy3 \\ hy4 \\ hy5 \\ hy6 \end{pmatrix}$$

[where $(1/6)^{1/2} = 0.408$, $(1/2)^{1/2} = 0.707$, $(1/3)^{1/2} = 0.577$ and $(1/12)^{1/2} = 0.288$]. Following the same procedure as that developed for the tetrahedral sp^3 hybridization scheme:

$$\begin{pmatrix} hy1 \\ hy2 \\ hy3 \\ hy4 \\ hy5 \\ hy6 \end{pmatrix} = \begin{pmatrix} 0.408 & 0.707 & 0.000 & 0.000 & 0.577 & 0.000 \\ 0.408 & -0.707 & 0.000 & 0.000 & 0.577 & 0.000 \\ 0.408 & 0.000 & 0.707 & 0.000 & -0.288 & 0.500 \\ 0.408 & 0.000 & -0.707 & 0.000 & -0.288 & 0.500 \\ 0.408 & 0.000 & 0.000 & 0.707 & -0.288 & -0.500 \\ 0.408 & 0.000 & 0.000 & -0.707 & -0.288 & -0.500 \end{pmatrix} \begin{pmatrix} s \\ p_z \\ p_x \\ p_y \\ d_{z^2} \\ d_{x^2-y^2} \end{pmatrix} \quad (3)$$

The above two examples illustrate a very simple methodology for constructing the hybridized orbitals for spherical molecules where all the atoms are symmetry equivalent.

A more complicated example is the square anti-prismatic structure where the hybrids are no longer cylindrically symmetric. Although it was shown by Hultgren [12] that no more than six equivalent cylindrical hybridized orbitals can be constructed using *s–p–d* hybridization schemes, Racah [22] noted that eight equivalent non-cylindrical hybridized orbitals can be obtained. He utilized the rotation symmetry group to construct the eight equivalent non-cylindrical *s–p–d* hybridized orbitals for the square anti-prism. The methodology developed above provides a similar method for deriving the hybrid orbitals. From Eq. (1) it gives:

$$\begin{pmatrix} s \\ p_z \\ p_x \\ p_y \\ d_{xz} \\ d_{yz} \\ d_{xy} \\ d_{x^2-y^2} \end{pmatrix} = \begin{pmatrix} 0.354 & 0.354 & 0.354 & 0.354 & 0.354 & 0.354 & 0.354 & 0.354 \\ 0.354 & 0.354 & 0.354 & 0.354 & -0.354 & -0.354 & -0.354 & -0.354 \\ 0.500 & 0.000 & -0.500 & 0.000 & 0.354 & -0.354 & -0.354 & 0.354 \\ 0.000 & 0.500 & 0.000 & -0.500 & 0.354 & 0.354 & -0.354 & -0.354 \\ 0.500 & 0.000 & -0.500 & 0.000 & -0.354 & 0.354 & 0.354 & -0.354 \\ 0.000 & 0.500 & 0.000 & -0.500 & -0.354 & -0.354 & 0.354 & 0.354 \\ 0.000 & 0.000 & 0.000 & 0.000 & 0.500 & -0.500 & 0.500 & -0.500 \\ 0.500 & -0.500 & 0.500 & -0.500 & 0.000 & 0.000 & 0.000 & 0.000 \end{pmatrix} \begin{pmatrix} hy1 \\ hy2 \\ hy3 \\ hy4 \\ hy5 \\ hy6 \\ hy7 \\ hy8 \end{pmatrix}$$

[where $(1/8)^{1/2} = 0.354$ and $1/2 = 0.500$].
Therefore,

$$hy1 = 0.354s + 0.354p_z + 0.500p_x + 0.500d_{xz} + 0.500d_{x^2-y^2} \quad (4)$$

The other seven hybrids can also be obtained from the above matrix. The hy1 (4) is illustrated in Fig. 1(c) and its direction of maximum electron density makes an angle of 57.6° with respect to the tetragonal axis. The direction of maximum density is obtained by varying the θ, ϕ in Eq. (4), where the atomic orbitals *s*, *p* and *d* are the angular spherical harmonic functions $Y_{lm}(\theta, \phi)$. When $s - d_{z^2}$ mixing is considered the angle becomes larger and when the ratio of d_{z^2} to *s* coefficients is $-\tan(18°)$ the maximum density occurs at 60.90° in agreement with Racah's calculation [13].

In tetrahedral transition metal complexes, for example $Ni(CO)_4$, the hybridization scheme sp^3 is not obviously the most relevant since the nickel *d* orbitals could also participate in the σ-bonding. Since the d_{xy}, d_{yz} and d_{xz} orbitals have the same transformation properties as the p_z, p_x and p_y orbitals and the corresponding component pairs are:

$$p_z \leftrightarrow d_{xy}$$

$$p_x \leftrightarrow d_{yz}$$

$$p_y \leftrightarrow d_{xz}$$

The alternative extreme sd^3 hybridized orbitals are:

$$hy1 = 0.5s + 0.5(d_{xy} + d_{yz} + d_{xz})$$

$$hy2 = 0.5s + 0.5(d_{xy} - d_{yz} - d_{xz})$$

$$hy3 = 0.5s - 0.5(d_{xy} - d_{yz} + d_{xz})$$

$$hy4 = 0.5s - 0.5(d_{xy} + d_{yz} - d_{xz})$$

Since the *s* and *d* orbitals are both centrosymmetric the resultant sd^3 hybrids point simultaneously at opposite corners of a cube (see Fig. 2). Similarly sd^2 (*s*, d_{xy} and $d_{x^2-y^2}$) and *sd* (sd_{xy} or $sd_{x^2-y^2}$) hybrids point simultaneously at opposite vertices of a hexagon and square.

Clearly the alternative sp^m and sd^m (m = 1, 2 or 3) hybridization schemes represent extreme situations and in a molecule such as $Ni(CO)_4$ an intermediate hybridization scheme is relevant. Therefore a new set of hybrids for describing

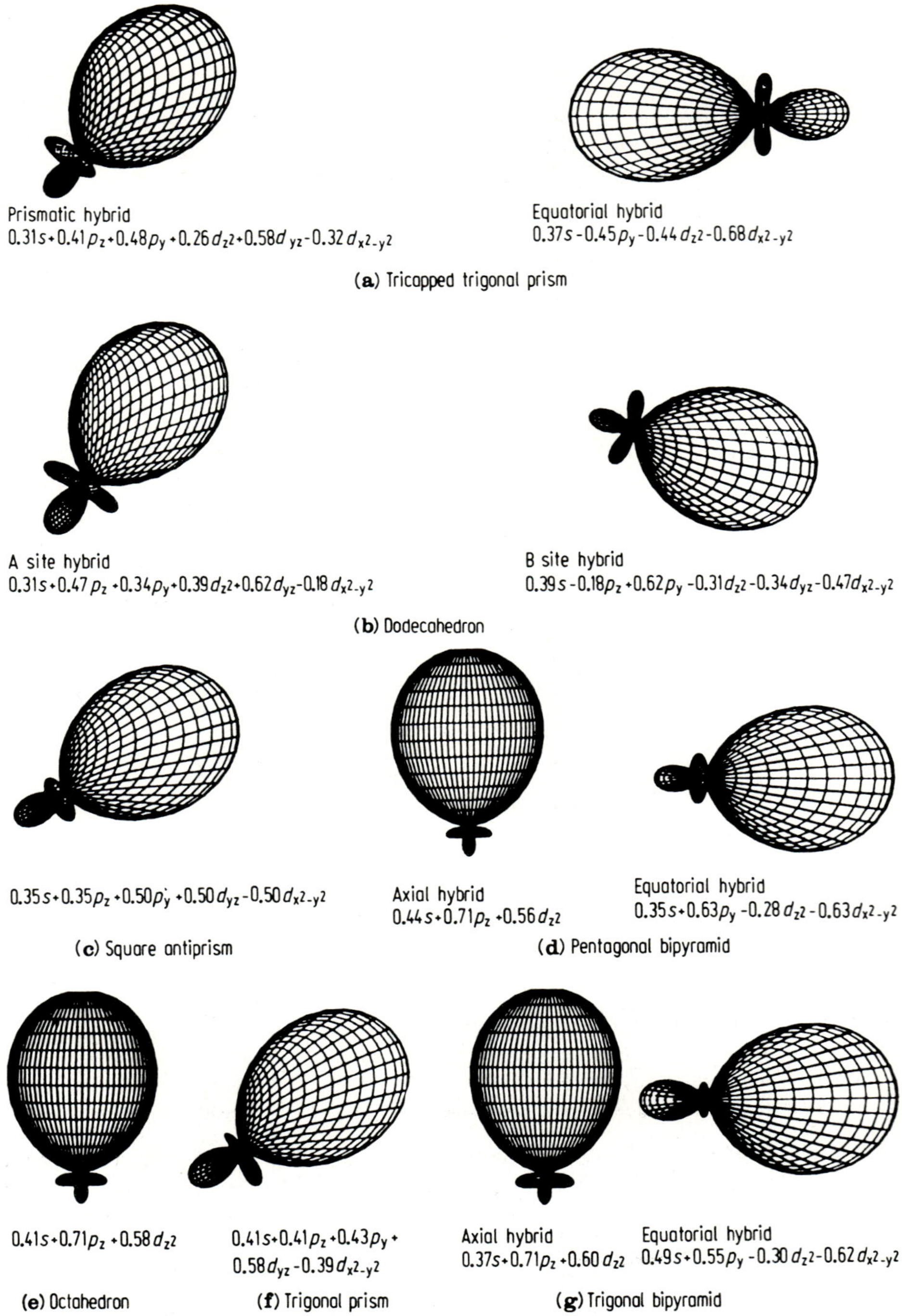

Fig. 1. The optimum hybrids for spherical co-ordination polyhedra

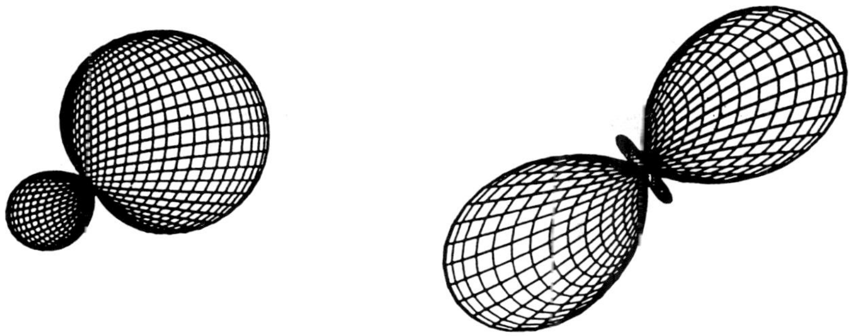

Fig. 2. The comparison between *sp* and *sd* hybrids

a tetrahedral transition metal complex can be obtained from Eq. (2) by replacing

$$p_z \rightarrow \cos A \; p_z + \sin A \; d_{xy}$$

$$p_x \rightarrow \cos A \; p_x + \sin A \; d_{yz}$$

$$p_y \rightarrow \cos A \; p_y + \sin A \; d_{xz}$$

where the cos A and sin A indicate the relative contributions of the *p* and *d* components.

The inclusion of *d* orbitals into the hybridization scheme leads to no change in the maximum directions of the hybrids. The mixing of *d* character into the sp^3 hybrids enhances the directional character of the hybrids towards the four corners of a tetrahedral structure. The complementary (*d–p*) hybrid orbitals are:

$$\sin A \; p_z - \cos A \; d_{xy}$$

$$\sin A \; p_x - \cos A \; d_{yz}$$

$$\sin A \; p_y - \cos A \; d_{xz}$$

These three orbitals maximize their electron densities towards the directions of the four vacant corners of the cube and correspond to a rotation of 45° of the tetrahedral structure along the z axis (see (**1**)). Figure 3 illustrates the nodal characteristics of the p_z–d_{xy} hybrid orbital.

The *dp* mixing therefore creates four hybrids with superior directional characteristics in the tetrahedral bonding directions and three hybrids with inferior bonding characteristics. The latter become completely non-bonding when the

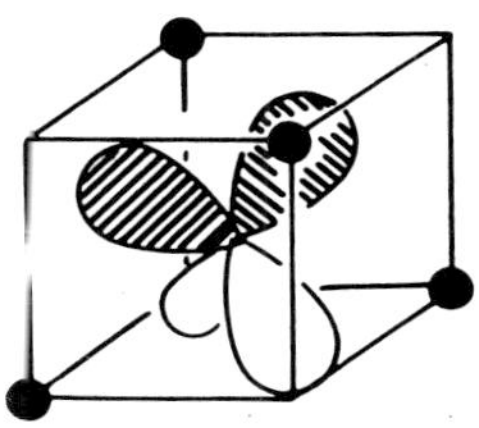

Fig. 3. The illustration of p_z–d_{xy} hybrid orbital

metal–ligand bonds coincide with their nodal planes since the metal–ligand overlap for this mixing coefficient is precisely zero. The following linear combinations:

$$0.790p_z - 0.612d_{xy}$$

$$0.790p_x - 0.612d_{yz}$$

$$0.790p_y - 0.612d_{xz}$$

have zero electron densities in the tetrahedral bond directions. Therefore the following linear combinations:

$$0.612p_z + 0.790d_{xy}$$

$$0.612p_x + 0.790d_{yz}$$

$$0.612p_y + 0.790d_{xz}$$

give maximum bonding in the tetrahedral bond directions.

These optimum $sp^{m-n}d^n$ hybridization schemes are summarized in Table 2 for 3 and 4 co-ordinate complexes. The hybridization for these co-ordination complexes are effectively $s(pd)^m$ (m = 2 or 3) with m *d–p* hybrids pointing away from the bond directions and essentially non-bonding. Although these hybrids are non-bonding as far as the σ-bonding framework they play an important role in metal-ligand π-bonding and this aspect will be discussed in more detail below. In summary, when alternative sd^m and sp^m hybridization schemes are possible, *p–d* mixing between orbitals which have the same symmetry transformation properties ensures the occurrence of optimum hybrids $s(pd)^m$ which maximize the overlap in

Table 2. The optimum $s(pd)^m$ (m = 2 and 3) hybridization schemes for 3 and 4 coordinate complexes

3 Coordination

σ-hybrids

$$hy1 = (1/3)^{1/2}s + (2/3)^{1/2}(0.667p_y - 0.745d_{x^2-y^2})$$

$$hy2 = (1/3)^{1/2}s - (1/6)^{1/2}(0.667p_y - 0.745d_{x^2-y^2}) + 1/2^{1/2}(0.667p_x - 0.745d_{xy})$$

$$hy3 = (1/3)^{1/2}s - (1/6)^{1/2}(0.667p_y - 0.745d_{x^2-y^2}) - 1/2^{1/2}(0.667p_x - 0.745d_{xy})$$

non-bonding hybrids

$$0.745p_y - 0.667d_{x^2-y^2}$$

$$0.745p_x - 0.667d_{xy}$$

4 Coordination

σ-hybrids

$$hy1 = 0.5s + 0.5[0.612(p_z + p_x + p_y) + 0.790(d_{xy} + d_{yz} + d_{xz})]$$

$$hy2 = 0.5s + 0.5[0.612(p_z - p_x - p_y) + 0.790(d_{xy} - d_{yz} - d_{xz})]$$

$$hy3 = 0.5s - 0.5[0.612(p_z - p_x + p_y) + 0.790(d_{xy} - d_{yz} + d_{xz})]$$

$$hy4 = 0.5s - 0.5[0.612(p_z + p_x - p_y) + 0.790(d_{xy} + d_{yz} - d_{xz})]$$

non-bonding hybrids

$$0.790p_z - 0.612d_{xy}$$

$$0.790p_x - 0.612d_{yz}$$

$$0.790p_y - 0.612d_{xz}$$

the bonding directions. The complementary *dp* hybrids are non-bonding because they have nodal planes in the bond directions.

3 Inequivalent Hybridized Orbitals

3.1 Construction of Inequivalent Hybrids

Molecules such as NH_3 and H_2O etc. are described in terms of an inequivalent hybridization scheme based on sp^3 in valence bond theory. The construction of hybridized orbitals in such molecules is different from that developed above. The tetrahedral molecule XAY_3 (**3**) provides a useful starting point. Since the hy1 is distinguished from hy2, hy3 and hy4, the symmetry-adapted linear combinations of these hybrids cannot be generated in terms of the spherical harmonic expansion in Eq. (1). But they can be derived as follows:

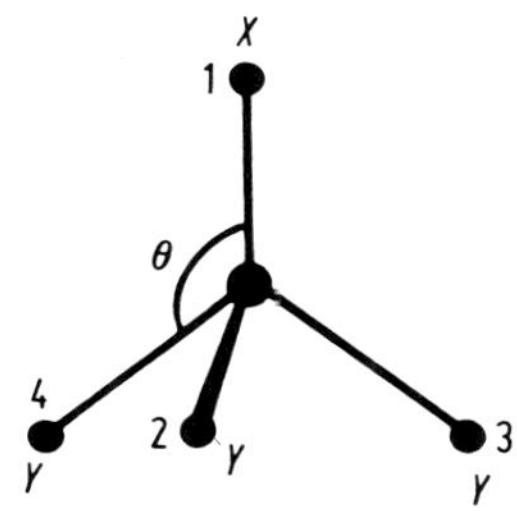

(**3**)

$$s: \quad \sin A\ \mathrm{hy}_1 + \cos A/(3)^{1/2}\ (\mathrm{hy}_2 + \mathrm{hy}_3 + \mathrm{hy}_4)$$

$$p_z: \quad \cos A\ \mathrm{hy}_1 - \sin A/(3)^{1/2}\ (\mathrm{hy}_2 + \mathrm{hy}_3 + \mathrm{hy}_4)$$

$$p_x: \quad 2/(6)^{1/2}\mathrm{hy}_2 - 1/(6)^{1/2}\ (\mathrm{hy}_3 + \mathrm{hy}_4)$$

$$p_y: \quad 1/(2)^{1/2}\ (\mathrm{hy}_3 - \mathrm{hy}_4)$$

The above four linear combinations are orthogonal to each other. Therefore, following the same method as developed above, we get

$$\mathrm{hy1}: \quad \sin A\ s + \cos A\ p_z$$

$$\mathrm{hy2}: \quad \cos A/(3)^{1/2}\ s - \sin A/(3)^{1/2}\ p_z + 2/(6)^{1/2}p_x$$

$$\mathrm{hy3}: \quad \cos A/(3)^{1/2}\ s - \sin A/(3)^{1/2}\ p_z - 1/(6)^{1/2}p_x + 1/(2)^{1/2}p_y$$

$$\mathrm{hy4}: \quad \cos A/(3)^{1/2}\ s - \sin A/(3)^{1/2}\ p_z - 1/(6)^{1/2}p_x - 1/(2)^{1/2}\ p_y$$

The hy2, hy3 and hy4 hybrids are equivalent. The bond angle between two hybrid orbitals can be calculated from the *s* and *p* characters in the hybrid orbital [23],

$$\cos \phi = -C_s^2/C_p^2 \quad (5)$$

where ϕ is the bond angle and C_s^2/C_p^2 is the relative *s*/*p* contribution. We can see how the *s* character in the hy1 orbital changes with the bond angle between the other three equivalent bonds. The results shown below indicate that the *s* character in the hy1 orbital increases with the decrease of bond angle between hy2-hy3-hy4. This illustrates the complementary nature of spherical electron density in these pyramidal molecules. Equation (5) shows that the angle ϕ decreases with the *s* character in the 3 equivalent hybrids. Consequently the *s* character increases in the hy1 hybrid from the normalization condition. In fact Eq. (5) results from the orthogonalization and normalization conditions. Therefore it can be concluded that the complementary nature of spherical electron density is a consequence of these two quantum mechanical principles.

	Orbital nature	Bond angle (°)	*s* character (%) in the hy1 orbital
$A = 0°$	p_z and sp^2 hybrids	120	0
$A = 30°$	sp^3 hybrids	109.47	25
$A = 45°$	1 hybrid + 3 hybrids	101.54	50
$A = 60°$	1 hybrid + 3 hybrids	95.22	75
$A = 90°$	*s* and three *p* orbitals	90	100

The decrease in the bond angles of $NH_3 \rightarrow PH_3 \rightarrow AsH_3$ pyramidal molecules results in an increase of *s* character in the hy1 orbital (non-bonding orbital). This leads to a more stable geometry since the energies of the atomic *s* orbitals relative to the *p* orbitals increase from $N \rightarrow P \rightarrow As$.

Another example is provided by the X_2AY_2 molecule (**4**). The orthogonal symmetry-adapted linear combinations of hy1, hy2, hy3 and hy4 may be written as follows:

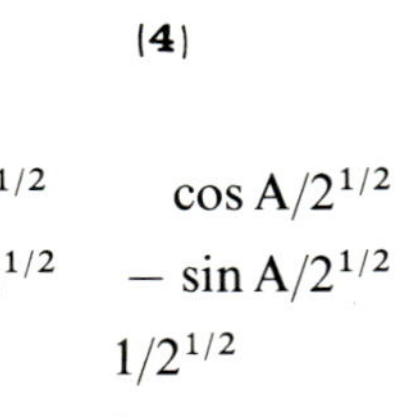

(**4**)

$$\begin{array}{l} s: \\ p_z: \\ p_x: \\ p_y: \end{array} \begin{pmatrix} \sin A/2^{1/2} & \sin A/2^{1/2} & \cos A/2^{1/2} & \cos A/2^{1/2} \\ \cos A/2^{1/2} & \cos A/2^{1/2} & -\sin A/2^{1/2} & -\sin A/2^{1/2} \\ 0 & 0 & 1/2^{1/2} & 1/2^{1/2} \\ 1/2^{1/2} & 1/2^{1/2} & 0 & 0 \end{pmatrix} \begin{pmatrix} \text{hy1} \\ \text{hy2} \\ \text{hy3} \\ \text{hy4} \end{pmatrix}$$

Therefore,

$$\mathrm{hy1} = \sin A/2^{1/2}\, s + \cos A/2^{1/2}\, p_z + 1/2^{1/2}\, p_y$$

$$\mathrm{hy2} = \sin A/2^{1/2}\, s + \cos A/2^{1/2}\, p_z - 1/2^{1/2}\, p_y$$

$$\mathrm{hy3} = \cos A/2^{1/2}\, s - \cos A/2^{1/2}\, p_z + 1/2^{1/2}\, p_x$$

$$\mathrm{hy4} = \cos A/2^{1/2}\, s - \cos A/2^{1/2}\, p_z - 1/2^{1/2}\, p_x$$

From these hybrids it can be seen that the angles of hy1-hy2 and hy3-hy4 are not independent. The bond angles between hy1 and hy2, hy3 and hy4 can be calculated from Eq. (5). The following results are obtained.

	angle (°) between hy1-hy2	*s* % in hy1 (or hy2)	angle (°) between hy3-hy4	*s* % in hy3 (or hy4)
A = 0°	90	0%	180	50%
A = 30°	98.21	12.5%	126.87	37.5%
Á = 45°	109.47	25%	109.47	25%
A = 60°	126.87	37.5%	98.21	12.5%
A = 90°	180	50%	90	0%

The relationship between θ_1 and θ_2 is illustrated in Fig. 4. Figure 4 also shows the experimental data [24] from main group molecules and some late transition metal complexes, for example X_2ZnY_2, which have a core-like d^{10} shell. The agreement between the theoretical results and experimental data is remarkably good.

In summary the normalization and orthogonalization conditions ensure a complementary distribution of electron density. If more *s* electron density is concentrated in one half of a molecule then the reduction in *s* character in the second half

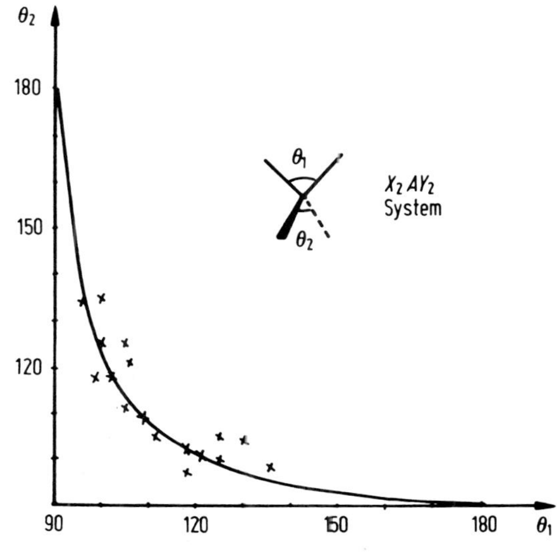

Fig. 4. The relationship between θ_1 and θ_2 in X_2AY_2 system

is compensated for by an increase in p orbital character. These ideas are of course central to the conclusions developed by Bent [25a] to describe the geometries and spectroscopic properties of a wide range of compounds.

In non-spherical coordination compounds, the spherical harmonic expansions of hyi are not orthogonal to each other. This means that there is no longer one-to-one mapping between ψ_{lm}(hy) and ϕ_{lm} atomic orbitals. For example the trigonal bipyramid (**5**) gives rise to S(hy), $P_{0,\pm 1}$(hy) and D_0(hy) linear combinations from (1). The S(hy) and D_0(hy) linear combinations belong to the same a_1' irreducible representation and correspond to admixtures of the s and d_{z^2} atomic orbitals. New linear combinations between S(hy) and D_0(hy) may produce a one-to-one mapping with the s and d_{z^2} atomic orbitals. However, there is no general way to obtain these new linear combinations.

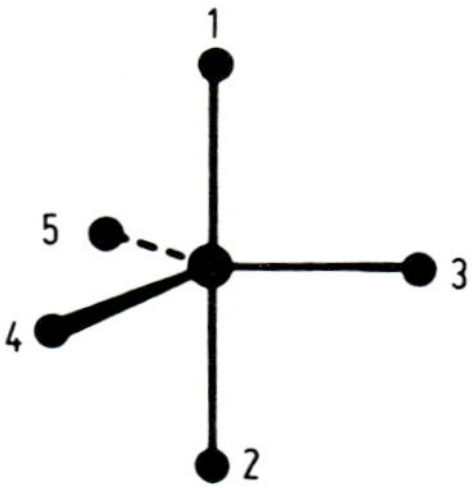

(**5**)

The two examples described above indicate that a set of linear combinations which are orthogonal to each other is needed to generate a set of hybridized orbitals. Since the spherical harmonic expansions can be approximately used for a pseudo-spherical structure, the expansions from (1) can be taken as the first set of linear combinations of hyi. Then we reorthogonalize them to get a new set of orthogonalized linear combinations which correspond to the atomic orbitals on the central atom. The construction of orthogonal functions from a set of n non-orthogonal and linear-independent functions ψ_{lm}(hy) can be achieved by using the Schmidt orthogonalization method [25b]. It is possible to construct a new set of orthogonal functions ψ_i' from the set ψ_{lm}(hy)'s which are assigned as ψ_i by means of the linear transformation:

$$\begin{aligned}
\psi_1' &= \psi_1 \\
\psi_2' &= C_{21}\psi_1 + C_{22}\psi_2 \\
\psi_3' &= C_{31}\psi_1 + C_{32}\psi_2 + C_{33}\psi_3 \\
&\vdots \\
\psi_n' &= C_{n1}\psi_1 + C_{n2}\psi_2 + C_{n3}\psi_3 + \ldots + C_{nn}\psi_n
\end{aligned}$$

The procedure is to choose $\psi_1' = \psi_1$, and first to orthogonalize ψ_2 to ψ_1' to give ψ_2', the ψ_3 to ψ_1' and ψ_2' to give ψ_3', and so on. The general formula is:

$$\psi_i' = \psi_i - \sum_{j=1}^{i-1} \psi_j' \left(\int \psi_j' \psi_i \, d\tau\right) / \left(\int \psi_j' \psi_j' \, d\tau\right) \qquad (i = 2, 3, \ldots n)$$

The resulting orthogonal functions may then be normalized. From the procedure above, it can be seen that the resulting functions depend on which $\psi_{lm}(hy)$ is chosen as ψ_1, which is ψ_2, and so on. Since the hyi's are not equivalent for a non-spherical structure, the S(hy), $(1/n)^{1/2}(hy1 + hy2 + hy3 + \ldots + hyn)$ which is evenly distributed, cannot be chosen as ψ_1. Therefore, $\psi_{lm}(hy)$'s are chosen as ψ_i's in decreasing order of l, and then m.

3.2 Some Representative Examples

In this section, some specific examples will be illustrated using the method developed above. The trigonal bipyramidal structure has the following spherical harmonic expansions:

$$
\begin{array}{lll}
S(hy): & 0.369(hy1 + hy2 + hy3 + hy4 + hy5) & \psi_5 \\
P_0(hy): & 0.707(hy1 - hy2) & \psi_4 \\
P_x(hy): & 0.707(hy4 - hy5) & \psi_3 \\
P_y(hy): & 0.408(2hy3 - hy4 - hy5) & \psi_2 \\
D_0(hy): & 0.301(2hy1 + 2hy2 - hy3 - hy4 - hy5) & \psi_1
\end{array}
$$

[where $(1/5)^{1/2} = 0.369$, $(1/2)^{1/2} = 0.707$, $(1/6)^{1/2} = 0.408$ and $(1/11)^{1/2} = 0.301$]. According to the Schmidt method, the resulting orthogonal functions are:

$$
\begin{pmatrix}
0.369 & 0.369 & 0.492 & 0.492 & 0.492 \\
0.707 & -0.707 & 0.000 & 0.000 & 0.000 \\
0.000 & 0.000 & 0.000 & 0.707 & -0.707 \\
0.000 & 0.000 & 0.816 & -0.408 & -0.408 \\
0.602 & 0.602 & -0.301 & -0.301 & -0.301
\end{pmatrix}
\begin{pmatrix} hy1 \\ hy2 \\ hy3 \\ hy4 \\ hy5 \end{pmatrix}
$$

Therefore,

$$
\begin{pmatrix} hy1 \\ hy2 \\ hy3 \\ hy4 \\ hy5 \end{pmatrix}
=
\begin{pmatrix}
0.369 & 0.707 & 0.000 & 0.000 & 0.602 \\
0.369 & -0.707 & 0.000 & 0.000 & 0.602 \\
0.492 & 0.000 & 0.000 & 0.816 & -0.301 \\
0.492 & 0.000 & 0.707 & 0.408 & -0.301 \\
0.492 & 0.000 & -0.707 & 0.408 & -0.301
\end{pmatrix}
\begin{pmatrix} s \\ p_z \\ p_x \\ p_y \\ d_{z^2} \end{pmatrix}
$$

These hybrids correspond to the conventional description of sp^3d hybridization scheme for a trigonal bipyramid.

Here it should be noted that the resulting orthogonal linear combinations depend greatly on the coordinates chosen since the orthogonalization scheme depends on the order of $\psi_{lm}(hy)$'s. Usually the definition of the principal axis of a molecule as the z axis gives a chemically reasonable result.

In the methodology developed above, it is necessary to know the locations of ligands, i.e. the polar coordinates (θ_i, ϕ_i), to generate the spherical harmonic expansions from (1). Then following the procedures above, we obtain the hybrids. However, the maximum electron densities of these hybrids obtained on the basis of the above methodology may not be located exactly in the direction of the (θ_i, ϕ_i) defined initially. For example an XAY_3 (**3**) molecule belonging to the C_{3v} point group can be used to illustrate the point. Figure 5 illustrates the relationship between the input θ and the calculated θ from the resulting hybrids. It can be seen from the Figure that the greater the deviation from a spherical arrangement, the worse the approximation. It is noteworthy that the overall trend is correct, i.e. the calculated θ increases with the input θ. Therefore, this approximation can be utilized in the discussion of the effect of the variation of *s*, *p* and *d* character of hybrids when the molecular geometry is distorted from spherical.

Following the methodology developed above, calculations on dodecahedron (**6**) and tricapped trigonal prism (**7**) were completed.

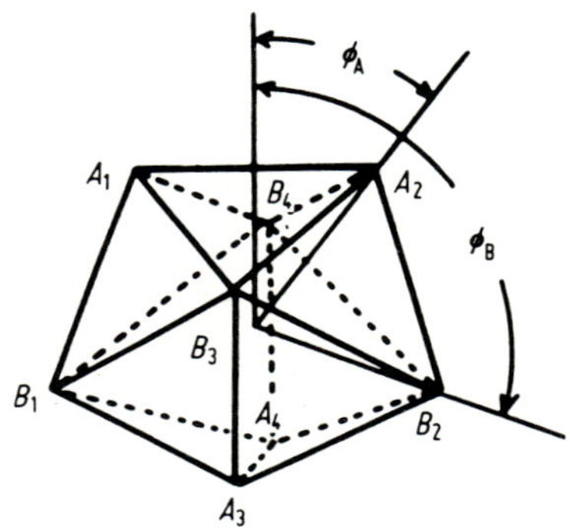

(**6**)

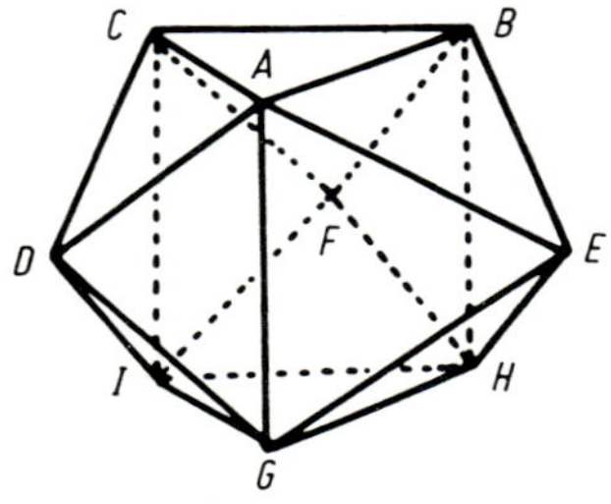

(**7**)

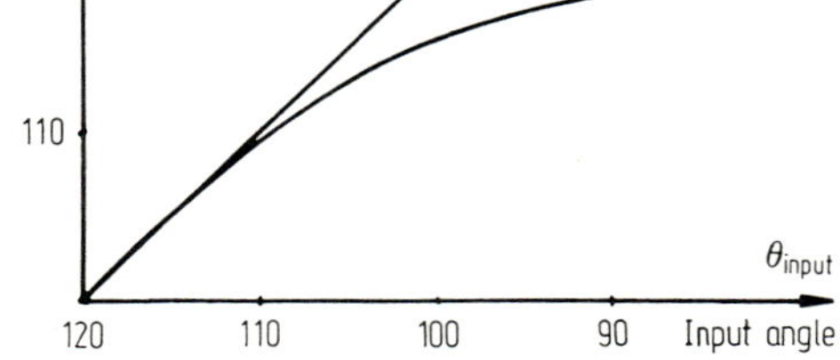

Fig. 5. The relationship between the input θ and the calculated θ from hybrids in XAY_3 system

For a tricapped trigonal prism, (**7**), the resulting hybrids with θ_A(input) = 43.2° are:

$$\text{hyA} = 0.31s + 0.41p_z + 0.48p_x + 0.26d_{z^2} + 0.58d_{xz} + 0.32d_{x^2-y^2}$$

$$\text{hyB} = 0.31s + 0.41p_z + 0.42p_y - 0.24p_x + 0.26d_{z^2} + 0.50d_{yz} - 0.29d_{xz} - 0.28d_{xy} - 0.16d_{x^2-y^2}$$

$$\text{hyC} = 0.31s + 0.41p_z - 0.42p_y - 0.24p_x + 0.26d_{z^2} - 0.50d_{yz} - 0.29d_{xz} + 0.28d_{xy} - 0.16d_{x^2-y^2}$$

$$\text{hyD} = 0.37s - 0.39p_y + 0.23p_x - 0.44d_{z^2} - 0.60d_{xy} - 0.34d_{x^2-y^2}$$

$$\text{hyE} = 0.37s + 0.39p_y + 0.23p_x - 0.44d_{z^2} + 0.60d_{xy} - 0.34d_{x^2-y^2}$$

$$\text{hyF} = 0.37s - 0.45p_x - 0.44d_{z^2} + 0.68d_{x^2-y^2}$$

$$\text{hyG} = 0.31s - 0.41p_z + 0.48p_x + 0.26d_{z^2} + 0.58d_{xz} + 0.32d_{x^2-y^2}$$

$$\text{hyH} = 0.31s - 0.41p_z + 0.42p_y - 0.24p_x + 0.26d_{z^2} - 0.50d_{yz} + 0.29d_{xz} - 0.28d_{xy} - 0.16d_{x^2-y^2}$$

$$\text{hyI} = 0.31s - 0.41p_z - 0.42p_y - 0.24p_x + 0.26d_{z^2} + 0.50d_{yz} + 0.29d_{xz} + 0.28d_{xy} - 0.16d_{x^2-y^2}$$

The maximum electron densities of A, B and C hybrids (or G, H and I) are located exactly in the direction of $\theta = 43.2°$ (or 136.8°). The two different hybrids are illustrated in Fig. 1(a). These angles are based on those determined experimentally in the crystal structure of $[ReH_9]^{2-}$ [17].

For the dodecahedron (**6**), the resulting hybrids with ϕ_A(input) = 36.0° and ϕ_B(input) = 106.0° are:

$$\text{hyA1} = 0.31s + 0.47p_z - 0.34p_y + 0.39d_{z^2} - 0.62d_{yz} - 0.18d_{x^2-y^2}$$

$$\text{hyA2} = 0.31s + 0.47p_z + 0.34p_y + 0.39d_{z^2} + 0.62d_{yz} - 0.18d_{x^2-y^2}$$

$$\text{hyA3} = 0.31s - 0.47p_z + 0.34p_x + 0.39d_{z^2} - 0.62d_{xz} + 0.18d_{x^2-y^2}$$

$$\text{hyA4} = 0.31s - 0.47p_z - 0.34p_y + 0.39d_{z^2} - 0.62d_{xz} + 0.18d_{x^2-y^2}$$

$$\text{hyB1} = 0.39s - 0.18p_z - 0.62p_y - 0.31d_{z^2} + 0.34d_{yz} - 0.47d_{x^2-y^2}$$

$$\text{hyB2} = 0.39s - 0.18p_z + 0.62p_y - 0.31d_{z^2} - 0.34d_{yz} - 0.47d_{x^2-y^2}$$

$$\text{hyB3} = 0.39s + 0.18p_z + 0.62p_y - 0.31d_{z^2} + 0.34d_{xz} + 0.47d_{x^2-y^2}$$

$$\text{hyB4} = 0.39s + 0.18p_z - 0.62p_y - 0.31d_{z^2} - 0.34d_{xz} + 0.47d_{x^2-y^2}$$

The calculated ϕ_A and ϕ_B from the above hybrids are 34.4° and 106.9° which are very close to the input angles. These angles are very close to the 34.55° and 107.22° angles obtained by Racah [22] although the methodologies are very different. The A site and B site hybrids are shown in Fig. 1(b). The total s and p character of the hybrids associated with the A and B sites are 43.26% and 56.89% respectively. This suggests that the stronger σ bonds will be formed between the central atom and B site ligands. This conclusion is in agreement with those based on extended Hückel calculations [27].

4 Mixing Between Alternative Hybridization Schemes

For the trigonal bipyramid the d_{xy} and $d_{x^2-y^2}$ orbitals transform according to the same representations of the D_{3h} point group as p_x and p_y. Therefore the limiting hybridization schemes are either sp^3d or spd^3. The optimum hybrids can be obtained using the same methodology as that developed above for the tetrahedron. The in-phase d–p hybrids

$$\sin A\ p_x + \cos A\ d_{xy}$$

$$\sin A\ p_y + \cos A\ d_{x^2-y^2}$$

are constructed which have nodal planes in the directions of the equatorial ligands. These are calculated to have $\sin A = 0.745$, consequently the optimum hybrids will have the following relative proportions of p and d orbital character:

$$0.667p_x - 0.745d_{xy}$$

$$0.667p_y - 0.745d_{x^2-y^2}$$

From the results in the preceding section, the following hybridized orbitals are obtained for the trigonal bipyramid:

$$\text{hy1} = 0.369s + 0.707p_z + 0.603d_{z^2}$$

$$\text{hy2} = 0.369s - 0.707p_z + 0.603d_{z^2}$$

$$\text{hy3} = 0.492s + 0.816(0.667p_y - 0.745d_{x^2-y^2}) - 0.301d_{z^2}$$

$$\text{hy4} = 0.492s + 0.707(0.667p_x - 0.745d_{xy}) + 0.408(0.667p_y - 0.745d_{x^2-y^2}) - 0.301d_{z^2}$$

$$\text{hy5} = 0.492s - 0.707(0.667p_x - 0.745d_{xy}) + 0.408(0.667p_y - 0.745d_{x^2-y^2}) - 0.301d_{z^2}$$

The methodology developed above has the advantage that the non-bonding and the hybridized σ bonding orbitals are determined simultaneously. As a result, it is very useful in the discussion of structural preferences in transition metal complexes.

The optimum hybridization schemes for coordination compounds are summarized in Table 3. In Fig. 1 the hybridization characteristics of spherical co-ordination polyhedra are summarized together with those for polyhedra with symmetry inequivalent bonds.

Table 3. Valence orbitals for a range of spherical or approximately spherical polyhedral complexes.

Coordination number and geometry	Hybrid schemes	Hybrid non-bonding orbitals	Pure atomic non-bonding orbitals
3 Trigonal plane	$s(pd)^2$	$e' \begin{cases} 0.745p_x + 0.667d_{xy} \\ 0.745p_y + 0.667d_{x^2-y^2} \end{cases}$	$a_1'(d_{z^2})^*$, $a_2''(p_z)$, $e''(d_{xz}, d_{yz})$
4 Tetrahedron	$s(pd)^3$	$t_2 \begin{cases} 0.790p_z + 0.612d_{xy} \\ 0.790p_x + 0.612d_{yz} \\ 0.790p_y + 0.612d_{xz} \end{cases}$	$e(d_{z^2}, d_{x^2-y^2})$
5 Trigonal bipyramid	$spd(pd)^2$	$e' \begin{cases} 0.745p_x + 0.667d_{xy} \\ 0.745p_y + 0.667d_{x^2-y^2} \end{cases}$	$e''(d_{xz}, d_{yz})$
6 Octahedron	sp^3d^2	none	$t_{2g}(d_{xy}, d_{xz}, d_{yz})$
6 Trigonal prism	$spd^2(pd)^2$	$e' \begin{cases} 0.670p_x - 0.740d_{x^2-y^2} \\ 0.670p_y + 0.740d_{xy} \end{cases}$	$a_1'(d_{z^2})$
7 Pentagonal bipyramid	sp^3d^3	none	$e''(d_{xz}, d_{yz})$
8 Square antiprism	sp^3d^4	none	$a_1(d_{z^2})$
8 Dodecahedron	sp^3d^4	none	$b_1(d_{xy})$
9 Tricapped trigonal prism	sp^3d^5	none	none

* For clarity the mixing of s character into d_{z^2} is ignored

5 Hybrids in Non-Spherical Polyhedra

In the previous sections, the hybrids for spherical or nearly spherical polyhedra were discussed. The hybrids for non-spherical polyhedra, such as *nido* and *arachno* structures, can also be constructed in a similar way to that developed above. The construction of hybrids for a square pyramidal structure provides a specific example for illustrating the methodology.

Since the d_{xy}, d_{xz} and d_{yz} orbitals have zero overlap with ligands, they are not involved in the 5 hybrids. The remaining s, p_z, p_x, p_y, d_{z^2} and $d_{x^2-y^2}$ orbitals should be taken into account when the hybrids are constructed. The construction of 5 hybrids from 6 atomic orbitals gives rise to one redundant orbital. This redundant orbital minimizes its overlap with ligands and is referred to as non-bonding orbital. Therefore, it can be defined by having its nodal cones coincident with the M–L bond directions. When the vertex in the $-z$ direction is removed from the octahedron, the non-bonding orbital has the following wave function:

$$as - bp_z + cd_{z^2}$$

The coefficients are determined by inserting $\theta = 0$ (axial ligand) and $\theta = 90°$ (equatorial ligands) into the following equation derived from the spherical harmonic functions (Y_{lm}) given in Table 1:

$$a - b \times 3^{1/2} \cos\theta + c \times 5^{1/2}/2(3\cos^2 \theta - 1) = 0$$

Combining this with the normalization condition, we obtain

$$0.456s - 0.791p_z + 0.408d_{z^2}$$

The other two remaining orthogonal linear combinations of s, p_z and d_{z^2} which are involved in the hybrids can be obtained by the Schmidt orthogonalization method. They are:

$$\phi_1 = 0.667s - 0.745d_{z^2}$$

$$\phi_2 = 0.589s + 0.612p_z + 0.527d_{z^2}$$

For a spherical polyhedron, the linear combinations of hyi's are obtained from Eq. (1), where $C_{l,m}$ are the normalized angular parts of pure atomic wavefunctions listed in Table 1. For a non-spherical polyhedron, the $C_{l,m}$ has to be revised since the atomic orbitals involved in the hybrids are now new linear combinations of atomic orbitals. For example, the hybrids for a square pyramid involve ϕ_1, ϕ_2, p_x, p_y and $d_{x^2-y^2}$ orbitals. Inserting these five new functions into Eq. (1) instead of $C_{l,m}(\theta, \phi)$, we have the following five normalized linear combinations:

	hy(+ z)	hy(+ x)	hy(+ y)	hy(− x)	hy(− y)
ϕ_1	0.316	0.474	0.474	0.474	0.474
ϕ_2	1.000	0.000	0.000	0.000	0.000
p_y	0.000	0.000	0.707	0.000	− 0.707
p_x	0.000	0.707	0.000	− 0.707	0.000
$d_{x^2-y^2}$	0.000	0.500	− 0.500	0.500	− 0.500

where the hy(+ z) indicates the hybrid in the $+z$ direction and similarly for the others. Following the methodology developed for spherical polyhedra, now we reorthogonalize the above matrix from bottom to top by the Schmidt method. The resulting orthogonal matrix is:

$$\begin{pmatrix} 0.000 & 0.500 & 0.500 & 0.500 & 0.500 \\ 1.000 & 0.000 & 0.000 & 0.000 & 0.000 \\ 0.000 & 0.000 & 0.707 & 0.000 & -0.707 \\ 0.000 & 0.707 & 0.000 & -0.707 & 0.000 \\ 0.000 & 0.500 & -0.500 & 0.500 & -0.500 \end{pmatrix}$$

Therefore the hybrids are:

$$\begin{pmatrix} \mathrm{hy}(+z) \\ \mathrm{hy}(+x) \\ \mathrm{hy}(+y) \\ \mathrm{hy}(-x) \\ \mathrm{hy}(-y) \end{pmatrix} = \begin{pmatrix} 0.000 & 1.000 & 0.000 & 0.000 & 0.000 \\ 0.500 & 0.000 & 0.000 & 0.707 & 0.500 \\ 0.500 & 0.000 & 0.707 & 0.000 & -0.500 \\ 0.500 & 0.000 & 0.000 & -0.707 & 0.500 \\ 0.500 & 0.000 & -0.707 & 0.000 & -0.500 \end{pmatrix} \begin{pmatrix} \phi_1 \\ \phi_2 \\ p_y \\ p_x \\ d_{x^2-y^2} \end{pmatrix}$$

$$= \begin{pmatrix} 0.589 & 0.612 & 0.000 & 0.000 & 0.000 & 0.527 \\ 0.333 & 0.000 & 0.000 & 0.707 & 0.500 & -0.373 \\ 0.333 & 0.000 & 0.707 & 0.000 & -0.500 & -0.373 \\ 0.333 & 0.000 & 0.000 & -0.707 & 0.500 & -0.373 \\ 0.333 & 0.000 & -0.707 & 0.000 & -0.500 & -0.373 \end{pmatrix} \begin{pmatrix} s \\ p_z \\ p_y \\ p_x \\ d_{x^2-y^2} \\ d_{z^2} \end{pmatrix}$$

The methodology developed by Murrell [19] is very similar to that described above. Instead of orthogonalizing the above matrix, he obtained new linear combinations of ϕ_1, ϕ_2, p_x, p_y and $d_{x^2-y^2}$ which give the maximum overlap with ligand orbitals. The results for a square pyramid based on his methodology are:

$$\begin{pmatrix} \mathrm{hy}(+z) \\ \mathrm{hy}(+x) \\ \mathrm{hy}(+y) \\ \mathrm{hy}(-x) \\ \mathrm{hy}(-y) \end{pmatrix} = \begin{pmatrix} 0.468 & 0.604 & 0.000 & 0.000 & 0.000 & 0.645 \\ 0.378 & 0.052 & 0.000 & 0.707 & 0.500 & -0.323 \\ 0.378 & 0.052 & 0.707 & 0.000 & -0.500 & -0.323 \\ 0.378 & 0.052 & 0.000 & -0.707 & 0.500 & -0.323 \\ 0.378 & 0.052 & -0.707 & 0.000 & -0.500 & -0.323 \end{pmatrix} \begin{pmatrix} s \\ p_z \\ p_y \\ p_x \\ d_{x^2-y^2} \\ d_{z^2} \end{pmatrix}$$

It can be seen that the results from both methods are very similar. The inclusion of the p_z orbital in the equatorial hybrids from Murrell's method indicates that the equatorial hybrids do not lie exactly in the xy plane.

When the four equatorial ligands are distorted from the xy plane, the mixing between (p_x, p_y) and (d_{xz}, d_{yz}) has to be taken into account as well as that between s, p_z and d_{z^2}. The same routine is used to find new linear combinations of p_x, p_y and d_{xz}, d_{yz} which are involved in the hybrids and then the new revised functions replace the p_x and p_y functions in Eq. (1). Finally the same procedures are applied to generate the hybrids for the distorted square pyramidal structure.

Using the procedures described above the hybrids for any kind of geometry can be generated. The geometric consequences for *nido* and *arachno* structures will be discussed in the next section.

6 Discussion

6.1 Spherical Polyhedra

6.1.1 General Nature of Valence Orbitals

The structures discussed in the sections 2–4 are referred to as spherical polyhedra since their hybridization schemes can be either exactly or very closely described in terms of the spherical harmonic methodology developed above. In summary the orbitals describing these polyhedra based on the *Valence Bond Theory* can be classified into three types:

(1) hybrid σ orbitals, pointing directly towards the ligands;
(2) hybrid non-bonding orbitals, usually *p–d* hybrids whose nodal planes coincide with the metal–ligand bond directions;
(3) pure atomic non-bonding orbitals, with nodal planes in the ligand directions.

The above three types of valence orbitals for a range of spherical polyhedra are summarized in Table 3. The hybrid σ orbitals which are not listed in the Table are shown in Fig. 1 and Table 2. These σ bonding orbitals are usually utilized to form M–L σ bonds. Therefore the occupation of the type (2) and (3) non-bonding orbitals leads to complexes which conform to the 18-electron rule. Since the *p* orbitals have high energies relative to the *s* and *d* orbitals for transition metal atoms either a stabilization of the hybrid non-bonding orbitals through π-interactions with π^* vacant orbitals from ligands, such as CO, CN^- etc., or a small *d–p* promotion energy is required for the stabilization and complete electron occupation of orbitals of type (2). The following examples illustrate this conclusion.

Tetrahedron	$Ni(CO)_4$	d^{10}
Trigonal bipyramid	$Fe(CO)_5$	d^8
Octahedron	$Cr(CO)_6$	d^6
Pentagonal bipyramid	$Mo(CN)_7^{5-}$	d^4
Square antiprism	$H_4W(CN)_8$	d^2
Dodecahedron	$K_4W(CN)_8$	d^2

Deviations from the inert gas rule are a consequence of partial occupations of the hybrid and pure atomic non-bonding orbitals. This often occurs because either the non-bonding orbitals are destabilized by π-donor ligands, such as halide and OR^-, or the metal has a large *d–p* promotion energy, e.g. the late transition metals.

6.1.2 *Site Preferences*

In the complexes where the M–L bonds are not equivalent, such as trigonal and pentagonal bipyramidal structures, the bonding abilities of these inequivalent hybrid σ orbitals depend on their *s*, *p* and *d* character. Since an *s* orbital has a larger overlap integral with ligand σ orbitals than the *p* and *d* orbitals, a greater *s* character in the hybrid σ orbital leads to a stronger bond with a ligand. Therefore the following conclusions can be made from the hybrid wavefunctions shown in Fig. 1.

Structure	Strong bond	Weak bond
Tricapped trigonal prism	Equatorial	Prismatic
Dodecahedron	B site	A site
Pentagonal bipyramid	Axial	Equatorial
Trigonal bipyramid	Equatorial	Axial

These conclusions are generally consistent with those derived from extended Hückel calculations [27–29]. In their calculations on trigonal bipyramidal complexes, Rossi and Hoffmann [28] noted that the axial bonds are stronger than the equatorial bonds when the electron configuration is d^8 for the central transition metal. Checking the experimental data (see Table 1 of Ref. [19]) for d^8 trigonal bipyramidal structures, we cannot find much evidence for such a site preference effect.

6.1.3 *Orientational Preferences for π-Acceptor Ligands Such as Ethylene*

The structure of tris(ethylene)nickel(0) has attracted a lot of interest from experimentalists and theoreticians [30]. A trigonal-planar arrangement (**8**) rather than perpendicular (**9**) has been explained using extended Hückel calculations [30]. In terms of a hybridization model, the σ interactions between Ni and ethylenes are based on the optimum $s(p-d)^2$ hybridization scheme given in Table 3. There are 6 non-bonding orbitals, one pure p_z(empty), two pure d(e″: d_{xz}, d_{yz}), two *d–p* complementary hybrids (e′) and one d_{z^2} (mixing with *s*) orbitals (see Table 3). Since ethylene is a π-acceptor ligand the preferred conformation is that which maximizes the back donation from metal to ethylene π*. Filled *d–p* hybrid non-bonding orbitals function most effectively in this sense since the relative orbital energies nd < (n + 1)s ≪ (n + 1)p ensures that such hybrids have a smaller ionization

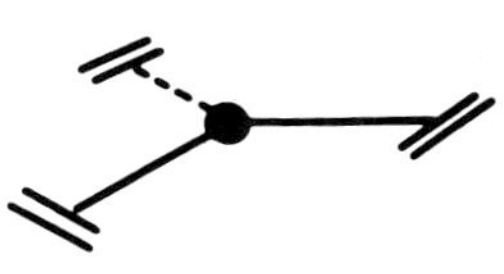

(**8**)

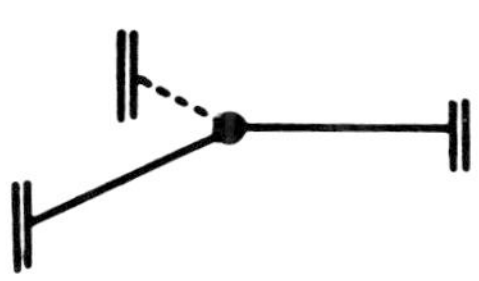

(**9**)

energy than pure *d* orbitals. The back donation interactions for (**8**) are between π^* and e′(*d*–*p* mixings) and for (**9**) π^* and e″ (pure *d* orbitals). A larger stabilization energy can therefore be expected for the trigonal-planar arrangement (**8**).

The X-ray structural studies of Howard, Spencer and Stone have established the platinum tris(olefin) complexes, Pt(olefin)$_3$ [31], do indeed have the anticipated planar arrangement. Furthermore, all the known PtL_2(olefin) [32] and PtL_2 (acetylene) complexes have the planar conformations shown in (**10**) and (**11**).

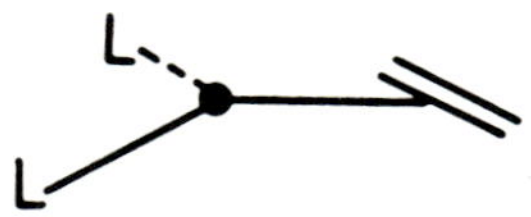

(**10**)

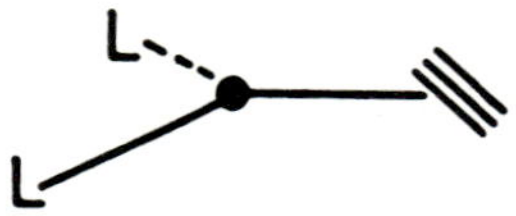

(**11**)

In trigonal bipyramidal ML_5 complexes the optimum $spd(pd)^2$ hybrids define the metal ligand σ-bonds. Two non-bonding *d*–*p* hybrids lie in the equatorial plane and the d_{xz} and d_{yz} orbitals lie perpendicular to this plane and by symmetry are non-bonding (see Table 3). Occupation of the complementary hybrids and the non-bonding *d* orbitals leads to a noble gas configuration for the metal. This of course depends on the absence of a very large *d*–*p* promotion energy. Where this occurs, e.g. platinum(II) and gold(III), the square planar geometry becomes energetically preferred. If one of the ligands in the trigonal bipyramidal complex is a non-cylindrical π-acceptor ligand, such as ethylene, then the more stable conformation is that which maximizes the back donation from the in-plane *d*–*p* hybrid orbitals, i.e. that shown in (**12**).

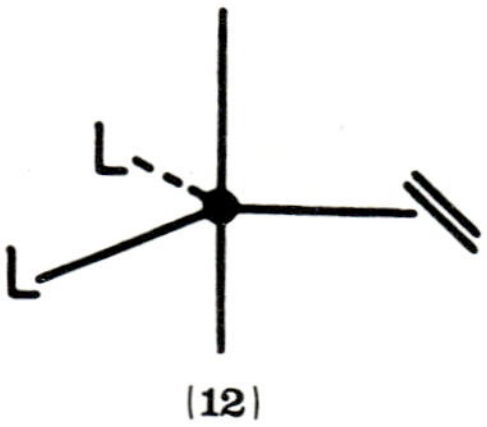

(**12**)

Tetrakis(ethylene)nickel(0), with 18 electrons, has a structure which can be either "quasidodecahedral" (**13**) or "quasicubical" (**14**), both with idealized D_{2d} symmetry. The optimum $s(pd)^3$ hybridization scheme can be used to describe the four σ bond interactions. From Table 3, there remain 5 non-bonding orbitals which include two pure *d*(e: $d_{x^2-y^2}$, d_{z^2}) orbitals and three complementary t_2 *d*–*p* hybrid non-bonding orbitals. The stabilization energy of the t_2 *d*–*p* hybrid non-bonding orbitals by the π^* acceptor of ethylene is more significant than that of the e set of pure *d* orbitals. The three (*d*–*p*) hybrid non-bonding orbitals have b_2 + e symmetries in the D_{2d} point group. The four π^* acceptor orbitals transform as $a_1 + b_2 + e$ for (**13**) and $a_2 + b_1 + e$ for (**14**). Therefore, greater stabilization

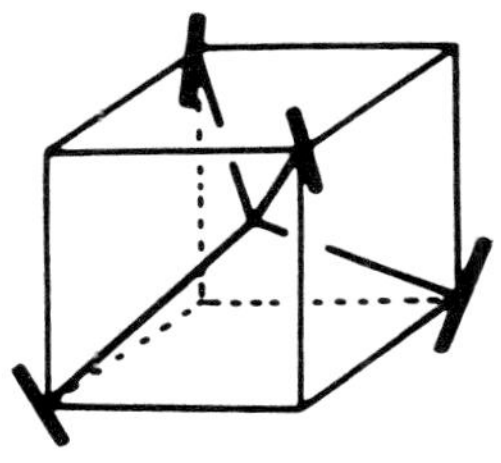

(13)

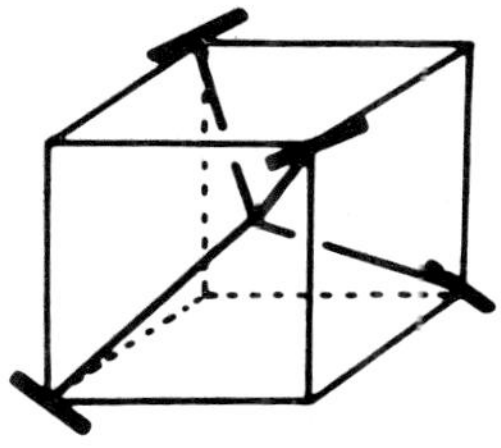

(14)

energy is expected for the "quasidodecahedral" structure (**13**) by symmetry consideration. This conclusion is confirmed by Hoffmann's MO calculations [30]. The "quasidodecahedral" structure of $[Cr(O_2)_4]^{3-}$ [33] can also be understood from the above discussion. For the O_2^{2-} ligands the interaction between the empty (d–p) hybrid non-bonding orbitals and the π-donor orbitals of O_2^{2-} are of primary importance. Therefore the resultant conclusion is the same.

6.1.4 Square Antiprism and Dodecahedron

Since the M–L bonds are equivalent in a square anti-prismatic ML_8 complex, the hybridized orbitals can be constructed as described in Sect. 2. The maximum densities of those hybridized orbitals make an angle of 57.6° with the tetragonal axis. Therefore the sp^3d^4 hybridization scheme with a pure d_{z^2} non-bonding orbital (i.e. d_{z^2} is excluded from the hybridized orbitals) leads to a geometry with a cone angle of 57.6° in square anti-prismatic ML_8 complexes. From the method developed above, we can see that any deformation from the angle of 57.6°, i.e. increase or decrease, results in the mixing of s character into the d_{z^2} non-bonding orbital and the participation of d_{z^2} in the hybridized σ orbitals. Consequently the above argument indicates that the d^2 complexes prefer the structure having an angle close to 57.6° and d^0 complexes tend to have cone angles which deviate from this value. Since a square anti-prism based on equal edge lengths makes an angle of 59.26° with the tetragonal axis, the ligands in the d^0 complexes tend to adopt this spherical arrangement. Specific examples [24] include $H_4W(CN)_8 \cdot 6H_2O$ (d^2, $\theta = 57.6°$), $[Nd(ONC_5H_5)_8]$ $(ClO_3)_3$ (d^2, $\theta = 56.1°$), $[Sr(H_2O)_8]$ $(AgI_2)_2$ (d^0, $\theta = 58.2°$) and Na_3TaF_8 (d^0, $\theta = 59°$) and support the argument developed above. From extended Hückel calculations [27], Hoffmann and coworkers found an energy-minimum at $\theta = 57° (d^2)$ and $\theta = 59° (d^0)$ for ML_8 square antiprismatic complexes.

In the dodecahedral structure ML_8 (7), with D_{2d} symmetry, the non-bonding orbital is a pure d_{xy} (b_1) orbital for all θ_A and θ_B angles. The most favoured structure is the one in which the ligands are evenly distributed on the surface of a spherical shell for both d^0 and d^2 complexes. The hybrids make θ_A and θ_B angles of 34.4° and 106.9° respectively. Extended Hückel calculations [27] have shown that for both d^0 and d^2 systems minimum-energy geometries are located at $\theta_A = 36°$ and $\theta_B = 106°$. Experimental data [24] include $K_4Mo(CN)_8$ (d^2) which has $\theta_A = 36.0°$ and $\theta_B = 107.1°$, and $(Bu_4N)_3Mo(CN)_8$ (d^1) has $\theta_A = 37.2°$ and

$\theta_B = 107.5°$. These results emphasize that the conclusions derived from extended Hückel and the hybridization schemes are almost identical, because they both depend greatly on the nodal properties of the spherical harmonic functions.

6.2 Capped Polyhedra

Coordination polyhedra can be described as mono-, bi- and tri-capped polyhedra if they can be represented as a spherical polyhedron with one, two and three ligands located on faces. Since the methodology developed above is based on the spherical harmonic functions, the capped polyhedra are best described in terms of their parent polyhedra which are spherical or approximately spherical. The capping ligand can be either a σ-acceptor, i.e. L^{2+}, or a σ-donor when the parent polyhedron has empty non-bonding orbitals. The orbital interactions for capped polyhedra arise from the hybrid *d–p* or pure atomic non-bonding orbitals on the central atom and the capping ligand orbital. As indicated above stronger interactions occur for hybrid *d–p* non-bonding orbitals than the pure *d* orbitals. Since the capping ligand orbital interacts primarily with the hybrid and pure *d* non-bonding orbitals which have no *s* character, the following general points can be made:

(1) The metal-capping ligand bonds are weaker than the other metal–ligand bonds because they have less *s* character.
(2) The competition of the capping ligand for *s* character leads to a distortion of the geometry of the parent polyhedron.
(3) Distortion of the parent polyhedron also results in a mixing of *p* character into pure *d* non-bonding orbitals and consequently a destabilization of the non-bonding orbitals.

The majority of capped structures are based on the tetrahedron, the octahedron and the trigonal prism. In the following section these capped polyhedra are discussed in terms of the valence orbitals listed in Table 3.

6.2.1 Capped Tetrahedra

From Table 3 it can be seen that there are three t_2 (*d–p*) hybrid and two e pure *d* non-bonding orbitals for a tetrahedral complex. The discussion in Sect. 2 showed that the three non-bonding t_2 (*d–p*) hybrid orbitals maximize their electron densities towards the triangular faces of the tetrahedron and the $e(d_{x^2-y^2}, d_{z^2})$ orbitals point towards the edges. Therefore the triangular faces are preferred for capping ligands because of the mixing of *p* character into the t_2 non-bonding orbitals. Examples of bicapped tetrahedral structures, $H_2Fe[P(C_6H_5)(OC_2H_5)_2]_4$ and $(AuPR_3)_2[Fe(CO)_4]$, and the mono-capped tetrahedral stuctures, $HCo(PF_3)_4$ and $HRh[P(C_6H_5)_3]_4$, where the H and $AuPR_3$ are capping ligands, support the arguments above [34, 35].

In ML_4L' and $ML_4L'_2$ complexes, where L and L′ have approximately equal electronegativities, the ligands have similar bonding capabilities and compete equally for s character on the central atom. This ensures that a rehybridization occurs from the $s(pd)^3$ optimum scheme of a tetrahedral structure to sp^3d and sp^3d^2. Therefore such compounds are characterized by trigonal bipyramidal and octahedral structures. For examples, $[Ni(PMe_3)_4Br]^-$ and $[Ni(PMe_3)_4CH_3]^-$ are trigonal bipyramidal structures [36] and $[Rh(H_2O)_4Cl_2]^+$ is octahedral [37]. When L and L′ have very different electronegativities, e.g. $AuPR_3$ and PR_3, it can be predicted that the more electropositive ligands prefer the capping positions because, according to (1) above, weaker bonds to the central atom are formed at these positions. In the examples discussed above H and $AuPR_3$ are located in capping positions.

6.2.2 Capped Octahedra

In the sp^3d^2 hybridization scheme of an octahedral complex, the three t_{2g} non-bonding orbitals defined with respect to the C_3 axis are:

$$t_{2g}(O_h)\begin{cases} d_{z^2} & a_1(C_{3v}) \\ (2/3)^{1/2}d_{x^2-y^2} + (1/3)^{1/2}d_{xz} & \multirow{2}{*}{e(C_{3v})} \\ (2/3)^{1/2}d_{xy} - (1/3)^{1/2}d_{yz} & \end{cases}$$

The capped ligand then interacts with the $d_{z^2}(a_1)$ orbital of these non-bonding orbitals. It was concluded from MO calculations [29] that $\theta_2(=\theta_3=\theta_4)$ and $\theta_5(=\theta_6=\theta_7)$ (see **(15)** for the definition of θ) both decrease as the number of d electrons is increased from 0 to 4. This can be understood simply in the following context. In the d^0 complexes where the $e(C_{3v})$ non-bonding orbitals are not occupied, the ligands tend to arrange themselves evenly on a spherical surface. Therefore, the deformation from a regular capped octahedron towards a spherical arrangement results in the increase of the θ_2 and θ_5. In the d^4 complexes, the occupation of the two non-bonding orbitals leads to a structure which maximizes the d character in the $e(C_{3v})$ non-bonding orbitals. This can be achieved by a geometrical change which would lead back to the regular capped octahedron. One example of a capped octahedron is $[NbTe_{10}]^{3-}$, which was considered as a cluster compound in the original paper [38]. Alternatively it can be described as a seven coordination compound $[Nb(Te_3)_3Te]^{3-}$ with d^0 configuration, where Te_3^{2-} is the polytelluride ion.

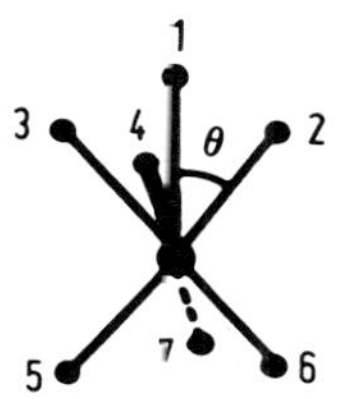

(15)

When the second capping ligand is introduced, it interacts with the remaining non-bonding orbitals of e symmetry. Therefore a *trans*-bicapped octahedral complex is unfavourable because the remaining available non-bonding orbitals have zero electron density along the *z* axis.

6.2.3 Capped Trigonal Prisms

In a trigonal prismatic transition metal complex (**16**), in addition to the six σ hybrids which form M–L bonds there are three non-bonding orbitals which include one pure d_{z^2} and two e′ *d–p* hybrid non-bonding orbitals (see Table 3). As discussed above, it can be concluded that capping the square faces is preferred to the triangular faces in the mono- and bi-capped trigonal prismatic transition metal complexes because of the two *d–p* hybrid non-bonding orbitals which maximize their electron density towards the square faces. The existence of structures such as $[Mo(CNR)_6I]^+$ (mono-capped) [39] and ZrF_8^{4-} (bicapped) [40] supports the conclusions above.

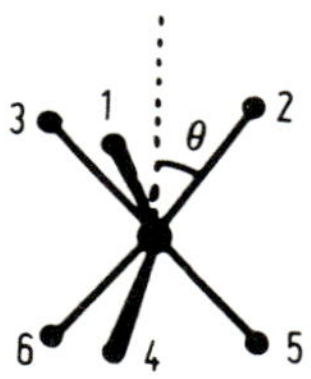

(**16**)

Capping two ligands on two of three square faces of the prism results in a great decrease in the θ cone angle (see (**16**)) compared to the parent trigonal prism. The decrease in the θ cone angle leads to a mixing of *s* orbital character into the non-bonding d_{z^2} orbital. Consequently, an as + bd_{z^2} hybridized orbital is involved in the hybridized σ orbitals of the trigonal prism and a rehybridization occurs when two ligands are capped onto the square faces of a trigonal prism. The complementary component bs − ad_{z^2}, which maximizes its electron density in the equatorial plane, is available for the third capping ligand. Therefore capping the third square face rather than the triangular faces is preferred when the third capping ligand is introduced.

6.3 Nido and Arachno Structures

Co-ordination polyhedra can be described as *nido*, *arachno* and *hypho* if they can be represented as fragments of spherical polyhedra with one, two and three vertices missing respectively. The grossly non-spherical nature of these polyhedra means that their hybrid orbitals are no longer adequately described in terms of the methodology developed above. It might appear that it is possible to describe the hybrids associated with the missing vertices in terms of the parent polyhedron. This is correct for the tetrahedron, where for example in AH_3 molecules the lone pair orbital can be represented as a sp^3 hybrid as long as the H-A-H angle is 109.47°. However, for spherical polyhedra which require d orbitals to define the hybrids the nodal cones of the hybrids do not coincide with the ligand positions. For example, in an octahedron the relevant d^2sp^3 hybrid has the form

$$0.408s + 0.707p_z + 0.577d_{z^2}$$

This hybrid has nodal cones at 38° and 99° with respect to the four fold symmetry axis (see (**17**)) and do not coincide with the axes of the octahedron. Therefore, it is hardly surprising that in transition metal square pyramidal complexes the angle between axial and equatorial ligands is always greater than 90°. Indeed extended Hückel calculations [28] on d^8 PtL_5 complexes have resulted in an energy minimum at $\theta = 98°$ almost precisely in the nodal cone of a d^2sp^3 hybrid. The experimentally determined angle in the related d^8 complex $[Ni(CN)_5]^{3-}$ is 100° [41]. In a d^6 low spin complex the hybrid orbital is empty and the geometry approximates more close to that of an octahedron. For example, $Mo(CO)_5$ has a θ angle of 91° [8a]. This emphasizes that undoubted success of approximate molecular orbital methods such as the extended Hückel lies in the fact that they are reproducing the important properties of the angular parts of the wavefunctions and defining the angles which maximize the metal ligand overlaps and placing the ligands in the nodal cones of the remaining orbitals.

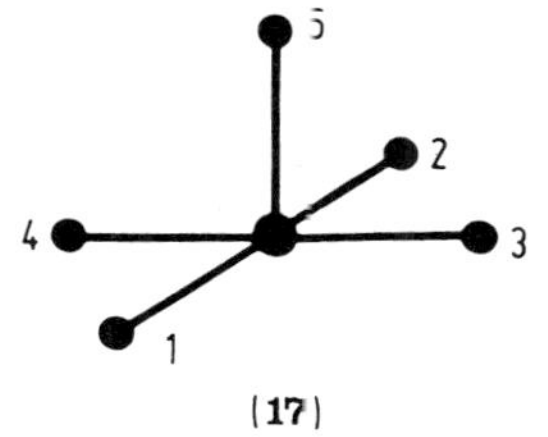

(**17**)

In more general terms one can explore the nodal characteristics of the non-bonding a_1 orbital in a square pyramidal complex by defining it initially as:

$$as - bp_z + cd_{z^2}$$

The coefficients are then sought which lead to a hybrid with zero overlap with the ligand σ-orbitals of the square pyramid. Inserting the angular parts of the spherical harmonic functions this leads to:

$$a - b \times 3^{1/2} \cos\theta + c \times 5^{1/2}/2(3\cos^2\theta - 1) = 0$$

When the ligand angular co-ordinates are inserted, i.e. when $\theta = 0$ for the axial ligand and $\theta = 90°$ for the equatorial ligands, the a, b and c parameters can be determined. The Table below summarizes the variation in s, p and d character as a function of the θ angle

	a(*s*)	b(p_z)	c(d_{z^2})
$\theta = 80°$	.565	.755	.332
$\theta = 85°$	.518	.773	.367
$\theta = 90°$	.456	.791	.408
$\theta = 95°$	.377	.806	.456
$\theta = 100°$	.273	.816	.510

For a d^8 complex where the energy minimum is close to 100° the non-bonding hybrid orbital defined above is similar to a d^2sp^3 hybrid but not identical to it. It has a higher proportion of *p* orbital character and less *s* and *d* character. Nonetheless, the nodal characteristics of the d^2sp^3 hybrid give a good indication of the magnitude and sense of the anticipated distortion. It follows that the bonding in *nido, arachno* and *hypho* fragments can to a first approximation be described in terms of the hybrid orbitals of the parent polyhedron. If the hybrid orbitals are vacant then the geometry of the fragment closely approximates to that of the parent spherical molecules. However, if the hybrid(s) is occupied then the geometry will distort somewhat so that the ligands are located in the nodal cones of the idealized hybrid orbitals.

The following series of matrix isolated molecules [8a] $Mo(CO)_5$, $Mo(CO)_4$ and $Mo(CO)_3$ illustrated these ideas since they all have geometries based on the parent octahedron. $Mo(CO)_5$ is square pyramidal with a θ angle of 91°, $Mo(CO)_4$ has a *cis* divacant octahedral structure with OC–Mo–CO (axial) = 174° and OC–Mo–CO (equatorial) = 107° and $Mo(CO)_3$ has a *cis* trivacant C_{3v} octahedral structure with OC–Mo–CO = 93.2°. Since these fragments also have approximate d^2sp^3 hybrids pointing towards the missing vertices their bonding capabilities are readily defined. The presence of these out-pointing hybrid orbitals has important consequences for describing the bonding in clusters derived from $M(CO)_n$ fragments and forms the basis of the *isolobal* analogy.

Although for d^6 fragments the geometries are referred to the octahedron for metals with fewer *d* electrons the fragments are referred to higher co-ordination number polyhedra. For example, a d^4 $M(CO)_4$ fragment can be referred to a trigonal-base tetragonal base co-ordination number 7 polyhedron (**18**). The computed C_{4v} geometry is consistent with the formation of d^3sp^3 hybrids to the four basal ligands and three hybrids towards the missing vertices. The four ligands lie in the nodal cone of d_{z^2} and the nodal planes of d_{xy} leading to two non-bonding orbitals localized on the metal. The occurrence of such non-bonding orbitals with maximum *d* character contributes significantly the stabilization of the geometry in view of the orbital energy ordering n*d* > (n + 1)*s* ≫ (n + 1)*p*. Similarly, the

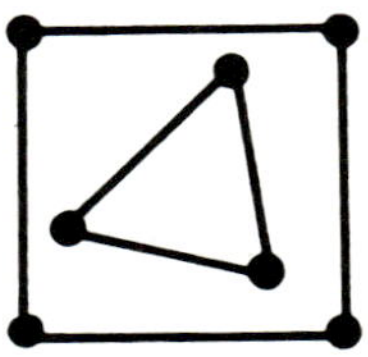

(**18**)

d^2 $M(CO)_4$ fragment can be referred to a square antiprism with four outpointing hybrids, four bonding hybrids and a d_{z^2} non-bonding orbital. Extended Hückel calculations [8a] on $M(CO)_4$ fragments predict C_{4v} pyramidal structures for $d^2(\theta = 117.5°)$ and d^4 $(\theta = 122.5°)$. The nodal cone of d_{z^2} occurs at $\theta = 125.27°$.

Since the position of the nodal cones in hybrids is relevant to the stereochemical problems described above their positions are summarized in Fig. 6. In those cases where the hybrids are not axially symmetric the nodal lines occur at different sides of the hybrid.

When the square pyramid is distorted from the idealized octahedral geometry the relative proportions of *s*, *p* and *d* character change according to whether $\theta > 90°$ or $< 90°$. For $\theta < 90°$ the *s* character in the non-bonding hybrid is increased at the expense of *p* and *d*, whereas when $\theta > 90°$ the *s* character decreases relative to *d* and *p*. Therefore, it is possible to account for the different stereochemical properties of lone pairs in transition metal and main group ML_5 complex. For a molecule such as BrF_5 $\theta < 90°$ [42] because the geometry maximizes the *s* character in the non-bonding hybrid orbital, and this is energetically favourable because of the atomic level ordering $ns > np \gg nd$. In the transition metal d^8 complexes the level ordering $nd > (n+1)s \gg (n+1)p$ leads to a geometry which maximizes the *d* character.

6.4 *Anisotropic π-Bonding Effects*

In the examples cited above it has been assumed that the ligands are bonding approximately equally and that σ-bonding effects are significantly more important than π-bonding effects. There are numerous complexes, however, where there is a unique ligand which is either a π-donor or π-acceptor. Some typical examples are illustrated in **(19)** [43] and **(20)** [44] where the unique ligand is a π-donor or acceptor.

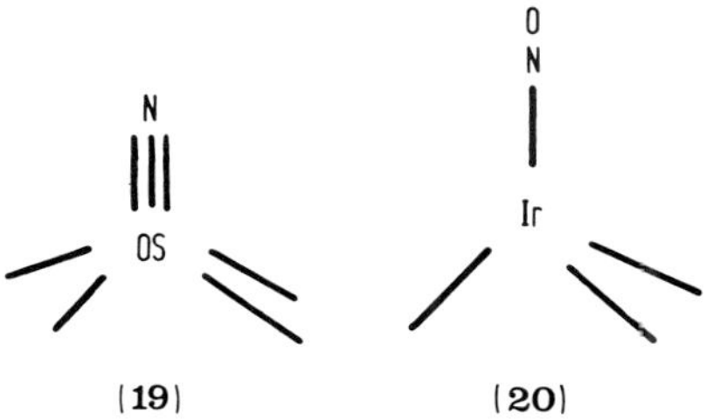

(19) **(20)**

In a square pyramid the change in θ angle from the idealized octahedral angle not only changes the character of the hybrid pointing towards the vacant site, but also causes a mixing of (p_x, p_y) and (d_{xz}, d_{yz}). In a hybridized sense the extent of mixing can be estimated by defining the resultant hybrid in such a way that it has a nodal cone coincident with the equatorial ligand directions. The results of such calculations for $\theta = 80$–$100°$ are summarized over the page:

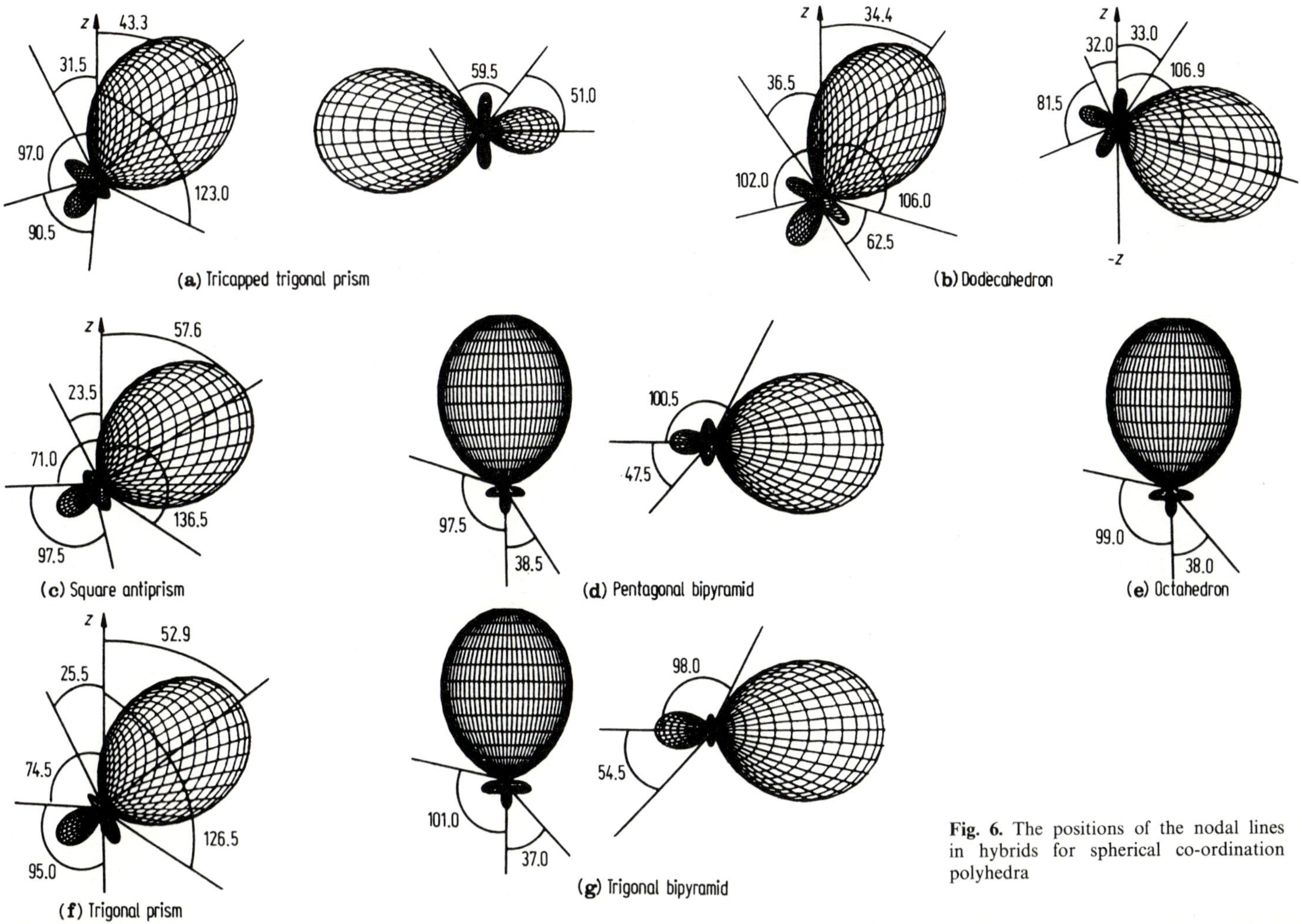

Fig. 6. The positions of the nodal lines in hybrids for spherical co-ordination polyhedra

	p_x	d_{xz}
$\theta = 80°$	-0.362	0.932
$\theta = 85°$	-0.191	0.982
$\theta = 90°$	0.	1.000
$\theta = 95°$	0.191	0.982
$\theta = 100°$	0.362	0.932

It is noteworthy that although the mixing coefficients for θ angles of $90 \pm x$ are equal, their signs are opposite. This difference is represented schematically in (**21**) and (**22**). In each case the hybrid points away from the direction of distortion. If the axial ligand is a good π-donor or acceptor then a superior π-interaction is achieved with the metal *d–p* hybrid if the θ is larger than 90°. The square pyramidal π-donor complexes $[OsNBr_4]^-$ (θ = 104.3°) [43], $[RuNBr_4]^-$ (θ = 104.2°) [45] and $ReNCl_2(PPh_3)_2$ (θ = 109.7°) [46] provide examples of such geometrically distorted situations. The effect is not limited to square pyramidal complexes and for example $Pt(CO)(PR_3)_3$ (θ = 113°) [47] and $Ir(NO)(PPh_3)_3$ (θ = 117°) [44] have distorted tetrahedral geometries. Despite the fact that there are bulky ligands present, the distortion increases the steric repulsions between the phosphine ligands.

(**21**)

(**22**)

If the situation is reversed such that the majority of ligands are good π-acceptors and the unique ligand is a σ-donor then similar geometric effects are observed. For example, $[Mn(CO)_5H]$ [48] has an octahedral structure but the OC(axial)-Mn-CO(equatorial) angle is 97°. Similarly $Co(CO)_4H$ [49] has a geometry intermediate between capped tetrahedral and trigonal bipyramidal with θ = 100°.

The octahedral 18 electron compounds $[MNL_5]$ and $[MOL_5]$ [50] provide particularly interesting examples of π-driven distortions. These complexes generally have a long metal–ligand bond *trans* to the unique multiply bonded *nitrido* or *oxo* ligand. In addition the N-M-L bond angles are significantly larger than 90°, e.g. in $[ReN(NCS)_5]^{2-}$ (θ = 96°) [51]. Clearly the stronger multiple bond leads to rehybridization of *d* and *p* associated with an increase in θ. However, this angular change also rehybridizes the hybrid pointing in the direction of the *trans* ligand. The decreased s character in this hybrid leads to a significantly longer metal–ligand *trans* bond.

7 Summary

This review has provided a general methodology for deriving the hybrid orbitals of transition metal co-ordination compounds. The important types of hybridization which can occur are summarized in Figs. 7–11, together with an indication of the location of their nodal lines and cones. For *s–p* hybrids (Fig. 7) the nodal cone moves from 70.5° from the $-z$ direction for $0.500s + 0.866p_z$ to 54.5° for $0.707s + 0.707p_z$. In contrast a *dsp* hybrid has two nodal cones (see Fig. 8), one in the $+z$ and the other in the $-z$ direction. The former approaches 90° as the percentage of d_{z^2} character is increased and the latter decreases from 38.5° to 37°. The $as + bd_{z^2}$ and $as - bd_{z^2}$ hybrids are illustrated in Fig. 9. The former concentrates electron density in the $+z$ and $-z$ directions and has two symmetrical nodal cones. The angles between the cones decreases as the percentage of *s* character is increased. In contrast the $as - bd_{z^2}$ hybrids concentrate electron density in the xy plane and their nodal cones make smaller angles with $+z$ and $-z$ axes. These hybridization modes are particularly important for discussing the bonding in linear d^{10} complexes of gold and mercury [52].

The hybridization schemes described above are relevant for describing σ-bonding effects in molecular compounds. Hybrids which maximize π-bonding effects can be constructed from atomic orbitals which have a nodal plane in the metal–ligand bond direction and are described as $p^{\pi}d^{\pi}$ hybrids. For example the $ap_y + bd_{yz}$ hybrids illustrated in Fig. 10 can be used to maximize the bonding with a π-donor or acceptor ligand located in the $+z$ direction. The nodal planes associated with these hybrids make an angle of 75° for $0.500p_y + 0.866d_{yz}$ with the $-z$ axis and this angle decreases as the percentage of *p* character is increased.

In Fig. 11 an alternative hybridization mode based on $p_z + d_{x^2-y^2}$ mixing is illustrated. With respect to the z axis the *p* orbital is not noded and therefore has σ-pseudo symmetry. In contrast $d_{x^2-y^2}$ is doubly noded and has δ-pseudo symmetry. The resulting hybrid has the effect of increasing the directionality of the

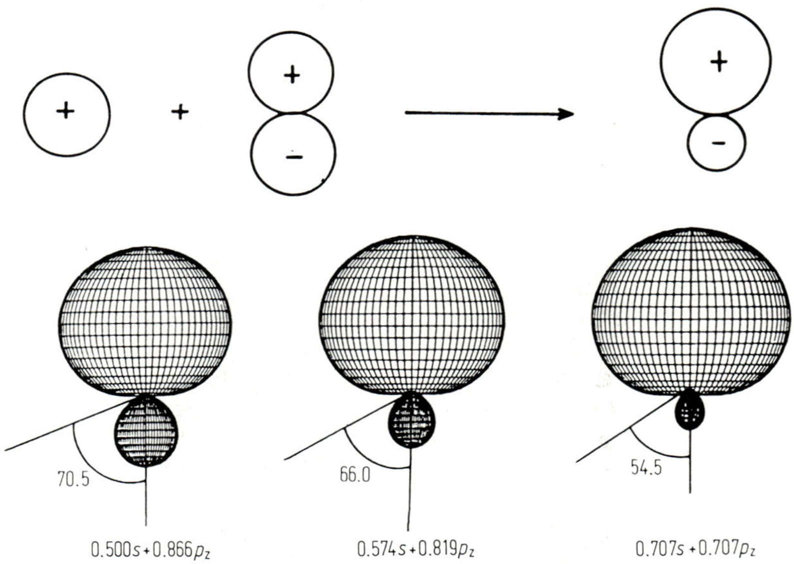

Fig. 7. *sp* hybrid orbitals

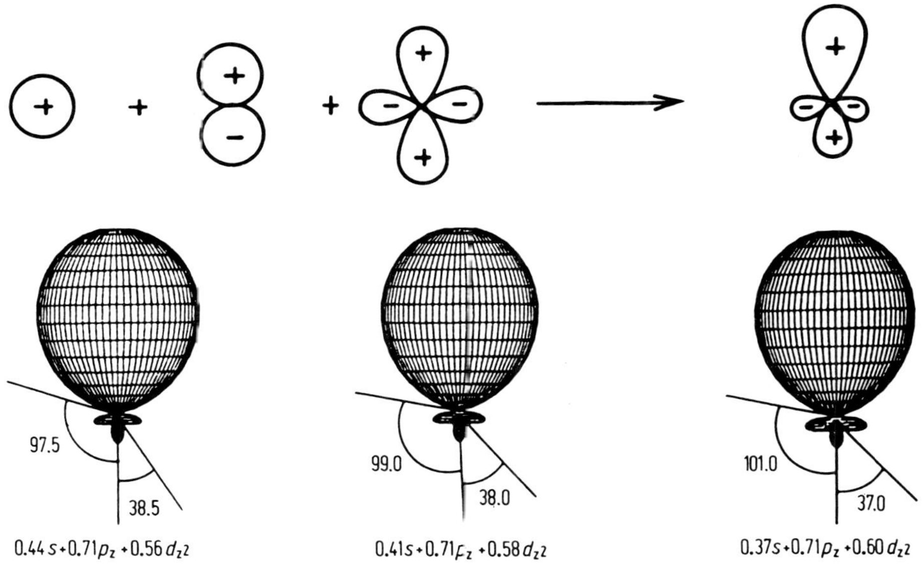

Fig. 8. *spd* hybrid orbitals

+ lobes of $d_{x^2-y^2}$ in the z direction and − lobes in the − z direction. The relevant nodal lines for a range of *p* and *d* character are illustrated in Fig. 11. This type of $p^\sigma d^\delta$ hybridization is particularly important when the bonding in tetrahedral co-ordination compounds is discussed as in a previous section of this review.

For spherical co-ordination compounds the equivalent σ-hybrids can be readily derived using spherical harmonic expansions. For spherical co-ordination compounds with ligands occupying two orbits it is necessary to re-orthogonalize the wave functions using, for example, Schmidt orthogonalization procedures. The characters of the relevant hybrid orbitals are summarized in Fig. 1. Where the ligands lie on separate orbits the relative proportions of *s*, *p* and *d* character vary significantly and can be used to estimate ligand site preferences. Generally, better σ-donor ligands will interact most strongly with those hybrids with a higher proportion of s character.

In the spherical co-ordination compounds ML_n it is possible to define n hybrid orbitals which have their maxima in the metal–ligand bond directions as long as the ligand linear combinations can be defined using *s*, *p* and *d* spherical harmonics. The remaining (9-n) orbitals are non-bonding either because the ligands lie in their nodal planes or because they are *d–p* hybrids which have nodal lines pointing in the metal–ligand directions. The characteristics of these hybrids are summarized in Table 3. The occupation of these orbitals and the hybrid σ-orbitals leads to compounds which conform to the noble gas rule. The conformational preferences of non-axially symmetric π-acceptor and donor ligands are determined primarily by the nodal characteristics of the non-bonding *d–p* hybrid orbitals.

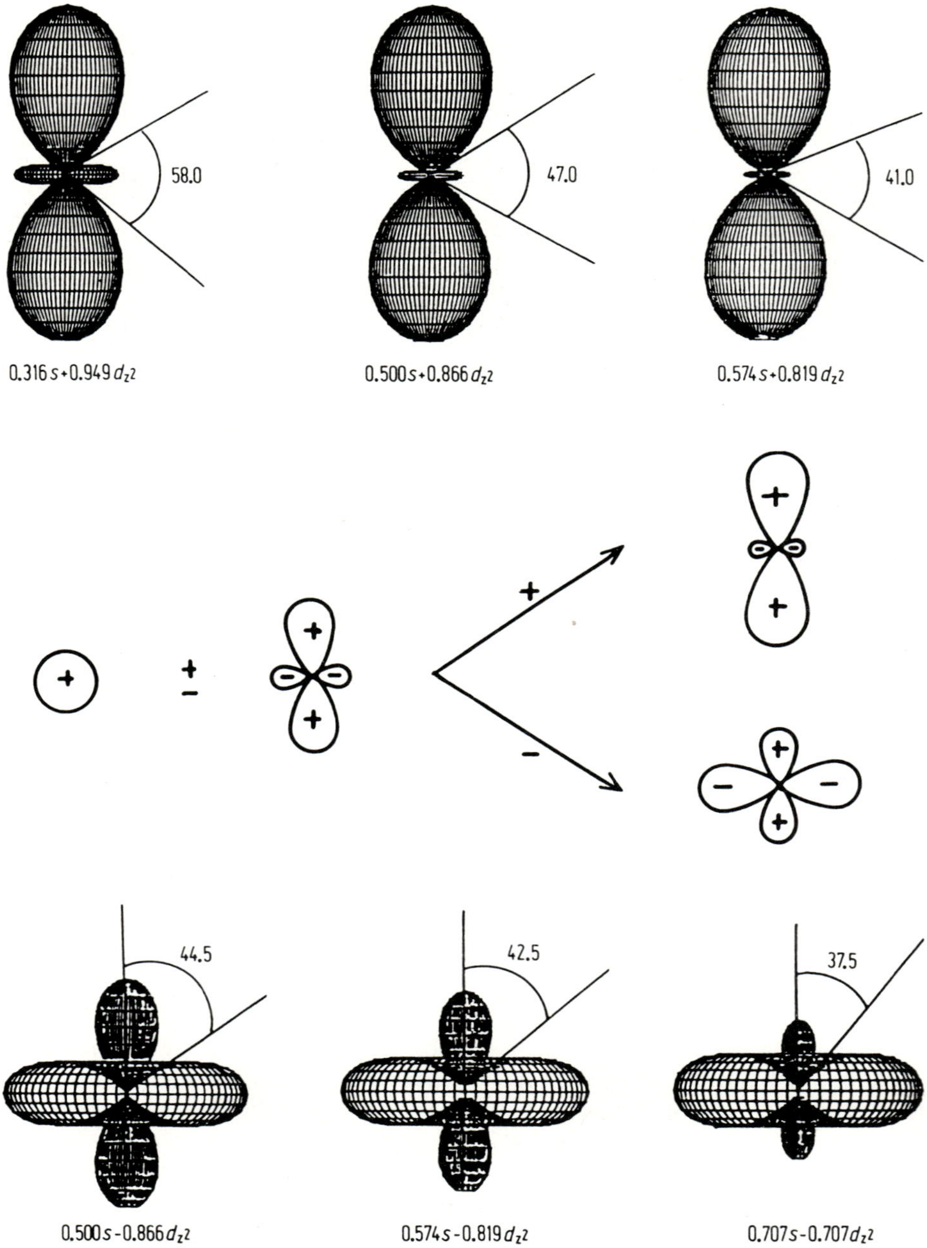

Fig. 9. $s + d_{z^2}$ hybrid orbitals

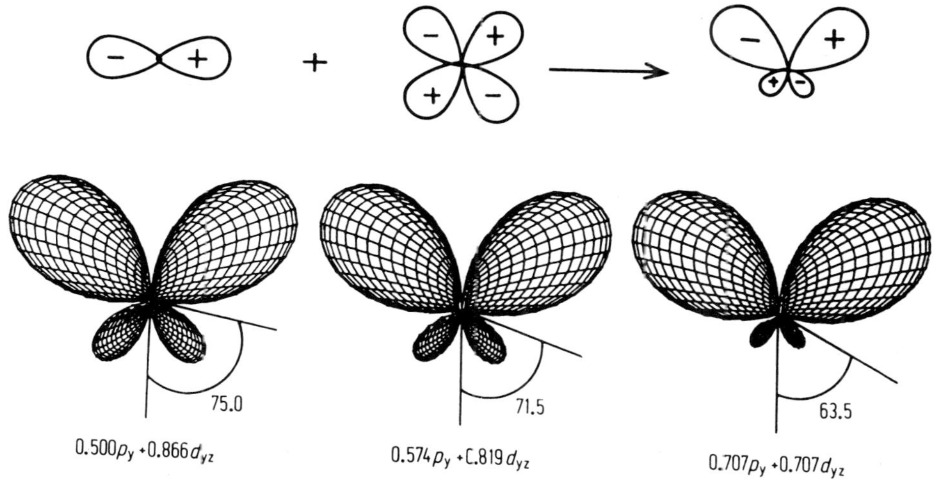

Fig. 10. $p^{\pi} d^{\pi}$ hybrid orbitals

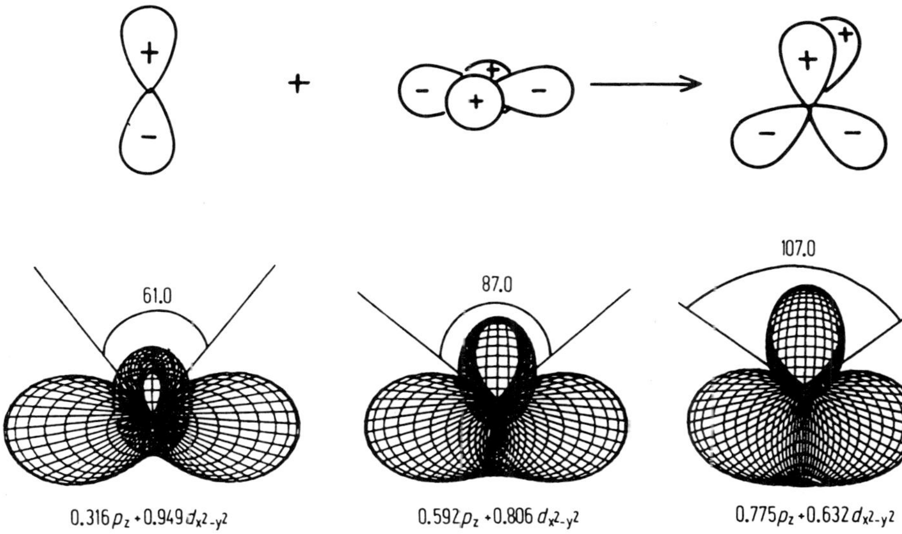

Fig. 11. $p^{\sigma} d^{\delta}$ hybrid orbitals

For ML_n co-ordination compounds with fewer than 18 electrons two alternative situations are generally observed. (a) Spherical co-ordination polyhedra are adopted and there are electron holes in the non-bonding *d* manifold. (b) When the ligands are strong π-acceptors, then *nido*, *arachno* and *hypho* non-spherical polyhedra are adopted. These fragments have hybrids pointing towards the missing vertices of the polyhedron. The location and number of these hybrids leads to the *isolobal* analogies and plays an important role in influencing the conformational preferences of ML_n (η-polyene) complexes.

The discussion presented in this review has assumed that the radial parts of the wavefunctions are relatively unimportant and the important geometric features associated with transition metal co-ordination compounds are decided primarily by the angular parts of the wavefunctions. For transition metal atoms where the n*d*, (n + 1)*s* and (n + 1)*p* valence orbitals are known to have very different radial characteristics [8] some consideration needs to be given to this problem. For spherical co-ordination compounds the number of angular variables is very limited and the hybridization model results presented above give geometries very similar to those predicted by molecular orbital methods which take into account differences in the radial distribution functions. In these co-ordination compounds the geometries are decided primarily by the nodal characteristics of non-bonding *d* orbitals. Since these do not involve admixtures of *d*, *s* and *p* differences in the radial parts are relatively unimportant. In non-spherical co-ordination compounds, *dsp* hybrids do play an important stereochemical role and a comparison between these hybrids and the results of molecular orbital calculations is more significant. For example, in a square-pyramidal MH_5 complex with $\theta = 90°$ the out-pointing hybrid orbital is calculated to have the following wave function:

$$\psi_{nb} = 0.22s - 0.40p_z + 0.81d_{z^2} + \text{hydrogen atom contributions}$$

The calculated hybrid in contrast has the following wave function:

$$\mathrm{Hy(nb)} = 0.46s - 0.79p_z + 0.408d_{z^2}$$

It tends to overemphasize the *s* and *p* character in the non-bonding hybrid and underemphasize their contribution to the bonding hybrids. This occurs because the methodology developed for generating the hybrids does not take into account the superior overlap integrals between the *s* and *p* orbitals with the ligand orbitals. Nevertheless, the relative signs of the orbital mixings are reproduced as are the trends associated with the variation of the nodal characteristics as the *d*, *s* and *p* contributions are varied.

The other assumption associated with the hybridization model is that the n*d*, (n + 1)*s* and (n + 1)*p* valence orbitals have the same valence state ionization energies. Molecular orbital calculations using the extended Hückel approximation have indicated that the changes in the valence state ionization energies do not cause changes in the non-bonding wavefunctions.

Acknowledgement: The S.E.R.C. and the Chinese Academy of Sciences are thanked for their financial support. We also received helpful comments on the manuscript from Prof. Pauling's associate Dr. Z.S. Herman.

8 References

1. Heilter W, London F (1927) Z. Physik. 44: 455
2. Pauling L (1931) J. Amer. Chem. Soc. 53: 1367
3. Pauling L (1960) Nature of the chemical bond, 3rd edn, Cornell University Press, Ithaca, New York
4. Ballhauson CJ, Gray HB (1964) Molecular orbital theory, Benjamin, New York

5. Figgis BN (1966) Introduction to ligand fields, Wiley Interscience, New York
6. (a) McWheeney R (1986) Nature 323: 666
(b) Cooper DL, Gerratt J, Raimondi M (1986) Nature 323: 699
(c) Pauling L (1987) Nature 325: 396
7. (a) Walsh AD (1953) J. Chem. Soc. 2260
(b) Gimarc BM (1974) Acc. Chem. Res. 7: 384
(c) Buenker RJ, Peyerimhoff SD (1974) Chem. Rev. 74: 127
8. (a) Elian M, Hoffmann R (1975) Inorg. Chem. 14: 1058
(b) Elian M, Chen MM-L, Mingos DMP, Hoffmann R (1976) Inorg. Chem. 15: 1148
9. Mingos DMP (1977) Adv. Organomet. Chem. 15: 1
10. Herman ZS (1983) Int. J. Quantum Chem. 23: 921
11. Kimball GE (1951) Ann. Rev. Phys. Chem. 2: 177
12. Hultgren R (1932) Phys. Rev. 40: 891
13. Pauling L (1975) Proc. Nat. Acad. Sci. USA. 72: 3799, 4200; (1976) 73: 274, 1403, 4290; (1978) 75: 12, 569; (1984) 81: 1918
14. Pauling L (1978) Acta. Cryst. B34: 746
15. Pauling L (1978) Canadian Mineralogist 16: 447
16. Pauling L, Herman ZS (1984) J. Chem. Edu. 61: 582
17. Herman ZS, Pauling L (1984) Croat. Chemica Acta. 57: 765
18. Kimball GE (1940) J. Chem. Phys. 8: 188
19. Murrell JN (1960) J. Chem. Phys. 32: 767
20. Yang C (1988) J. Mol. Struct. (Theochem.) 169: 1
21. Pauling L, Herman ZS, Kamb BJ (1982) Proc. Nat. Acad. Sci. USA. 79: 1361
22. Racah G (1943) J. Chem. Phys. 11: 214
23. Kutzelnigg W (1984) Angew. Chem. Int. Ed. Engl. 23: 272
24. Kepert DL (1982) Inorganic Stereochemistry, Springer, Berlin Heidelberg New York
25. (a) Bent HA (1961) Chem. Rev. 61: 275
(b) Steiner E (1976) The determination and interpretation of molecular wave functions, Cambridge University Press, London, p 40
26. Abrahams SC, Ginsberg AP, Knox K (1964) Inorg. Chem. 3: 558
27. Burdett JK, Hoffmann R, Fay RC (1978) Inorg. Chem. 17: 2553
28. Rossi AR, Hoffmann R (1975) Inorg. Chem. 14: 365
29. Hoffmann R, Beier, BF, Muetterties, EL, Rossi AR (1977) Inorg. Chem. 16: 511
30. Rosch N, Hoffmann R (1974) Inorg. Chem. 13: 2656
31. Stone FAG (1975) J. Organomet. Chem. 100: 257
32. Russel DR, Tucker PA: J. Chem. Soc. Dalton Trans. 1975: 1752
33. Swalen JD, Ibers J (1962) J. Chem. Phys. 37: 17
34. Frenz BA, Ibers JA (1971) In: Muetterties EL (ed) Transition metal hydrides, Marcel Dekker, New York, p 133
35. Puddephatt RJ (1987) In: Wilkinson G (ed) Comprehensive coordination chemistry, Pergamon, Oxford, vol 5 p 905
36. Kepert DL (1987) In: Wilkinson G (ed) Comprehensive coordination chemistry, Pergamon, Oxford, vol 1 p 31
37. Cotton FA, Wilkinson G (1972) Advanced inorganic chemistry 3rd edn John Wiley, New York, p 1024
38. Flomer WA, Kolis JW (1988) J. Amer. Chem. Soc. 110: 3682
39. Lewis DF, Lippard SJ (1972) Inorg. Chem. 11: 621
40. Kojic-Prodic B, Scavnicars, Matkovic B (1971) Acta. Cryst. B27: 638
41. Holmes RR (1985) Prog. Inorg. Chem. 32: 119
42. Mingos DMP, Hawes JC (1985) Structure and Bonding 63: 1
43. Collison D, Garner CD, Mabbs FE, Salthouse JA, King TJ: J. Chem. Soc. Dalton Trans. 1981: 1812
44. Albano VG, Bellon PL, Sansoni M: J. Chem. Soc. A 1971: 2420
45. Collison D, Garner CD, Mabbs FE: J. Chem Soc. Dalton Trans. 1981: 1820
46. Doedens RJ, Ibers JA (1967) Inorg. Chem. 6: 204
47. Albano VG, Ricci, GMB, Bellon PL (1969) Inorg. Chem. 8: 2109
48. Laplaca SL, Hamilton WC, Ibers JA, Davison A (1969) Inorg. Chem. 8: 1928
49. McNeil EA, Scholer FR (1977) J. Amer. Chem. Soc. 99: 6243
50. Jean Y, Lledos A, Burdett JK, Hoffmann R (1988) J. Amer. Chem. Soc. 110: 4506 and references therein
51. Carrondo MAA, De CT, Shahir R, Skapski AC: J. Chem. Soc. Dalton Trans. 1978: 844
52. Orgel LE (1960) An introduction to transition metal chemistry: Ligand field theory, London

The ^{1}H NMR Parameters of Magnetically Coupled Dimers—The Fe_2S_2 Proteins as an Example

Lucia Banci[1], Ivano Bertini[1] and Claudio Luchinat[2]

[1] Department of Chemistry, University of Florence, Via G. Capponi 7, 50121 Florence, Italy
[2] Institute of Agricultural Chemistry, University of Bologna, Viale Berti Pichat 10, 40127 Bologna, Italy

The theoretical approach to the understanding of the NMR parameters in paramagnetic exchange coupled dimers is discussed and simplified formulas are provided. The effect of magnetic coupling on electron relaxation times in magnetic coupled heterodimetallic systems is also discussed. The ^{1}H NMR spectra of two oxidized and reduced Fe_2S_2-ferredoxins are interpreted and structural information is obtained.

Other systems encountered in the literature are briefly reviewed and the guidelines for the interpretation of the ^{1}H NMR spectra of Fe_4S_4 systems are given as an extension of the above model.

1 Introduction

The theoretical principles necessary for understanding the NMR parameters in magnetically coupled dimers are a relatively complicated matter and their application is not straightforward even to the researchers in the field of NMR of paramagnetic molecules [1–7]. It is the aim of this article to summarize the principles and to discuss theoretical and physical aspects of the coupling between resonating nuclei and magnetically coupled electrons. We also discuss the effect of magnetic coupling on electron relaxation in dimers and its consequences on the NMR parameters. As a representative example, such a theory will be applied to the NMR spectra of two iron–two sulphur proteins [8] (ferredoxins) which contain the cluster shown in Scheme I. The complete ^{1}H NMR spectrum of the reduced protein, containing iron(II) and iron(III), isolated from spinach, has already been reported [9]. We have now measured the spectrum of the reduced ferredoxin isolated from red algae in order to ascertain the generality of the ^{1}H NMR features of this class of compounds; we have also measured the spectra of the oxidized species containing two iron(III) ions and the nuclear relaxation times of all species. Finally, we comment about previously investigated systems of pertinence here.

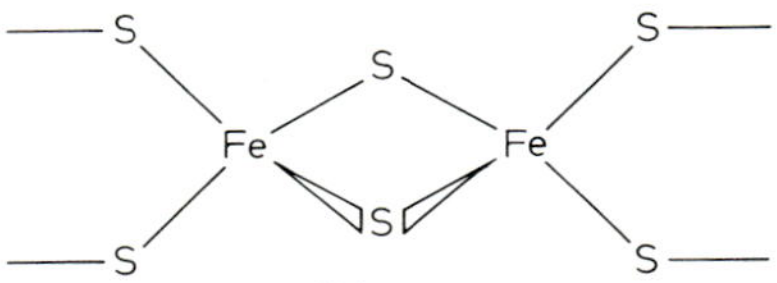

Scheme I

2 The Interaction Energy Between Nuclei and Magnetically Coupled Electrons

2.1 The General Approach

The contact contribution to the isotropic shift, $\left(\frac{\Delta\nu}{\nu_0}\right)^{con}$, experienced by a nucleus I interacting with a single S manifold is proportional to the hyperfine coupling constant A_c according to the following relationship [1, 10]

$$\left(\frac{\Delta\nu}{\nu_0}\right)^{con} = -\frac{1}{\hbar\gamma_I B_0} A_c \langle S_z \rangle \tag{1}$$

where $\hbar$ is the Planck constant and γ_I is the nuclear magnetogyric ratio. The expectation value of S_z, $\langle S_z \rangle$, is given by

$$\langle S_z \rangle = \frac{\sum\limits_{S,\,M_S} \exp(-E_{S,M_S}/kT)\langle S, M_S|S_z|S, M_S\rangle}{\sum\limits_{S,\,M_S} \exp(-E_{S,M_S}/kT)} \tag{2}$$

where $|S, M_S\rangle$ are the spin eigenfunctions with E_{S, M_S} eigenvalues. Since Eq. (1) holds for a single S manifold, Eq. (2) reduces to

$$\langle S_z \rangle = -S(S+1)\frac{g_e \mu_B B_0}{3kT} \tag{3}$$

Let us now consider the case in which a nucleus interacts with two electron spin moments, S_1 and S_2, of two metal ions. With no interaction between the two electron spins, it is intuitive that the two contributions to the isotropic shift are additive:

$$\left(\frac{\Delta\nu}{\nu_0}\right)^{con} = -\frac{1}{\hbar\gamma_I B_0}(A_{c1}\langle S_{1z}\rangle + A_{c2}\langle S_{2z}\rangle) \tag{4}$$

Equation (4) is just the sum of two contributions of the same type as Equation (1).

We are interested in the changes occurring when the two S moments are magnetically coupled. We consider here the case of an isotropic coupling situation, described by the additional spin Hamiltonian term

$$\mathscr{H} = JS_1 . S_2 \tag{5}$$

where J is the isotropic coupling constant. This isotropic coupling situation, besides simplifying our reasoning, turns out to be a good assumption, since anisotropic contributions to the magnetic exchange coupling are usually small.

Because of magnetic exchange coupling, new energy levels originate, each described by a different S'_i. The S'_i's range between $|S_1 + S_2|$ and $|S_1 - S_2|$, and the relative energies are given by [11].

$$E(S'_i) = \tfrac{1}{2}J[S'_i(S'_i + 1) - S_1(S_1 + 1) - S_2(S_2 + 1)] \tag{6}$$

Schemes of the type of that shown in Fig. 1 are obtained. In that scheme J is positive (antiferromagnetic coupled systems). With $J < 0$ (ferromagnetic coupled systems) the resulting energy levels are reversed. From now on, we will neglect the sign of J, since it does not affect the general treatment, except when discussing real cases. The $|S'_i, M_{S'_i}\rangle$ eigenfunctions of Hamiltonian (5) are expressed as linear combinations of $|S_1, S_2, M_{S_1}, M_{S_2}\rangle$ functions relative to the two metal ions

$$|S'_i, M_{S'_i}\rangle = \sum_{M_{S_{1,2}}} C_{S_1 S_2 S'_i M_{S_1} M_{S_2} M_{S'_i}} |S_1, S_2, M_{S_1}, M_{S_2}\rangle \tag{7}$$

where $C_{S_1 S_2 S'_i M_{S_1} M_{S_2} M_{S'_i}}$ are the coefficients with which each $|S_1, S_2, M_{S_1}, M_{S_2}\rangle$ function contributes to a given $|S'_i, M_{S'_i}\rangle$ function. These coefficients are obtained as a result of application of Hamiltonian (5).

The effect of magnetic exchange coupling in the analysis of NMR parameters cannot be neglected everytime J is larger than A_{c1} and A_{c2}, and larger than $\hbar\tau_s^{-1}$, where τ_s is the relaxation time of the electron spin system.[1] The first condition

[1] In this case $\tau_s = T_{2e}$; generally, however, in this article we refer to $\tau_s = T_{1e}$ It is possible, however, that $T_{1e} \simeq T_{2e}$ for slowly relaxing systems as a result of rotation independent electronic relaxation mechanisms.

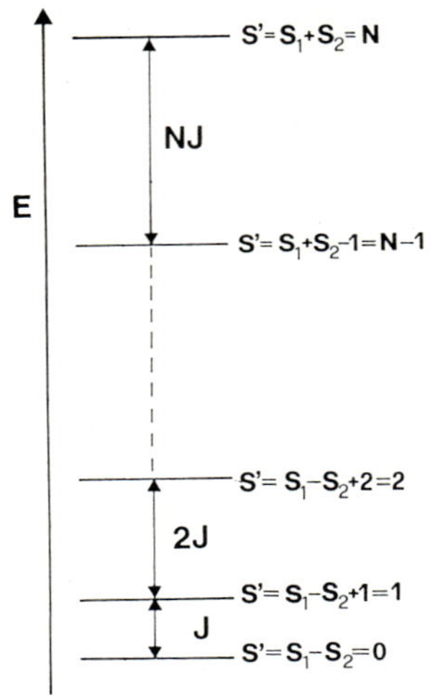

Fig. 1. Energy levels diagram for two magnetically coupled spin momenta $S_1 = S_2$

arises from the fact that, if the interaction energy between the electron spins is smaller than each electron–nuclear coupling energy, the former can be neglected at least in a first approximation. The second condition requires that the energy separations among the S′ levels (Fig. 1) be larger than their energy uncertainty. As long as the exchange coupling is isotropic, J does not need to be larger than the Zeeman term.

Equation (4) holds also for magnetically coupled dimers as long as the $\langle S_{jz}\rangle$ values are evaluated over all the S'_i levels. Therefore

$$\langle S_{jz}\rangle = \frac{\sum\limits_{S'_i, M_{S'_i}} \exp(-E_{S'_i, M_{S'_i}}/kT)\langle S'_i, M_{S'_i}|S_{jz}|S'_i, M_{S'_i}\rangle}{\sum\limits_{S'_i, M_{S'_i}} \exp(-E_{S'_i, M_{S'_i}}/kT)} \tag{8}$$

with j = 1 or 2. Here S_{jz} operates only on the S_j-containing part of the total S'_i functions. It can be shown that each of the $\langle S_{jz}\rangle$ values in (8) tends to the value given by Eq. (3) in the high temperature limit. The correct estimate of the isotropic shift is achieved as a function of the two original A_{cj} (j = 1, 2) and of the new $\langle S_{jz}\rangle$ values:

$$\left(\frac{\Delta\nu}{\nu_0}\right)^{con} = -\frac{1}{\hbar\gamma_I B_0}\sum_j \frac{A_{cj}\sum\limits_{S'_i, M_{S'_i}} \exp(-E_{S'_i, M_{S'_i}}/kT)\langle S'_i, M_{S'_i}|S_{jz}|S'_i, M_{S'_i}\rangle}{\sum\limits_{S'_i, M_{S'_i}} \exp(-E_{S'_i, M_{S'_i}}/kT)} \tag{9}$$

The physical meaning of this treatment is to project out of the total S'_i functions the contributions from S_1 and S_2, and evaluate each of them separately for each and all S_i' levels. We therefore obtain the total contribution to the $\langle S_{1z} \rangle$ and $\langle S_{2z} \rangle$ in the coupled system. The major drawback of this approach lies in the cumbersome calculations that need to be performed, and in the fact that they must be performed on a case-by-case basis, i.e. there is no general formula that one can calculate once and for all. Dunham and Palmer [2] have developed this theory just for Fe_2S_2 proteins. However, this approach has been ignored by most of the researchers who have subsequently been dealing with NMR of magnetically coupled systems.

It should be pointed out that this procedure also holds for systems larger than dimers, as long as Hamiltonians of the type of Eq. (5) can be used.

2.2 The Shortcut

Let us now suppose that in a given isolated metal ion more than one electronic level is thermally populated, each with its own S. Equation (1) becomes

$$\left(\frac{\Delta\nu}{\nu_0}\right)^{con} = -\frac{1}{\hbar\gamma_I B_0}\sum_i A_{c_i}\langle S_z\rangle_i \quad (10)$$

where $\langle S_z\rangle_i$ is separately evaluated on each i multiplet. In other words, a hyperfine coupling constant for each populated level is needed to correctly define the coupling with the nucleus I. This is a consequence of the fact that A_c reflects a specific distribution of unpaired spin density over the occupied molecular orbitals, so that S multiplets belonging to excited states have different unpaired spin density distributions.

In magnetic coupled dimers there are several S' levels, and the isotropic shift of a nucleus interacting with the coupled pair should accordingly be given by

$$\left(\frac{\Delta\nu}{\nu_0}\right)^{con} = -\frac{1}{\hbar\gamma_I B_0}\sum_i A'_{c_i}\langle S'_z\rangle_i \quad (11)$$

where A'_{c_i} are the hyperfine coupling constants for each S_i' level.

Equation (11), as it is, is of little use in analyzing the NMR parameters of exchange coupled systems, because as many A'_c values as the different S_i' multiplets seem to be required. There have been many attempts in the literature, even in recent years, to analyze NMR data using Eq. (11) or equivalents [4, 12, 13]. The next step is to realize that the A'_c values relative to each S_i' level are not independent of one another, as they are in the case described by Eq. (10). As long as all the S_i' levels originate from exchange coupling interactions between two pure S_1 and S_2 states, the A'_{c_i} values of Eq. (11) are related to the A_{c_1} and A_{c_2} values of Eq. (4). To find out this relationship the magnetic exchange coupling is treated in the total spin formalism, in which the spin Hamiltonian is reshaped in such a way to operate only on the total spin functions $|S_i' M_{S_i'}\rangle$. The exchange coupling term can be written as

$$\mathscr{H} = \tfrac{1}{2}J[S'\cdot S' - S_1(S_1+1) - S_2(S_2+1)] \quad (12)$$

from which the energy of the S_i' states can be calculated as shown before (Fig. 1 and Eq. (6)). The A'_{c_i} values in Eq. (11) are related to the A_{cj} values in (4) through coefficients C_{ji} [14], where j refers to the metal ion and i to the S'_i level. The C_{ji} value is 1/2 for every homodimer [15], whereas different coefficients are obtained in the case of heterodimers [14, 16]. Such coefficients are only functions of S'_i, S_1 and S_2 and can thus be calculated once for all. They are reported in Appendix I. They give the contribution of the isolated spin j to the ith spin level of the pair.

These coefficients were calculated more than a decade ago [14] and routinely used since then in EPR of homo- and hetero-dimetallic complexes [17, 18]. According to this treatment it is possible to express the isotropic shift in a coupled dimetallic pair as

$$\left(\frac{\Delta\nu}{\nu_0}\right)^{con} = -\frac{1}{\hbar\gamma_I B_0}\left(A_{c1}\sum_i C_{1i}\langle S'_z\rangle_i + A_{c2}\sum_i C_{2i}\langle S'_z\rangle_i\right) \tag{13}$$

where the $\langle S'_z\rangle_i$ refer to the expectation values of S_z for each particular S'_i multiplet. Note that Eq. (13), rearranged as

$$\left(\frac{\Delta\nu}{\nu_0}\right)^{con} = -\frac{1}{\hbar\gamma_I B_0}\sum_i (A_{c1}C_{1i} + A_{c2}C_{2i})\langle S'_z\rangle_i \tag{14}$$

is the exact equivalent of Eq. (11), where now each of the unknown A_{c_i} values is expressed in terms of a product between A_{c_1} (or A_{c_2}) and a known coefficient.

Equation (14), with the explicit expression of $\langle S'_z\rangle_i$, written in analogy to Eq. (8), takes the form:

$$\left(\frac{\Delta\nu}{\nu_0}\right)^{con} = -\frac{1}{\hbar\gamma_I B_0}\sum_j \frac{A_{cj}\sum_{S'_i, M_{S'_i}} C_{ji}\exp(-E_{S'_i, M_{S'_i}}/kT)\langle S'_i, M_{S'_i}|S'_z|S'_i, M_{S'_i}\rangle}{\sum_{S'_i, M_{S'_i}} \exp(-E_{S'_i, M_{S'_i}}/kT)} \tag{15}$$

In a more explicit form we finally obtain

$$\left(\frac{\Delta\nu}{\nu_0}\right)^{con} = -\frac{g_e\mu_B}{\hbar\gamma_I 3kT}\sum_j \frac{A_{c_j}\sum_i C_{ji}S'_i(S'_i+1)(2S'_i+1)\exp(-E_{S'_i}/kT)}{\sum_i (2S'_i+1)\exp(-E_{S'_i}/kT)} \tag{16}$$

The equivalence between Eq. (15) and Eq. (9) allows us to evaluate the C_{ij} coefficients and accounts for their physical meaning: they indicate how much S_j and M_{Sj} are contributing to each S'_i wavefunction of the couple. Equation (16) is easily applicable in all cases. As a practical example of use of Eq. (16) Appendix II shows the numerical calculation of the high temperature limit. It can be noted that the temperature dependence of the isotropic shifts is often opposite to that observed in monomeric complexes (anti-Curie behavior). In principle, fitting of experimental temperature dependent data to Eq. (16) can give a measure of J. Attempts to estimate J from temperature dependence were previously reported in a number of cases [4, 13, 19]; however, equations of the type of (11) were used, with the unjustified assumption that the A_{c_i} values were all the same and equal to that of the monomer.

3 The Electronic and Nuclear Relaxation Times

When one metal ion, let us say copper(II) without loss of generality, senses another copper(II) ion, under the condition $J < kT$ and $J > A$ and $\hbar\tau_s^{-1}$, no obvious physical mechanism can be conceived that can increase the electron relaxation rate of the former copper ion. To first approximation, the electronic relaxation is not affected by magnetic coupling in homodinuclear complexes [20–23].

In heterodinuclear systems one of the two metal ions has a shorter electronic relaxation time than the other. The magnetic coupling allows the slow relaxing metal ion to relax via the mechanisms of the other metal ion, so that its electronic relaxation rate results increased [21–23]. The question lies in determining the limits within which these mechanisms are operative. It is reasonable to set the lower limit as $J > \hbar\tau_{s_1}^{-1}$ where 1 refers to the slow relaxing metal ion [23]. This means that magnetic coupling produces a splitting of the electron spin levels larger than the uncertainty of their energy. Therefore the probability of the electronic spin transitions, operative for relaxation, is changed by the presence of magnetic coupling. As long as J is still much smaller than $\hbar\tau_{s_2}^{-1}$, the new values of electronic relaxation rates can be estimated through an equation analogous to that derived for contact nuclear relaxation [24], since the Hamiltonian $A\ \mathbf{I}\cdot\mathbf{S}$ is analogous to that of $J\ \mathbf{S}_1\cdot\mathbf{S}_2$

$$\tau_{s_1}^{-1}(J) = \tau_{s_1}^{-1}(0) + \tfrac{2}{3}(J^2/\hbar^2)S_2(S_2+1)\frac{\tau_{s_2}}{1+\omega_{s_2}^2\tau_{s_2}^2} \tag{17}$$

Equation (17) lacks experimental verification. Qualitatively, it would predict the effect of magnetic coupling on $\tau_{s_1}^{-1}$ to be observable when $J/\hbar \simeq \sqrt{\tau_{s_1}^{-1}(0)\tau_{s_2}^{-1}}$.

It is reasonable to believe that, when $J \gg \hbar\tau_{s_2}^{-1}$, the actual $\tau_{s_1}(J)$ will tend to become τ_{s_2} [21, 23].

Provided that we may have τ_s different from those of the uncoupled system, we should still take into consideration that the interaction energy between the unpaired electron on a metal ion and a nucleus sensing it changes if the metal ion is magnetically coupled to another metal ion, as discussed in the previous section. Nuclear relaxation depends on the square of the above energy [1]. For a monomeric system the contribution to nuclear relaxation due to the presence of unpaired electron(s) can be described by the general equation:

$$T_{1M}^{-1} = KA^2\langle S^2\rangle f(\tau_c, \omega) \tag{18}$$

where A is the hyperfine coupling and can be dipolar or contact in origin and $\langle S^2\rangle$ is evaluated applying the S^2 operator on the electronic spin levels. In the case of a dinuclear system the nuclear relaxation rates of a nucleus in the presence of only one metal ion (j) in the couple is given by [3]:

$$T_{1M}^{-1} = KA_{cj} f(\tau_{cj}, \omega_j) \sum_i \langle S_j^2 \rangle_i \tag{19}$$

where $\langle S_j^2 \rangle$ is calculated over all the S_j components of the S_i' levels, as shown in Eq. (7), but applying S_j^2 instead of S_{jz}. Of course if a nucleus senses both paramagnetic centers at different distances, then T_{1M}^{-1} will be the sum of the two contributions.

In the "shortcut" approach, i.e. in the total spin representation, we have to calculate the hyperfine coupling for each new electronic spin level, using the coefficients reported in Appendix I; then the squares of the hyperfine coupling must be calculated. The $\langle S'^2 \rangle_i$ value has to be evaluated by operating with the S'^2 operator over each new S_i' level and then performing the sum on all the spin levels, taking into account their population distribution. The general equation for nuclear relaxation expressed in terms of total spin numbers takes therefore the form:

$$T_{1M_j}^{-1} = \frac{2}{15}\left(\frac{\mu_0}{4\pi}\right)^2 \frac{\gamma_N^2 g_e^2 \mu_B^2}{r_{M_j-H}^6} \frac{\sum_i C_{ji}^2 S_i'(S_i'+1)(2S_i'+1)\exp(-E_{S_i'}/kT)}{\sum_i (2S_i'+1)\exp(-E_{S_i'}/kT)} \left(\frac{7\tau_{c_j}}{1+\omega_{s_j}^2\tau_{c_j}^2} + \frac{3\tau_{c_j}}{1+\omega_I^2\tau_{c_j}^2}\right) \tag{20}$$

When $J \ll kT$, i.e. all the electronic spin levels are equally populated, Eq. (20) is simplified in:

$$T_{1M_j}^{-1} = \frac{2}{15}\left(\frac{\mu_0}{4\pi}\right)^2 \frac{\gamma_N^2 g_e^2 \mu_B^2}{r_{M_j-H}^6} \frac{\sum_i C_{ji}^2 S_i'(S_i'+1)(2S_i'+1)}{\sum_i (2S_i'+1)} \left(\frac{7\tau_{c_j}}{1+\omega_{s_j}^2\tau_{c_j}^2} + \frac{3\tau_{c_j}}{1+\omega_I^2\tau_{c_j}^2}\right) \tag{21}$$

This equation is equal to that for the uncoupled system (besides the variation in τ_c) times a coefficient that depends only on the total spin numbers of the coupled system and of the two single metal centers. These coefficients, that should multiply the equation of the uncoupled system in the case of coupled systems in which $J \ll kT$, are reported in Appendix III [5, 6].

When a nucleus senses both the metal ions a straightforward extension of Eqs. (20) or (21) must be used.

4 The Fe_2S_2 Case

Two-iron two-sulfur (Fe_2S_2) clusters are present in several ferredoxins. In the oxidized state they contain two high spin antiferromagnetically coupled iron(III) ions with J values in the range 180–200 cm^{-1} [25]. In the reduced state one high spin iron(III) and one high spin iron(II) ion are antiferromagnetically coupled with J values of about 100 cm^{-1} [25]. It should be reminded that $S = 5/2$ ions have long electronic relaxation times unless the zero field splitting is large enough [26, 27], as in Fe(III) porphyrins [28]. Long electronic relaxation times provide broad ^{1}H NMR signals. For example, oxidized rubredoxin, with the Fe(III)S_4 chromophore, does not show any detectable isotropically shifted ^{1}H NMR signal [29].

The magnetic coupling, as already stated, does not help to reduce the electronic relaxation times when it occurs between the same metal ions. In fact, oxidized magnetic coupled ferredoxin systems only give rise to two broad signals, that contain all the peaks from the eight β-CH_2 and the four α-CH protons, as shown in Fig. 2 for oxidized red algae and spinach ferredoxins.

Iron(II) compounds have short electronic relaxation times and the ^{1}H NMR spectra are well resolved. Reduced rubredoxin containing a Fe(II)S_4 system has beautiful ^{1}H NMR spectra [29]. In the reduced Fe_2S_2 system it is expected that the fast relaxing iron(II) (from the electron point of view) drives the electronic relaxation times of iron(III) to similar values, J being larger than $\hbar\tau_{s(Fe(II))}^{-1}$. Sharp ^{1}H NMR signals are expected and indeed observed [8, 9] for reduced Fe_2S_2 proteins (Fig. 3).

4.1 *The Oxidized Fe_2S_2 Case*

As already mentioned, both ^{1}H NMR spectra of red algae and spinach oxidized ferredoxins show a broad absorption at about 34 ppm downfield and a sharper

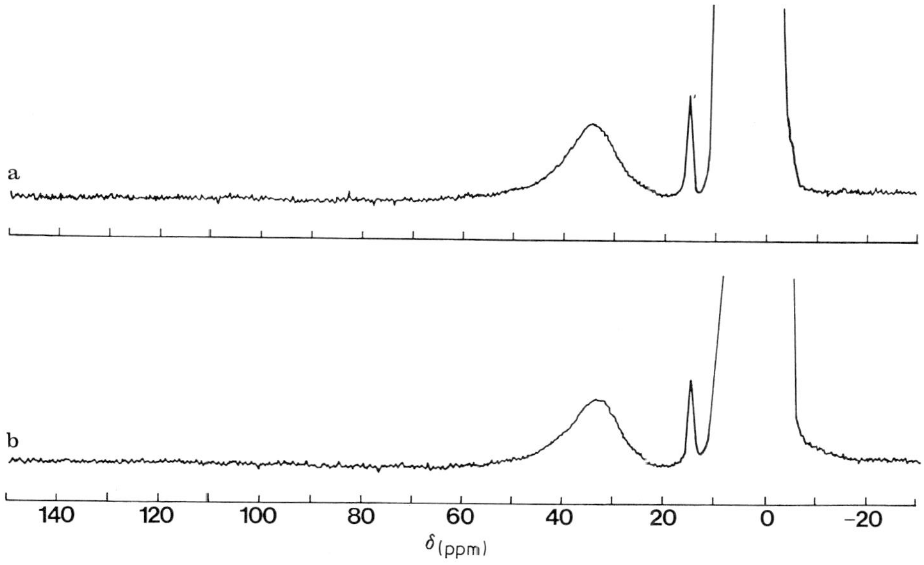

Fig. 2a, b. 303 K 200 MHz ^{1}H NMR spectra of oxidized Fe_2S_2 ferredoxins solutions (in 50 mM phosphate buffer, pH 7.4) from: **a**) spinach; **b**) red algae *Porphyra umbilicalis*

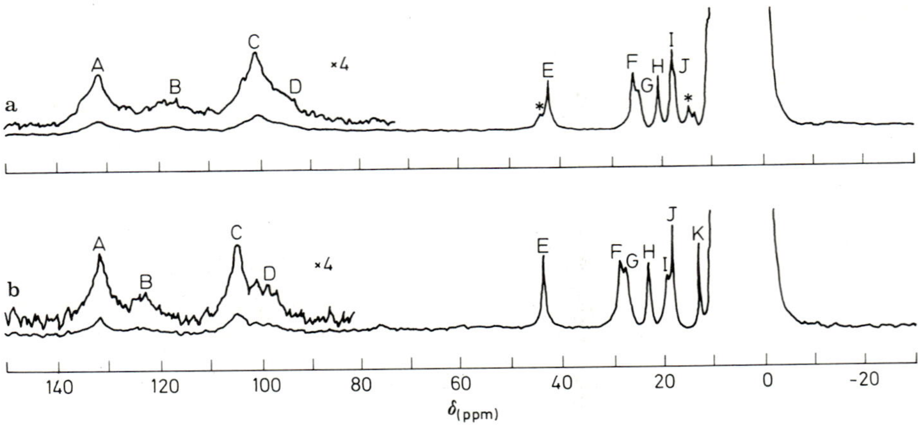

Fig. 3a, b. 303 K 200 MHz ^{1}H NMR spectra of reduced Fe_2S_2 ferredoxins from: **a**) spinach; **b**) red algae *Porphyra umbilicalis* (same experimental conditions as in Fig. 2)

signal at about 15 ppm (Fig. 2). The isotropic shifts are 31 and 10 ppm respectively. The former signal is probably due to βC–H_2's and the latter to αC–H's. The J values are in the range 180–200 cm^{-1} [25]. The energy levels are shown in Fig. 4A. By using eq. (12) with $S_1 = S_2 = 5/2$, the energies of Fig. 4A with $J = 180$–$200\ cm^{-1}$, $A_c/h = 1.0$ MHz (= 300 ppm) and $C_{1S'_i} = C_{2S'_i} = 1/2$ for every S'_i level (see Appendix II), we calculate an isotropic shift at 303 K of 28–31 ppm. The choice of $A_c/h = 1.0$ MHz is arbitrary in principle although it is consistent

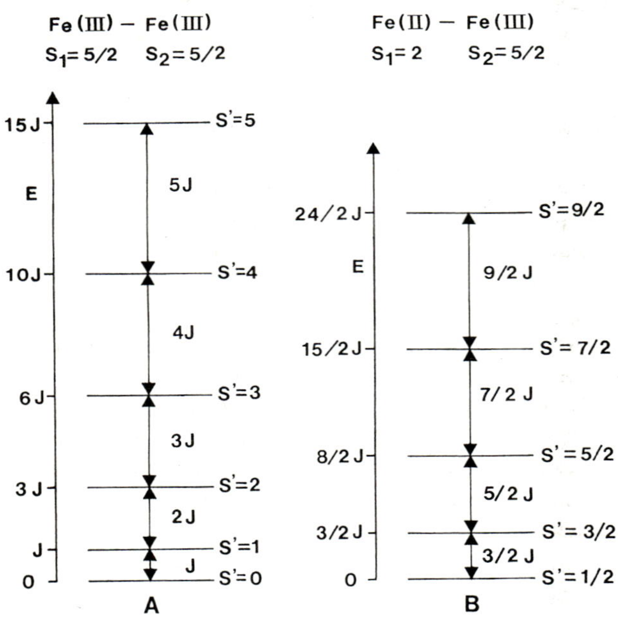

Fig. 4. Energy levels diagram for two magnetically coupled spin momenta: **a**) $S_1 = S_2 = 5/2$; **b**) $S_1 = 5/2$, $S_2 = 2$

with the reduced case and with the known cobalt–cysteine systems [30, 31]. We learn therefore that: 1) the ^{1}H NMR shifts are accounted for; 2) the linewidth indicates a short nuclear T_2 which is the result of the coupling with a long τ_s. The magnetic coupling in homodimers is not expected to change τ_s which is then set around 10^{-11} s.

4.2 The reduced Fe_2S_2 Case

If we assume that the hyperfine coupling of protons sensing iron(II) and iron(III) is equal, the isotropic shifts can be calculated for protons sensing each of the two metal ions as expressed in Eq. (16). The energy levels for the couple $S_1 = 5/2$ and $S_2 = 2$ are shown in Fig. 4B. The values of the isotropic shift, δ, as a function of temperature calculated as reported in Appendix II are shown in Fig. 5A. We used $J = 100\ cm^{-1}$, the $E_{S_i'}$ and S_i' values of Fig. 4B and the coefficients of Appendix I for each set of S_i', S_1 and S_2 values. This assumption is not a loss of generality since we are interested in trends rather than in absolute values. In Fig. 5B are reported the isotropic shift values calculated at 303 K as a function of the J coupling constant. Palmer et al. [2] also included a parameter describing the zero field splitting of the $S = 2$ of iron(II) with analogous results. If we take a reasonable A_c/h value of 1.0 MHz (giving rise to a shift of 300 ppm in the oxidized monomeric protein) we can assign the signals in the range 140–100 ppm to the β-CH_2 protons of cysteines bound to iron(III) and the signals below 30 ppm to the β-CH_2 of cysteines bound to iron(II). In the 30 ppm range the α-CH of cysteines bond to iron(III) could also be present.

The temperature dependence of the isotropic shifts could be a check for this assignment. As shown in Fig. 5A, the shift values of the β-CH_2 protons of cysteines sensing iron(III) decrease with temperature, whereas those of β-CH_2 of cysteines sensing iron(II) increase with increasing temperature. This is what is indeed found for both the two ferredoxins (Fig. 6). It appears therefore that the signals above 100 ppm belong to the β-CH_2 of Cys residues sensing Fe(III) and the signals at 45 and 25 ppm to the α-CH; signals F, G, H and I are assigned to β-CH_2 of Cys residues of the Fe(II) domain.

Analogous treatment can be performed for the T_1 values. The experimental values at different temperatures are reported in Table 1. In Table 2 are reported the T_1 values calculated at 303 K for β-CH_2's and α-CH's protons, using Eq. (20), assuming only dipolar interaction with both metal ions to be operative. Unfortunately, there is not much temperature dependence of the experimental values and therefore no much information is added for signal assignment and structural purposes.

From the X-ray structure we know that the arrangement of the cysteines bound to the metal ions is highly distorted, in particular one cysteine is in a completely distorted position with respect to the others and one β-CH_2 proton is very near to both iron ions. Using these structural data we can calculate the proton T_1 values reported in Table 2 by assuming a purely point-dipolar metal-centered model. For reasons we are going to discuss the experimental values cannot be reproduced.

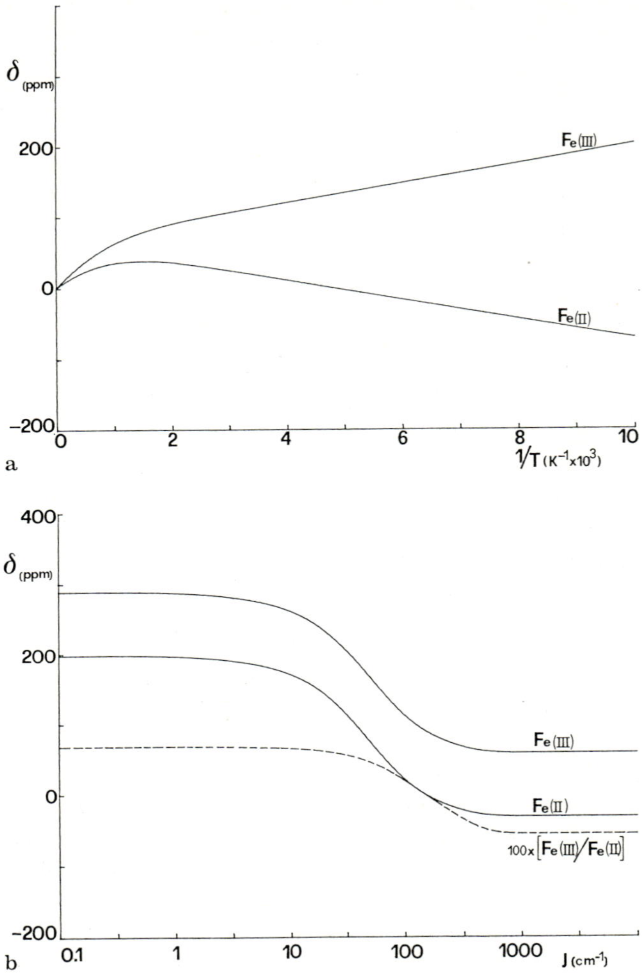

Fig. 5. Temperature dependence of the isotropic shift values calculated in a Fe(II)–Fe(III) magnetic coupled system, with a J coupling constant of 100 cm^{-1} (see text); **b**) Calculated dependence of the isotropic shift values in a Fe(II)–Fe(III) magnetic coupled system, and their ratio (*broken line*) as a function of the J value. Temperature is 303 K

Indeed, in the present case, where sulphur tends to host electrons from the metal ion, the quantitative analysis of the NMR data require a full account of spin delocalization. This is outside the scope of this review.

Finally NOE measurements might provide information on the inter-proton distances and add further information on the structure of the metal cluster and on the assignment of the NMR spectra.

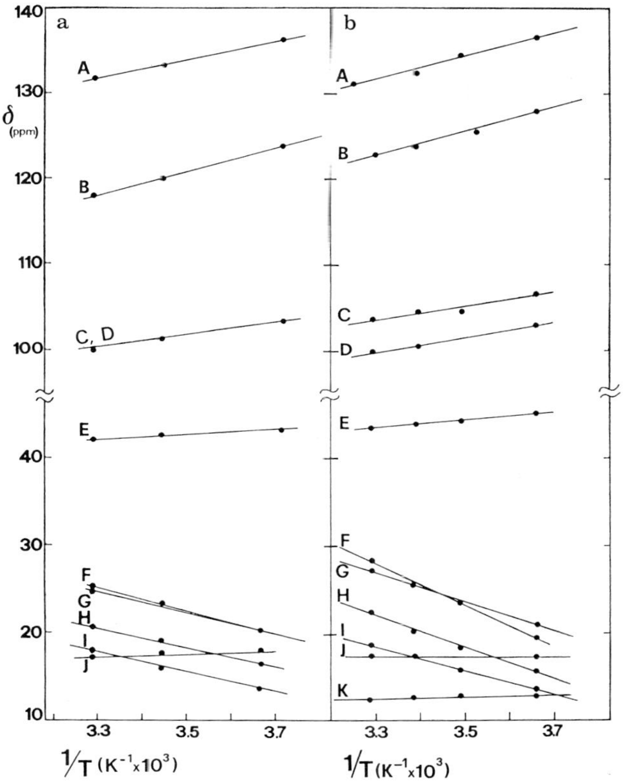

Fig. 6a, b. Experimental temperature dependence of the isotropic shift values for the reduced Fe_2S_2 ferredoxins from: **a**) spinach; **b**) red algae *Porphyra umbilicalis* (Same experimental conditions as in Fig. 2; signals' labelling as in Fig. 3)

5 Concluding Remarks and Perspectives in Dimetallic Systems

The isotropic shifts experienced by the cysteine protons belonging to both iron(II) and iron(III) have been accounted for on the basis of the magnetic coupling scheme discussed previously. In principle the differences in shifts experienced by the β-CH_2 protons of the cysteines coordinated to iron(III), and the differences experienced by the β-CH_2 protons of the cysteines bound to iron(II) could be ascribed to their geometrical properties around the two metal centers.

The first consequence of the present assignment is that the two oxidation states are well defined. The possibility that for each metal ion there is a distribution of population between oxidation number 2 and 3 under rapid exchange conditions is disfavored by the temperature dependence of the shifts: if such a distribution were

Table 1. Experimental T_1 values for reduced Fe_2S_2 ferredoxins at different temperatures

A) Spinach ferredoxin

Signal	T_1 (ms) Temp (K)	273	290	303
A		1	1.5	2.5
B		—	—	—
C		0.8	1.1	1.5
D		—	—	—
E		3.9	4.9	5.8
F		4.9	5.0	7.0
G				5.5
H		5.0	7.7	8.5
I		5.4	6.6	5.5
L		3.2	3.6	5.5

B) Red algae

Signal	T_1 (ms) Temp (K)	273	286	295	303
A		0.8	1.2	1.4	1.9
B		—	—	—	—
C		0.7	0.9	1.2	1.4
D		—	—	—	—
E		3.7	4.4	5.3	5.6
F		5.2	4.6	4.9	6.1
G		3.6			4.4
H		6.9	7.0	—	8.1
I		3.6	3.3	3.4	3.0
J		4.7	5.2		7.5
K		4.6	6.2	6.2	7.8

50%, the antiCurie behavior of the signals would have been minimal and different from that observed. Therefore we believe that there is neither fast electron exchange between the two metal centers nor an intermediate valence situation. Electron exchange could be slow; however, if the possibility existed of inversion of the two oxidation states, a second ^{1}H NMR spectrum different from the actual one and superimposed on it could be observed. Since this is not the case, the amount of this hypothetical species must be below detectability.

The next step in the analysis of two-iron two-sulphur proteins is a closer attempt to relate the NMR parameters with the structure of the protein by taking advantage of the X-ray data obtained on a similar protein from *Spirulina Platensis* [31] (which has about 75% homology with the present red algae ferredoxin [32] and about 65% homology with the spinach one [33]) and a ^{1}H NMR spectrum in the 50–0 ppm region very similar to those discussed here [34]. The shift values are related to the equatorial-axial nature of the protons of the CH_2 groups and the T_1

Table 2. Calculated T_1 values for the protons in the Fe_2S_2 cluster[a, b]

Proton	$r_{H-Fe(II)}$	$r_{H-Fe(III)}$	T_1 (ms)
HCβ Cys 41	3.0	5.5	5.7
HCβ Cys 41	4.1	6.3	18.1
HCα Cys 41	4.7	6.9	34.8
HCβ Cys 46	2.8	4.9	3.2
HCβ Cys 46	4.0	5.7	10.6
HCα Cys 46	5.1	6.4	24.9
HCβ Cys 49	3.8	2.4	0.1
HCβ Cys 49	4.1	2.9	0.3
HCα Cys 49	6.4	5.0	6.0
HCβ Cys 79	5.1	3.2	0.4
HCβ Cys 79	5.8	4.0	1.6
HCα Cys 79	6.9	5.0	5.8

[a] T_1 values calculated using Eq. (20), keeping into account the interaction with both metal ions, and using the structural data reported for ferredoxin from *Spirulina platensis* [31]. Cys 41 and 46 were assumed bound to Fe(III) and Cys 49 and 79 to Fe(II). Similar results are obtained for the reverse choice. Their difference is not large enough to discriminate between the two possibilities.
[b] Temperature is 303 K, $J = 100\ cm^{-1}$, $\tau_{s(Fe(III))} = 6 \times 10^{-12}$ s, $\tau_{s(Fe(II))} = 2 \times 10^{-13}$ s

values are essentially dipolar in origin, and therefore related to the proton-unpaired electron distance. At this point, however, other factors should be taken into account. It is generally true that the unpaired electrons are never localized on the metal ions. In the case of cobalt(II)-substituted superoxide dismutase, in which a copper-cobalt couple is present, nuclear relaxation indicated largely delocalized spin density [6]. Despite the effect of an unpaired electron on a carbon atom being small, it is very near to the protons and this makes it the predominant effect. Nevertheless, a full analysis on this system where magnetic coupling between cobalt and copper occurs has to start from the correct equation derived from the above treatment and requires the use of the Eqs. (20) or (21).

We would like to stress that any analysis should start using the correct equations derived for the magnetically coupled system. This treatment is quite general and can be applied to both homo- and heterodimers. In the former class we can mention μ-oxo diiron(III) porphyrins [4] and dinuclear copper(II) complexes [19]. The relatively sharp line of these systems is due to the large and negative value of J which reduces the overall magnetic susceptibility and then $\langle S_z \rangle$. The electronic relaxation rates are not increased by the magnetic coupling, as already discussed in Sect. 3. The NMR parameters can be analyzed by using Eqs. (16) and (20) for shifts and relaxation times, respectively. Among the heterodimers we may mention are Cu_2Co_2 superoxide dismutase [6], Cu_2Co_2 alkaline phosphatase [35] and proteins containing iron(II) and iron(III) centers as, for example, hemerythrin, acid phosphatase, ribonucleotide reductase [36]. For small complexes we should call attention to the possibility that the rotational correlation time τ_r is the dominant or a contributing term to the correlation time for the electron–nucleus interaction. In this case, in the present analysis of nuclear relaxation rate we should consider τ_r when it is relevant in Eqs. (20) and (21), whereas the analysis of the contact shifts still holds.

6 Inferences for Iron-Sulfur Systems of Higher Molecular Complexity

When the number of atoms increases, the calculations become more complex but the general philosophy still holds. Once the energy levels are calculated with their S_i' values and wavefunctions, the C_{ji} coefficients, which represent the ratio between $\langle S_{jz} \rangle_i$ and $\langle S_z' \rangle_i$, can be evaluated and used to compute the shifts and their temperature dependence with equation 16. Analogously, the C_{ij} coefficients squared can be used for the analysis of nuclear relaxation. Of course, also the analysis of the effects of magnetic coupling on the electronic relaxation rates of the single ions become more difficult. An analysis of the ^1H NMR spectra of a Co_4S_{11} cluster is available [37].

The cases of iron sulfur proteins with more than two iron atoms are presently a subject of international debate. For example, several oxidized proteins contain the cluster $Fe_4S_4^{3+}$ with three iron(III) and one iron(II) ions [1]. A theoretical treatment is available [38] which allows for the possibility of having two J values – among the six possible pairwise couplings among four ions – different from one another and from the other four. It is assumed that one of the J values within a Fe^{3+}–Fe^{3+} pair differs by an amount ΔJ_{12}, and one between the other iron(III) ion and the iron(II) ion differs by an amount ΔJ_{34} (Figure 7). The energies are given by the following equation [38].

$$E(S_i', S_{12i}', S_{34i}') = (J/2)[S_i'(S_i' + 1)] + (\Delta J_{12}/2)[S_{12i}'(S_{12i}' + 1)] + (\Delta J_{34}/2)[S_{34i}'(S_{34i}' + 1)] \quad (22)$$

where the energies are labeled according not only to the total S_i' value but also to the spin quantum numbers of the subpairs 1–2 and 3–4. Analogously to dimers, S_{12}' varies from $|S_1 - S_2|$ to $S_1 + S_2$ and S_{34}' from $|S_3 - S_4|$ to $S_3 + S_4$. S' varies from $|S_{12} - S_{34}|$ to $S_{12} + S_{34}$.

The coefficients are [38]:

$$C_{i1} = C_{i2} = (\alpha\gamma/\Delta_1),\ C_{i3} = (\delta_1\varepsilon/\Delta_2) \text{ and } C_{i4} = (\delta_2\varepsilon/\Delta_2) \quad (23)$$

where

$$\alpha = [S_{12i}'(S_{12i}' + 1) + S_1(S_1 + 1) - S_2(S_2 + 1)]/2$$

$$\gamma = [S'(S' + 1) + S_{12i}'(S_{12i}' + 1) - S_{34}'(S_{34i}' + 1)]/2$$

$$\Delta_1 = S_{12i}'(S_{12i}' + 1)S(S + 1)$$

$$\varepsilon = [S'(S' + 1) + S_{34i}'(S_{34i}' + 1) - S_{12i}'(S_{12i}' + 1)]/2$$

$$\delta_1 = [S_{34i}'(S_{34i}' + 1) + S_3(S_3 + 1) - S_4(S_4 + 1)]/2$$

$$\delta_2 = [S_{34i}'(S_{34i}' + 1) + S_4(S_4 + 1) - S_3(S_3 + 1)]/2$$

It is reasonable to take $J = 150\ cm^{-1}$, $\Delta J_{12} = 50\ cm^{-1}$ and $\Delta J_{34} = -50\ cm^{-1}$, in agreement with the J values found in Fe_2S_2 clusters. The result of the chosen J values is that the Fe^{3+}–Fe^{2+} pair is forced to be ferromagnetically coupled with $S_{34}' = 9/2$ ground state and the pair as a whole is antiferromagnetically coupled to

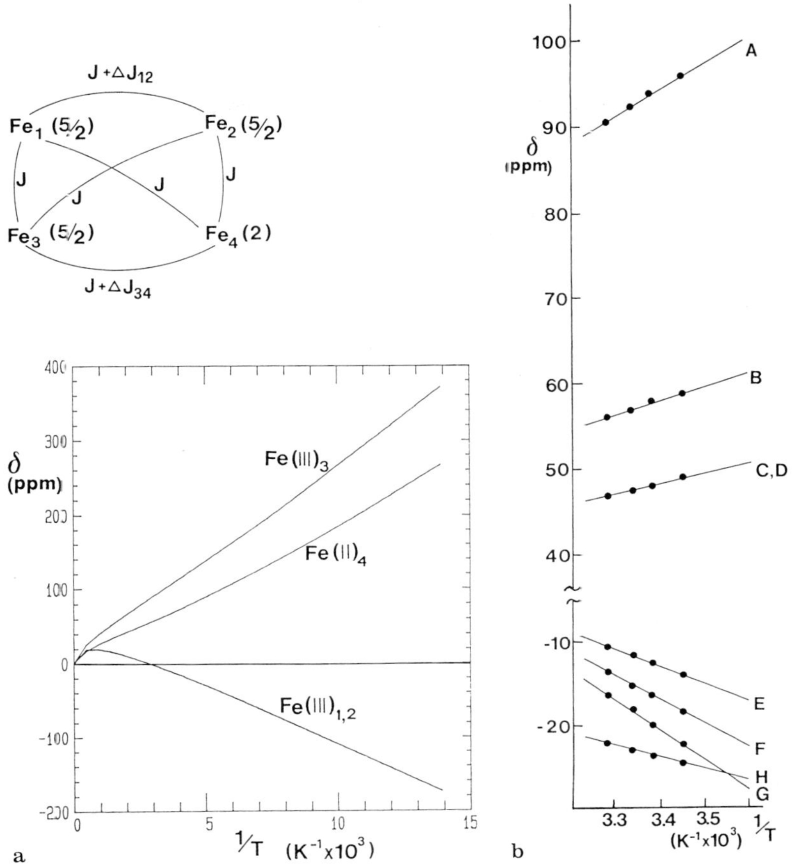

Fig. 7a. Temperature dependence of the isotropic shift values in a $Fe_4S_4^{3+}$ system with $J = 150$, $J_{12} = 50$, $J_{34} = -50\,cm^{-1}$ (see text); **b** Experimental temperature dependence of the isotropic shift values for the $Fe_4S_4^{3+}$ cluster in oxidized HiPIP II from E. halophila [41].

the 1–2 pair. In order to account for the Mössbauer data which suggest that two iron ions have a $+2.5$ oxidation state [39], a double exchange term within the 3–4 pair has been added in the spin Hamiltonian [38]. The importance of this term has been previously stressed [38, 40]. However, we like to underline that most of the effects on the eigenfunctions and eigenvalues due to the delocalization term can be obtained by using different J values (i.e. lower symmetry) in the clusters as mentioned above. We therefore drop such a term which is not needed at least for accounting for the gross features of the NMR data. The isotropic shifts will have different temperature dependence according to whether a proton senses the Fe^{2+}–Fe^{3+} pair or an Fe^{3+} ion, in a fashion similar to that reported in Figure 5 for the $Fe_2S_2^+$ pair. The ¹H NMR spectra of the $Fe_4S_4^{3+}$ system display signals both up and downfield, and could never be understood [1]. We propose that both kinds of far shifted signals belong to cysteines β-CH_2's. The signals upfield are due to negative $\langle S_{12z}\rangle$ of the iron(III) ions, just like $\langle S_{jz}\rangle$ can be negative for iron(II) in Fig. 5. The temperature dependences of the $\langle S_{jz}\rangle$ values for a $Fe_4S_4^{3+}$ system calculated with equation 22 and the above J and ΔJ values are shown in Fig. 7a; the experimental temperature dependence of the isotropic shifts of the oxidized HiPIP II from *E. Halophila* [41] is reported in Fig. 7b. More complicated patterns like the protons of three CH_2's experiencing downfield shifts and those of the fourth experiencing upfield shifts, observed for other HiPIP's [41, 42] can be accounted for by setting the two iron(III) inequivalent. We expect also that the protons sensing the iron(III) ion with positive $\langle S_{1(2)z}\rangle$ and being shifted downfield will have an anticurie behavior.

In the case of Fe_3S_4, it has been shown that there is a mixed valence Fe^{2+}–Fe^{3+} pair with $S' = 9/2$ ground state and a Fe^{3+} ion antiferromagnetically coupled to it [43]. The theoretical description of the system is obtained with three equal J values and a large delocalization term [40]. Again, the same $S' = 9/2$ ground state for the Fe^{2+}–Fe^{3+} pair can be obtained by reducing the coupling constant between the latter two ions. We predict that the protons sensing the above pair will behave like those sensing iron(III) in the Fe_2S_2 cluster, and the others will behave as the protons sensing iron(II). Some preliminary ^1H NMR spectra are available [44].

Besides the difficulties in evaluating eigenvalues and eigenstates for every particular case, the fitting of the ^1H NMR parameters will shed light into the details of the electronic structure of the iron-sulphur clusters.

7 Appendices

Appendix I

C_{ji} coefficients relating the hyperfine coupling of an S_1 or S_2 spin system in a monomer to that in a coupled system, for each S'_i spin level.

$$C_{1(2)i} = [S'_i(S'_i + 1) \pm S_1(S_1 + 1) \mp S_2(S_2 + 1)]/[2S'_i(S'_i + 1)] \qquad \text{(AI.1)}$$

$S_1 = 1/2$

S_2	1/2	1	3/2	2	5/2
S′	1	3/2	2	5/2	3
C_1	1/2	1/3	1/4	1/5	1/6
C_2	1/2	2/3	3/4	4/5	5/6
S′	0	1/2	1	3/2	2
C_1		− 1/3	− 1/4	− 1/5	− 1/6
C_2		4/3	5/4	6/5	7/6

$S_1 = 1$

S_2	1/2	1	3/2	2	5/2
S'	3/2	2	5/2	3	7/2
C_1	1/3	1/2	2/5	1/3	2/7
C_2	2/3	1/2	3/5	2/3	5/7
S'	1/2	1	3/2	2	5/2
C_1	− 1/3	1/2	4/15	1/6	4/35
C_2	4/3	1/2	11/15	5/6	31/35
S'		0	1/2	1	3/2
C_1			− 2/3	− 1/2	− 2/5
C_2			5/3	3/2	7/5

$S_1 = 3/2$

S_2	1/2	1	3/2	2	5/2
S'	2	5/2	3	7/2	4
C_1	1/4	2/5	1/2	3/7	3/8
C_2	3/4	3/5	1/2	4/7	5/8
S'	1	3/2	2	5/2	3
C_1	− 1/4	4/15	1/2	13/35	7/24
C_2	5/4	11/15	1/2	22/35	17/24
S'			0	1/2	1
C_1				− 1	− 3/4
C_2				2	7/4

$S_1 = 2$

S_2	1/2	1	3/2	2	5/2
S'	5/2	3	7/2	4	9/2
C_1	1/5	1/3	3/7	1/2	44/99
C_2	4/5	2/3	4/7	1/2	55/99
S'	3/2	2	5/2	3	7/2
C_1	− 1/5	1/6	13/35	1/2	26/63
C_2	6/5	5/6	22/35	1/2	37/63
S'		1	3/2	2	5/2
C_1		− 1/2	1/5	1/2	12/35
C_2		3/2	4/5	1/2	23/35
S'			1/2	1	3/2
C_1			− 1	1/2	2/15
C_2			2	1/2	13/15
S'				0	1/2
C_1					− 4/3
C_2					7/3

$S_1 = 5/2$

S_2	1/2	1	3/2	2	5/2
S'	3	7/2	4	9/2	5
C_1	1/6	2/7	3/8	44/99	1/2
C_2	5/6	5/7	5/8	55/99	1/2
S'	2	5/2	3	7/2	4
C_1	− 1/6	4/35	7/24	26/63	1/2
C_2	7/6	31/35	17/24	37/63	1/2
S'		3/2	2	5/2	3
C_1		− 2/5	1/12	12/35	1/2
C_2		7/5	11/12	23/35	1/2
S'			1	3/2	2
C_1			− 3/4	2/15	1/2
C_2			7/4	13/15	1/2
S'				1/2	1
C_1				− 4/3	1/2
C_2				7/3	1/2
S'					0

Appendix II

We want first to perform a sample calculation to show that Eq. (16), which we report here for convenience

$$\left(\frac{\Delta\nu}{\nu_0}\right)^{\text{con}} = -\frac{g_e\mu_B}{\hbar\gamma_I 3kT}\sum_j \frac{A_{c_j}\sum_i C_{ji}S'_i(S'_i+1)(2S'_i+1)\exp(-E_{S'_i}/kT)}{\sum_i (2S'_i+1)\exp(-E_{S'_i}/kT)} \tag{AII.1}$$

reduces to Eq. (4) in the high temperature limit. If $E_{S'_i}/kT \ll 1$ the exponential terms in the numerator and denominator are all $\simeq 1$ and AII 1 becomes

$$\left(\frac{\Delta\nu}{\nu}\right)^{\text{con}} = -\frac{g_e\mu_B}{\hbar\gamma_I 3kT}\sum_j \frac{A_{c_j}\sum_i C_{ji}S'_i(S'_i+1)(2S'_i+1)}{\sum_i (2S'_i+1)} \tag{AII.2}$$

Let us take $S_1 = 2$ and $S_2 = 5/2$, as in the case of reduced ferredoxins. From Appendix I we have

$S' = 1/2$	$S' = 3/2$	$S' = 5/2$	$S' = 7/2$	$S' = 9/2$
$C_{11} = -4/3$	$C_{12} = 2/15$	$C_{13} = 12/35$	$C_{14} = 26/63$	$C_{15} = 44/99$
$C_{21} = 7/3$	$C_{22} = 13/15$	$C_{23} = 23/35$	$C_{24} = 37/63$	$C_{25} = 55/99$

and the contribution from metal $1(S_1 = 2)$ is

$$\frac{-4/3\cdot 3/4\cdot 2}{2+4+6+8+10} + \frac{2/15\cdot 15/4\cdot 4}{2+4+6+8+10} + \frac{12/35\cdot 35/4\cdot 6}{2+4+6+8+10} + $$
$$+\frac{26/63\cdot 63/4\cdot 8}{2+4+6+8+10} + \frac{44/99\cdot 99/4\cdot 10}{2+4+6+8+10} =$$
$$= \frac{1}{30}(-2+2+18+52+110) = \frac{180}{30} = 6 = S_1(S_1-1)$$

while the contribution from metal 2 ($S_2 = 5/2$) is

$$\frac{7/3\cdot 3/4\cdot 2}{2+4+6+8+10} + \frac{13/15\cdot 15/4\cdot 4}{2+4+6+8+10} + \frac{23/35\cdot 35/4\cdot 6}{2+4+6+8+10} + $$
$$+\frac{37/63\cdot 63/4\cdot 8}{2+4+6+8+10} + \frac{55/99\cdot 99/4\cdot 10}{2-4+6+8+10} =$$
$$= \frac{1}{30}(7/2+13+69/2+74+275/2) = \frac{525}{60} = \frac{35}{4} = S_2(S_2+1)$$

Each is of course multiplied by the corresponding A_c. As we expected, these contributions are the same as those of the isolated ions.

We note that for both metal ions the stronger contributions come from the levels with higher S′, as it could be anticipated from the fact that these levels are "more paramagnetic". This increase in paramagnetic contribution with increasing S′ is, however, more pronounced for the ion with smaller S ($S_1 = 2$ in this example) than for the other. Moreover, for the ion with smaller S the contribution from the lowest level is negative. Therefore, with increasing J, the weight of the levels with higher S′ decreases, and this is reflected in a more marked decrease in the contribution of the ion with smaller S. Since in the ground S′ level the contribution from this ion is negative, an opposite temperature dependence of its contribution is expected. To illustrate this behavior the results of sample calculations are shown in Fig. 5. Figure 5A shows the temperature dependence of the contributions to the isotropic shifts from metal 1 ($S_1 = 2$) and metal 2 ($S_2 = 5/2$) for $J = 100\ cm^{-1}$. Metal 1 shows an opposite temperature dependence for any temperature below 10^3 K. Below room temperature the isotropic shift caused by metal 1 becomes negative. Figure 5B shows the isotropic shifts, as well as their ratios, as a function of J for T = 303 K.

Appendix III

Coefficients that should multiply the S(S + 1) term in the unpaired electron–nucleus coupling contribution to nuclear relaxation in a magnetically coupled system.

X_1 \ X_2	5/2	2	3/2	1	1/2
1/2	17/18 19/54	23/25 9/25	7/8 3/8	7/9 11/27	1/2 1/2
1	1049/1225 13/35	43/54 7/18	467/675 19/45	1/2 1/2	
3/2	107/144 173/432	113/175 379/875	1/2 1/2		
2	102031/165375 6257/14175	1/2 1/2			
5/2	1/2 1/2				

8 Acknowledgments

We are grateful to Drs. A. Bencini, G.N. La Mar, G. Martini, L. Noodleman, and J.J. Girerd for helpful discussions.

9 References

1. Bertini I, Luchinat C (1986) NMR of paramagnetic molecules in biological systems, Cummings, Boston
2. Dunham WR, Palmer G, Sands RH, Bearden AJ (1971) Biochim. Biophys. Acta 253: 373
3. Salmeen I, Palmer G (1972) Arch. Biochem. Biophys. 150: 767
4. La Mar GN, Eaton GR, Hom RH, Walker FA (1973) J. Am. Chem. Soc. 95: 63
5. a. Owens C, Drago RS, Bertini I, Luchinat C, Banci L (1986) J. Am. Chem. Soc. 108: 3298; b. Bertini I, Luchinat C, Owens C, Drago RS (1987) J. Am. Chem. Soc. 109: 5208
6. Banci L, Bertini I, Luchinat C, Scozzafava A (1987) J. Am. Chem. Soc. 109: 2328
7. Banci L, Bertini I, Luchinat C (1988) In: Que L Jr (ed) A.C.S. Symp. Ser., **372,** chap 4
8. a. Anderson RE, Dunham WR, Sands RH, Bearden AJ, Crespi HL (1975) Biochim. Biophys. Acta 408: 306; b. Poe M, Phillips WD, Glickson JD, McDonald CC, San Pietro A (1971) Proc. Natl. Acad. Sci. U.S.A. 68: 68; c. Salmeen I, Palmer G (1972) Arch. Biochem. Biophys. 150: 767; d. Nagayama K, Ozaki Y, Kyogoku Y, Hase T, Matsubara H (1983) J. Biochem. 94: 893
9. Bertini I, Lanini G, Luchinat C (1984) Inorg. Chem. 23: 2729
10. McConnell HM, Chesnut DB (1958) J. Chem. Phys. 28: 107
11. Abragam A, Bleaney B (1970) Electron paramagnetic resonance of transition ions, Clarendon, Oxford
12. NMR of Paramagnetic Molecules (1973) La Mar GN, Horrocks WDeW, Holm RH (eds) Academic Press, New York
13. Lauffer RB, Antanaitis BC, Aisen P, Que L Jr (1983) J. Biol. Chem. 258: 14212
14. Scaringe P, Hodgson DJ, Hatfield WE (1978) Mol. Phys. 35: 701
15. Slichter CP (1955) Phys. Rev. 99: 479
16. Chao CC (1973) J. Magn. Reson. 10: 1
17. Gatteschi D (1983) In: Bertini I, Drago RS, Luchinat C (eds) The coordination chemistry of metalloenzymes NATO-ASI C100, D. Reidel, Dordrecht, p 215
18. Gatteschi D, Bencini A (1985) In: Willett RD, Gatteschi D, Kahn O (eds) Magneto-structural correlations in exchange coupled systems, NATO-ASI C140, D. Reidel, Dordrecht, p 241
19. Byers W, Williams RJP: J. Chem. Soc., Dalton Trans. 1973: 555
20. Bertini I, Banci L, Brown RD III, Koenig SH, Luchinat C (1988) Inorg. Chem. 27: 951
21. Bertini I, Luchinat C, Mancini M, Spina G (1985) In: Willett RD, Gatteschi D, Kahn O (eds) Magneto-structural correlations in exchange coupled systems, NATO-ASI C140, D. Reidel, Dordrecht, p 421
22. Banci L, Bertini I, Luchinat C (1986) Magn. Res. Rev. 11: 1
23. Bertini I, Lanini G, Luchinat C, Mancini M, Spina GJ (1985) Magn. Reson. 63: 56
24. Solomon I, Bloembergen N (1957) J. Chem. Phys. 27: 575
25. Palmer G, Dunham WR, Fee JA, Sands RH, Izuka T, Yonetani T (1971) Biochim. Biophys. Acta (1971) 115: 711
26. Pyrz JW, Roe AL, Stern LJ, Que L Jr (1985) J. Am. Chem. Soc. 107: 614
27. Heinstand RH, Lauffer RB, Fikrig E, Que L Jr (1982) J. Am. Chem. Soc., 104: 2789
28. La Mar GN, Walker FA (1973) J. Am. Chem. Soc. 95: 6950
29. Werth MT, Kurtz DM, Moura I, LeGall J (1987) J. Am. Chem. Soc. 109: 273
30. Bertini I, Gerber M, Lanini G, Luchinat C, Maret W, Rawer S, Zeppezauer M (1984) J. Am. Chem. Soc. 106: 1826
31. Tsukihara T, Fukuyama K, Nakamura M, Katsube Y, Tanaka N, Kakudo M, Wada K, Hase T, aMatsubara H (1981) J. Biochem. 90: 1763
32. Takruri I, Haslett BG, Boulter D, Andrew PW, Rogers L (1978) J. Biochem. 173: 459
33. Nagayama K, Ozaki Y, Kyogoku Y, Hase T, Matsubara H (1983) J. Biochem. 94: 893
34. Matsubara H, Sasaki RM (1968) J. Biol. Chem. 243: 1732
35. Banci L, Bertini I, Luchinat C, Viezzoli MS, Wang Y (1988) Inorg. Chem. 27: 1442

36. Que L Jr, Scarrow RC (1988) In: Que L Jr (ed.) ACS Symp. Ser. 372: chap 8
37. Bertini I, Luchinat C, Messori L, Masak M (1989) J. Am. Chem. Soc. 111: 7300
38. Noodleman L (1988) Inorg. Chem. 27: 3677
39. Papaefthymiou V, Millar MM, Münck E (1986) Inorg. Chem. 25: 3010.
40. Münck E, Papaefthymiou V, Surerus KK, Girerd JJ (1988) in ACS Symp. Ser. 372, Que L Jr, Ed., Cap. 15
41. Krishnamoorthi R, Markley JL (1986) Biochemistry, 25: 60
42. Phillips WD, Poe M, McDonald CC, Bartsch RG (1970) Proc. Natl. Acad. Sci. U.S.A. 67: 682. Nettesheim DG, Meyer TE, Feinberg BA, Otvos GDJ (1983) Biol. Chem. 258: 8215. Sola M, Cowan JA, Gray HB (1989) Biochemistry, 28: 5261
43. Moura JJG, Xavier AV, Bruschi M, Le Gall (1977) J. Biochim. Biophys. Acta, 459, 278
44. Macedo AL, Moura I, Xavier AV, Le Gall J, Moura JJA (1989) J. Inorg. Biochem. 36: 254

Probing Metalloproteins by Voltammetry

Fraser A. Armstrong

Department of Chemistry, University of California, Irvine, California 92717, USA

Dynamic electrochemical methods, which have long held an important place among the techniques of the coordination chemist, have generally remained unexploited by those seeking to understand the complex and often elusive chemistry of metal centres in proteins. For a number of reasons, electron transfer between electrodes and proteins has been regarded as being too slow and irreversible to provide useful information. This article seeks to counter such a view and outlines the advances that have been made towards achieving and interpreting voltammetric responses from metal-containing active sites. The main theme, *exploitation*, is developed through discussion of several investigations that demonstrate the advantages that are now on offer from electrodes that "talk" to metalloproteins. Far from being restricted to measurements of stable equilibria, voltammetry can address reactive states; species which are thermodynamically inaccessible by normal chemical titration or which display interesting yet complicating dynamic properties such as structural change or catalytic activity. In such cases, the coupled processes are *visualised* and may be investigated quantitatively and under controlled conditions. The resolution of chemical activity which is thus afforded extends even to multi-site enzyme complexes.

Structure and Bonding 72

1 Introduction

Voltammetric techniques now feature so widely in the repertoire of coordination chemists that characterization of the redox properties of metal complexes has largely become a routine task. Consider, for example, the information that may be obtained from cyclic voltammetry (CV). In the familiar direct current (DC) experiment [1–4], the potential is swept linearly with time, forward and back, between two limits. The current due to induced oxidation or reduction of molecules interacting with the electrode is recorded as a function of potential. Starting from a single solution species, sequential redox couples may be probed across a potential range that is limited only by the stabilities of electrode and solvent. Most obviously, we may determine reduction potentials: in the general case, the formal reduction potential $E^{0'}$ is equal to or usually very close to the average value of reduction (cathodic) and oxidation (anodic) peak potentials, i.e. $(E_{pc} + E_{pa})/2$. We can generally determine the number of electrons transferred in the electrode reaction, and whether or not the reaction is diffusion-controlled or involves adsorbed species. We may be able to estimate the rate constant for the electrode reaction which will reflect the intrinsic electron-transfer reactivity of the couple involved. We can establish the chemical stability (or fate) of transformed species and determine if these possess catalytic properties. Furthermore, rate constants for all types of coupled reactions may be obtained by analysis of the voltammetric waveform and observed current. In short, the voltammogram provides a *dynamic* picture.

An armoury of powerful electrochemical methods is available. Potential step techniques such as differential pulse DP or square-wave SW voltammetry offer advantages in sensitivity and resolution. Hydrodynamic techniques involving use of rotating disc or rotating ring-disc electrodes allow the chemical steps of the electrode process to be separated from mass transport. Electrochemical transformations may be monitored optically with spectroelectrochemical methods. Even the electrode interface itself is amenable to study by *in situ* spectroscopic techniques. Detailed descriptions of these methods are to be found in appropriate texts [1–4].

In the past twenty years the field of "bioinorganic" chemistry has grown into a major area. Yet while those engaged in the syntheses and characterization of small active-site "model" complexes have freely exploited the techniques of *dynamic* electrochemistry, others pursuing the study of native metal sites contained within proteins have resorted to *static* techniques that, while excellent in terms of specificity and precision, are largely limited to *stable* systems *at equilibrium*. Indeed, potentiometric titrations in the presence of small-molecule mediators have long been standard procedure [5, 6] for manipulating and quantifying biological redox chemistry. The reason for this difference in approach lies in a general belief that metalloproteins tend not to give a reversible or even quasi-reversible electrode response even if the redox-site in question is an excellent agent for its own biological role. As widespread accounts now show [7–11], such a view is steadily losing ground.

The aim of this article is to bring to light the possibilities that voltammetric techniques now offer for discriminate study, manipulation, and exploitation of the active-site chemistry of metalloproteins. Examples of these aspects will be given in Sects 4 to 7. The relationship between proteins and electrodes is a very complex one, and I shall endeavour, in Sect. 3, to give an overview of investigations that shed light upon the somewhat critical conditions required for observing the electrochemical response and the mechanisms by which heterogeneous electron transfer may occur. Firstly, however, we shall examine, briefly, three of the often-quoted obstacles which, understandably or otherwise, have impeded progress.

2 The Feasibility of Protein Direct Electrochemistry

2.1 "Metal Redox Centres in Proteins Tend to Be Buried"

Drawing a simple comparison with "bare" complexes like ferrocene, we would expect that the electron-transfer activity of a metal centre enclosed or "buried" within a protein molecule should be considerably suppressed. Several investigators have addressed the problem of how electrons may move rapidly between fixed remote sites in proteins, and it is certain that both distance and the nature of the intervening medium are important [12–16]. At an electrode interface the electron may have to traverse some depth of polypeptide matrix and may also encounter strongly bound ions and solvent molecules. How much of a restriction might this impose? We may reason that two limiting situations will occur.

On the one hand, the protein may be "designed" for just this type of reaction or, more correctly, its biological equivalent. Small electron-transfer proteins, usually with molecular weights below 15000, convey electrons between larger, often membrane-bound, proteins in which the active sites are themselves buried or secluded. (This mediation is required because direct reaction between donor and recipient is not feasible, for example because of physical separation and effective immobility.) Those who have studied biological electron-transport systems are well aware that these processes are efficient in terms of the high turnover rate that may be achieved with a small thermodynamic driving force. Yet this is in spite of the need for two long-range *intermolecular* electron-transfer reactions to occur at each cycle. A useful perspective is drawn if we compare rate constants of electron self-exchange for a selection of inorganic redox reagents (each of which show reversible or quasi-reversible electrochemistry) with some values determined for proteins. These are given in Table 1.

There is no evidence here to suggest that "buriedness' of redox centre prevents electron transfer *between* these proteins. On the contrary, the respectably high self-exchange constants indicate the ease with which the metal centres may respond in the absence of a driving force, i.e. at zero overpotential. According to Marcus [17], the rate constant for electron self-exchange may be equated directly with that for the electrochemical process. We may thus argue that a reversible faradaic response should be obtainable for an electron-transfer protein just as it is for the small inorganic complexes.

Table 1. Experimental values of electron self-exchange rate constants for a selection of inorganic complexes and metalloproteins

Redox couple	Self-exchange rate constant/$M^{-1}s^{-1}$
$Ru(NH_3)_6^{3+/2+}$	820[a]
$Fe(C_5H_5)^{1+/0}$	6×10^6 [b]
$Fe(CN)_6^{3-/4-}$	1×10^4 [c]
$Co(phen)_3^{3+/2+}$	1.1 [d]
Cytochrome *c* ($Fe^{3+/2+}$)	1.4×10^5 [e]
Plastocyanin ($Cu^{2+/1+}$)	7×10^4 [f]
Azurin ($Cu^{2+/1+}$)	2.4×10^6 [g]

[a] Meyer TJ, Taube H (1968); Inorg. Chem. 7: 2369;
[b] Pladziewicz JR, Carney MJ (1982) J. Amer. Chem. Soc. 104: 3544
[c] Campion RJ, Deck CF, King P, Wahl AC (1967) Inorg. Chem. 6: 672
[d] Baker BR, Basolo F, Neuman HM (1959) J. Phys. Chem. 63: 371
[e] Ref. 107, rate constant is limiting value in presence of ATP or $Co(CN)_6^{3-}$;
[f] Ref. 106, rate constant in presence of 8 mM $Co(NH_3)_6^{3+}$.
[g] Groeneveld CM, Dahlin S, Reinhammar B, Canters GW (1987) J. Amer. Chem. Soc. 109: 3247

Since mitochondrial cytochrome *c* was available commercially (horse heart muscle being the most common source) and could readily be purified to a high level, it formed the basic subject for most of the pioneering studies. Many ideas concerning the electrochemical mechanism, in particular, the mode of interaction with the electrode, have developed around the considerable wealth of information that is available [14, 18] on the structure and properties of the protein molecule. The extent to which the metal centre is "buried" is illustrated well in Fig. 1 which shows the 3D structure [19] of yeast (iso-1) cytochrome *c* and a view of the exposed active site. The major function of cytochrome *c* is as electron donor to cytochrome *c* oxidase (Complex IV), the membrane-bound enzyme that is the terminus of the aerobic respiratory chain and a site for proton translocation. Another physiological oxidant of cytochrome *c* (in yeasts) is cytochrome *c* peroxidase, a soluble enzyme whose crystal structure is known (see Sect. 7). The most important reductant of cytochrome *c* is the cytochrome c_1 component of the membrane-bound bc_1 complex (Complex III), but others (see Sect. 6, Scheme 5) include cytochrome b_5, sulfite oxidase, and flavocytochrome b_2 (lactate dehydrogenase, found in yeasts).

The molecular weight of cytochrome *c* is approx. 12500, and the active site is a heme (Fe porphyrin) group that is attached to the polypeptide chain at two cysteines through thioether bridges. The Fe atom itself lies in the plane of the ring, and the two axial positions are occupied by methionine sulfur (Met-S) and histidine nitrogen (His-N) donors. Except for a small part of the edge, the heme group is almost entirely buried within the protein. The environment "tunes" the potential of the redox couple (Eq. 1) to approx. 260 mV vs. SHE as measured at 25° with ionic strength I = 0.1 M, pH = 7. The Fe is low-spin in each redox state

$$\text{Cytochrome } c(III)(\text{Fe(III)}) \longleftrightarrow \text{Cytochrome } c(II)(\text{Fe(II)}) \quad (1)$$

Crystal-structure studies [18, 19] show that the molecular shape is that of a prolate ellipsoid with dimensions of the order of 25 × 25 × 37 Å. High-resolution

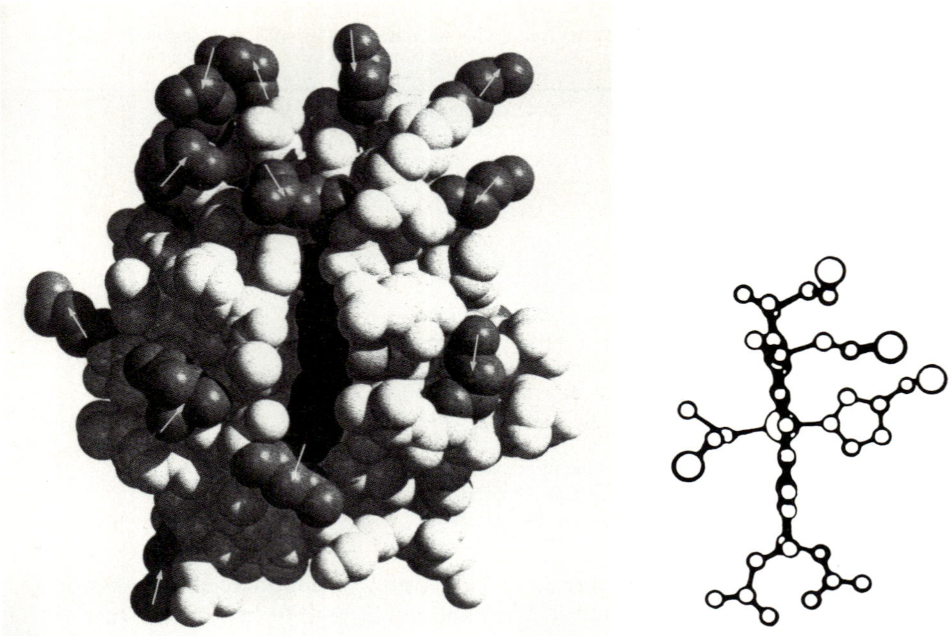

Fig. 1. Space-filling drawing of yeast *iso*-1-cytochrome *c* (see Ref. 19). Atoms coloured black form a part of the heme prosthetic group (shown exposed alongside). Atoms coloured grey (arrowed) belong to lysine and arginine side chains on the protein surface. Atoms coloured white represent those not falling into either of these categories. This illustration was kindly provided by Michael Murphy and Gary Brayer (University of British Columbia)

^{1}H NMR studies have shown that the Fe(II) form is particularly stable and the native conformation is essentially retained up to 97°. Cytochrome *c(III)* has a more labile structure and the Fe coordination is more easily disrupted. It is known (see Sect. 4.2) that, upon exposure to higher pH, methionine-S is displaced from the Fe(III) by another (N) donor. Since the Fe donor–acceptor orbitals can mix with the porphyrin π^* system, the effective d-electron density is extended to the heme edge and thus to the molecular surface at which this is exposed. As evident from Fig. 1, the actual protein surface area taken up by the solvent-accessible heme is very small (amounting to less than 1% of the total) so that if electron transfer to redox partners involves this entity (as it is widely believed) then the orientation of the protein during the encounter will be critical. The electron-transfer activity of the Fe centre may thus be termed "anisotropic".

The net charge of horse heart cytochrome *c(III)* at pH 7 is + 9 based upon the amino-acid composition. This charge arises from a considerable excess of positively charged groups (mainly lysines, $-NH_3^+$) with respect to glutamate and aspartate, which confer negative charges at neutral pH. The distribution of the positive charge over the protein surface is asymmetric. In the vicinity of the exposed heme edge, in particular, are a number of lysines whose charge is largely uncompensated by negatively charged residues. This area is believed to be the primary *interaction domain* for physiological reaction partners.

In certain cases, the interactions of cytochrome *c* with protein reaction partners have been studied directly or modelled from X-ray data [20, 21]. We picture a relatively long-lived complex with the two proteins juxtaposed to optimise the rate of electron transfer between sites. So long as the formation of this complex is very fast so that it may be considered always to be in equilibrium with the separated reactants, the overall second-order rate constant k_{12} is a simple product of two terms, the formation equilibrium constant **K** and the first-order rate constant k_{et} describing the electron transfer between sites

$$k_{12} = k_{et} K \tag{2}$$

The two protein surfaces are held together by a multiple combination of hydrogen bonds, hydrophobic contacts and coulombic forces (salt bridges). The mobility of surface residues demands that our picture is not that of the docking of rigid structures, but rather the optimization of forces within a dynamic assembly. Since the electron-transfer activity of each buried redox centre may be termed "anisotropic", their chemistry can be addressed optimally only by redox partners that can associate in the correct manner. The protein–protein complexes thus envisaged view cytochrome *c* as recognizing and binding at a domain of groups whose properties complement those of its own interaction domain. Thus, typically, this complementary domain will feature negatively charged residues, i.e. aspartate or glutamate. The result is a system which exhibits a much greater degree of redox selectivity than is encountered among small molecules.

Ideally, proteins should prefer their natural redox partners and may not even be able to approach, in an intimate manner, those with which electrostatic interactions are strongly repulsive. This might indeed be expected for the self-exchange reactions and we should note, from Table 1, that the appreciable rates measured for two of the proteins are obtained through the inclusion of multi-charged anions or cations which may form ionic bridges or create considerable charge shielding between like-charged surfaces. In conclusion, the "buriedness of centre" argument requires qualification. "Buriedness" is not prohibitive. However, for an electron-transfer protein, it introduces anisotropy of reactivity and hence specificity. Electrochemistry may therefore be critically dependent upon how the protein interacts with the electric field at the electrode–solution interface. The answer lies, at least in part, upon an extension of the ideas of Hubbard and co-workers [22] who demonstrated "tailoring" of electrode surfaces to achieve specificity in reactions of small ionic complexes.

The other limiting situation concerns the larger redox proteins, most obviously those which are classified as enzymes. Here we need to re-state the "buriedness of centre" argument. The metal centre may again be buried beneath the molecular surface but, more importantly, this could lie within a "pocket" that prevents approach of all but small substrates or the relatively small electron-transfer proteins discussed above. The problem is more one of steric hindrance since the active site metal may actually be accessible to solvent. Under these circumstances, an electrode surface—in its ideal form a planar surface—might be a poor redox partner indeed.

Should steric hindrance be identified to be the problem, more elaborate modification either of the electrode or even of the protein itself becomes necessary. Modification of the protein might include introduction of electron "relays" that can enable electrons to reach otherwise inaccessible active sites starting from a geometrically simple surface contact. Here, however, the aim of having a technique to characterize the chemistry of *native* proteins falls aside. On the other hand, modification of the electrode surface means the creation of a suitable conducting microstructure that is capable of penetrating deep pockets.

2.2 "Diffusion Coefficients of Proteins are Too Small"

In most experiments, we are interested in obtaining a voltammetric response from protein molecules that undergo rapid exchange with those in the bulk solution. Here, in what I shall sometimes refer to as the "*Bulk Solution mode*", the current is limited ultimately by mass transport of reactant and product. In other experiments we may seek to address only protein molecules that are strongly adsorbed, i.e. bound tightly to the electrode surface. In this case (which I shall term the "*Adsorbed Film mode*") mass transport does not concern us further unless the protein of interest is an enzyme that yields a biocatalytic current from freely diffusing substrate molecules. Mass transport of a molecule or particle under the influence of a concentration gradient is described by a *diffusion coefficient* D that depends upon the molecule or particle size and hydrodynamic properties. For *bulk solution* electrochemistry, we must visualise the consequences of our redox system comprising *macro*molecules as opposed to small reagents. Since we are interested in a relative comparison, this can be done in the following idealised manner.

For a sphere of radius r, for which interactions with solvent molecules are negligible, the diffusion coefficient in a medium of viscosity η is given by the Stokes–Einstein equation—Eq. (3).

$$D = k_B T / 6\pi\eta r \tag{3}$$

If the sphere has mass m, with uniform density ρ, Eq. (3) can be rewritten as Eq. (4)

$$D = k_B T / \{11.7\eta(m/\rho)^{1/3}\} \tag{4}$$

Now, consider the reduction of solution species "O" at a planar electrode surface during a voltammetry experiment in which the potential is swept with time in a linear manner. This is, of course, equivalent to the first sweep of a cyclic voltammogram, and the faradaic current that we observe is ideally limited by the diffusion of "O" to the electrode surface. In this case the observed peak current i_p is typically described [1–4] by the *Randles-Sevcik* equation—Eq. (5).

$$i_p = 0.4463\, nFAC_0 \left(\frac{nF}{RT}\right)^{1/2} \upsilon^{1/2} D_0^{1/2} \tag{5}$$

in which n is the number of electrons transferred, A is the electrode surface area (in cm^2), C_0 is the concentration of component "O" (in moles cm^{-3}), υ is the scan rate in $Vsec^{-1}$, and F, R and T have their normal meanings.

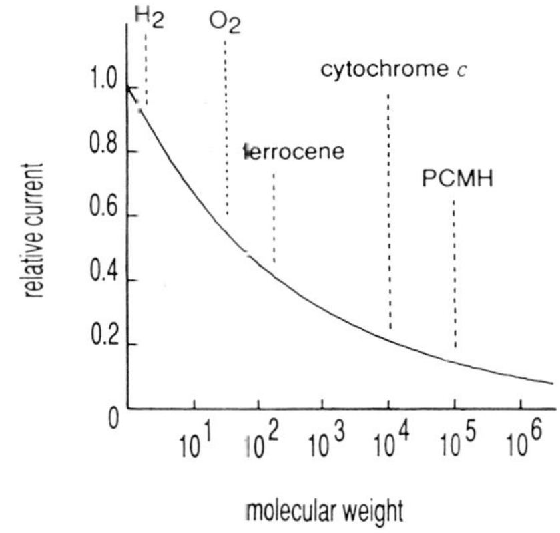

Fig. 2. Graph showing how voltammetric peak current is expected to diminish with increasing molecular weight in the case of redox-active spherical molecules diffusing to a planar electrode. PCMH = *p*-cresol-methylhydroxylase (see Sect 7.3)

If we extract the relevant parameters from the above equations, we arrive at the relationship $i_p \propto m^{-1/6}$. Our conclusion is that the diffusion-limited voltammetric current given by a redox species is surprisingly *insensitive* to its molecular weight *per se*. This is illustrated in Fig. 2 in which we see how i_p (and hence the effective sensitivity of voltammetric techniques) should vary with molecular weight (MW) in the limiting case of uniformly active spheres and linear diffusion. Remarkably, the diffusion restriction alone renders the current due to a macromolecule of MW 10^6 greater than 20% of that expected for a small complex of MW 100.

There are factors that we have not considered. The surface of a protein molecule generally comprises polar side chains that interact strongly with solvent molecules and ions. These interactions impede its mobility. Furthermore, we have neglected to take into account the anisotropy of reactivity that we outlined above. Diffusion to the electrode surface must be coupled to corrective rotational motion either on approach or as a "rolling" movement during encounter, otherwise contact may be restricted to an inactive area of protein surface. The question arises, "What is meant by a diffusion coefficient?". The value which is relevant to a voltammetric experiment, in which there is a dependence upon molecular orientation, must be lower than the value which is determined by a technique like ultrafiltration. The picture afforded by Fig. 2 is thus optimistic in that it compares the *maximum* faradaic responses that may be achieved.

In this article I shall adopt a practical view and describe as *diffusion-controlled* all cyclic voltammetry that adheres to a direct dependence of i_p upon $\upsilon^{1/2}$.

2.3 *"Proteins Usually Adsorb Strongly and Denature at Electrode Surfaces"*

The tendency of proteins to adsorb strongly and often denature at surfaces is well known [23–25]. The aspect of adsorption itself is central to discussions of the mechanism of electron transfer. In this article, I shall refer to just *two* electrochemical mechanisms for rapid oxidation and reduction of protein molecules in bulk solution. These mechanisms arise from a simple consideration of protein adsorption, particularly that occurring *without denaturation* i.e. in which the protein's structural and chemical integrity is preserved (*native* adsorption).

We have referred already to the need for a protein to associate at the electrode surface in a manner that will allow rapid electron transfer with its redox centre. In the sense that the protein–electrode interaction must be intimate enough to accomplish electron transfer, *weak* adsorption might be regarded as a minimum *prerequisite* for observation of a sharp and reversible voltammetric response [7]. We may discuss the options according to the scheme shown in Fig. 3.

First, if oxidation and reduction of freely diffusing molecules occurs *only* by *direct* electron transfer between electrode surface and the metal centre, then there are two limiting cases. At the one extreme, adsorption is weak. Both adsorption and desorption are rapid, but, during its brief association with the electrode surface, the protein molecule has been oriented so as to allow fast electron transfer to occur. The corresponding voltammetric response arises from molecules that are in rapid equilibrium with the bulk solution *via* direct exchange. This is the main theme of what I shall refer to as the *Rapid and Reversible Binding* (***RRB***) mechanism [7]. At the other extreme, adsorption may be so strong that the voltammetric response stems from molecules that do not, within the experimental time domain, exchange with those in the bulk solution. In the intermediate case, the response may arise from molecules both adsorbed and freely diffusing, and either may be addressed preferentially by varying the experimental time constant. Thus a

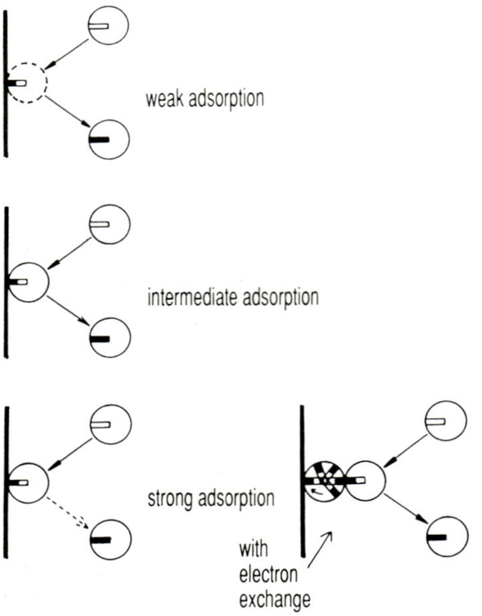

Fig. 3. Schematic representation of modes of electron transport from an electrode surface to protein molecules adsorbed and diffusing freely in solution (see text)

fast modulation will give emphasis to adsorbed species, while slow modulation will allow a relatively greater degree of exchange to occur with freely diffusing molecules.

In the Second mechanism for rapid oxidation and reduction of solution species, electrons are transferred *indirectly* between the electrode and bulk protein molecules by passage through a layer of strongly adsorbed protein [26]. This will be referred to as the *Adsorbed Protein Electron Exchange* (***APEE***) mechanism. The question is, "How does electron transfer to bulk species occur?" We have already considered that protein electron-transfer activity should be anisotropic; consequently, much is now demanded of the adsorbed protein molecule in terms of its dynamic versatility. Two factors may facilitate electron exchange. On the one hand, adsorbed molecules are not necessarily *immobilized*. Electron transfer to free protein molecules may occur via rapid rotation and consequent translocation of the redox centre and interaction domain. On the other hand, the protein may contain multiple redox sites so that a relay system can operate provided intramolecular electron transfer is rapid. Both modes of exchange are likely to be blocked if the adsorbed protein undergoes denaturation. In either case, and by analogy with the homogeneous self-exchange process, the *transient* interaction occurs between two protein molecules rather than between protein and electrode.

RRB and ***APEE*** mechanisms represent extreme cases. The ***RRB*** mechanism is to be favoured in cases in which strong protein adsorption is not apparent. The ***APEE*** mechanism, on the other hand, is to be favoured wherever diffusion-controlled voltammetry is observed despite strong adsorption with saturative electrode surface coverage. For a heterogeneous electrode surface, it is quite likely that both mechanisms can operate simultaneously with ***RRB*** and ***APEE*** occurring, respectively, at "coolspots" and "hotspots".

More generally, for characterization of the intrinsic properties of a protein, the detailed mechanism of electron transfer is not of importance. More relevant is the question, "How can I obtain useful information in the most direct manner with a minimal amount of sample?" We can distinguish two types of configuration. In the usual "bulk solution mode" experiment, the action is directed toward freely diffusing protein molecules, and the electrochemical perturbation is propagated by mass transport. Most reported work on proteins has been carried out in this mode, including, of course, bulk electrolysis. Thin-layer electrochemical cells offer several advantages including easy coulometric analysis and economy in sample usage [27]. The "adsorbed film mode" goes a stage further since the action is directed at or limited entirely to *adsorbed* species. With strong adsorption, the presence of protein molecules in solution is not required; and transferable films, even sub-monolayers of pre-adsorbed material, can be studied. This situation makes possible the characterization of proteins for which only very small quantities are available. The magnitude of the current will depend upon the surface coverage of molecules. Furthermore, if an *enzyme* is adsorbed in an electro-active native state, its catalytic activity towards freely diffusing substrate, observed as the "biocatalytic current", can be probed. Thus adsorption, far from being a prohibitive factor in protein electrochemistry, offers itself as an exploitable asset. The real problem, as we shall now discuss briefly, is denaturation.

If voltammetric techniques are to be useful for characterizing metalloproteins and their reactions, it is clear that denaturation must be prevented. Data obtained are invalid if the protein molecules responsible for the voltammetric response incur reconformation or inactivation whilst at the electrode surface. Natural electron-transfer pathways may become inoperative, thus leading to poor kinetics. Artifactual voltammetric waves may arise from degraded active sites.

Denaturation occurs because of a change in the balance of forces which normally favour the native conformation. The effects of high temperature, extremes of electrolyte concentration (including pH), and addition of solvents or other reagents are well known. At an electrode surface also, a protein molecule is subject to disruptive influences [28]. Distortion may occur as a result of the large electric field that is generated across the double layer. Part of the protein surface will be in contact with the electrode, and the normal ionic and hydration shell may be broken. There may be a strong tendency to make more extensive contact with the electrode surface by unfolding and "reburying" the hydrophobic interior with expulsion of numbers of water molecules—the so-called "hydrophobic effect". Furthermore, there may be specific chemical reactions illustrated, for example, by the affinity of metals such as mercury (Hg) for sulfur ligands as presented by cysteine, methionine, or the bridging sulfur atoms of Fe–S clusters.

Since proteins differ so greatly, there can be no general formula for controlling the adsorption process to minimise denaturation. The tendency of proteins to denature depends upon what we may term their "rigidity". In studies of protein adsorption on negatively charged (sulfonato) polystyrene latex particles, Lyklema and co-workers compared isotherms for human plasma albumin (HPA) and bovine pancreas ribonuclease (RNase) [23]. They found that, even under conditions in which the protein is negatively charged and the net coulombic interaction is thus repulsive, adsorption could occur if it was accompanied by denaturation. This is easy for the conformationally labile HPA, but much less so for RNase, which is far more "rigid". Thus "tough" proteins, for example those that have extensive β-sheet structure, should be more able to "survive" adsorption. Electron-transfer proteins normally have relatively robust structures. Thus we should not expect denaturation to be a prohibitive factor, particularly if low temperatures are used. Another interesting feature to emerge was that non-denaturing adsorption of protein on a like-charged surface is promoted by co-adsorption of counter ions, particularly those with multiple charge. This we may rationalize as follows: the resulting high charge density that would otherwise accumulate at the interface would either prevent adsorption altogether (coulombic repulsion in the manner responsible for the stabilization of colloids) or trigger structural changes to remediate the adverse electrostatics. An electrode is more complicated still, since its surface charge depends not only upon the chemical entities present, but also upon the potential that is applied by the potentiostat.

In conclusion, it is difficult to predict whether denaturation may or may not prove to be a problem in any particular case. To *control* the adsorption process by providing the correct combination of surface interactions is a vague, yet appropriate, description of the task in hand.

3 Electrodes and Interfaces for Proteins

3.1 Electrochemistry at Hg Electrodes

Early investigators of the direct electrochemistry of proteins made particular use of polarography (voltammetry at the dropping Hg electrode). The theory and instrumentation for this technique were well established [1–4]. Moreover, since measurements were made at a reproducible and continually *renewed* surface, there lay an answer, at least in principle, to the problem of electrode fouling by strongly adsorbed, denatured protein. But proteins do adsorb rapidly and irreversibly at Hg, and various studies have addressed the question of whether or not this is necessarily accompanied by denaturation. Reduction and reoxidation of disulfide bridges in contact with the Hg surface is electrochemically reversible, and this reaction has been exploited to determine numbers of accessible cystines and to investigate conformational changes, reversible and irreversible, that occur as these groups are reduced [29]. The activity of adsorbed urease, which depends upon the integrity of a disulfide bridge, could actually be modulated by the applied electrode potential [30]. Here it was clear that the process of adsorption *per se* did not cause inactivation of the enzyme.

By contrast, the Hg electrode has not fared well in studies of electron-transfer metalloproteins. In almost all cases, the electrochemical response has been dominated by undesirable features: signals arising from adsorbed molecules, which do not reflect the accepted redox chemistry of the metal centres, and irreversible waves due to sluggish reduction and reoxidation of protein molecules in bulk solution. Nevertheless, the studies that have been made are informative and some will be mentioned briefly here. All potentials given in this article have been corrected to correspond with the Standard Hydrogen Electrode (SHE).

Significantly, the best examples of direct electrochemistry at Hg have been studies on the robust tetraheme proteins termed cytochrome c_3. These are low-potential electron carriers (MW 14000) that occur in sulfate-reducing bacteria (*Desulfovibrio*) [31, 32]. Each of the four heme groups are relatively exposed to solvent and inter-heme distances are short, suggesting the likelihood of mutual interaction and facile intramolecular electron exchange. The electrochemistry of cytochrome c_3 from *Desulfovibrio vulgaris* (Strain Miyazaki) provides [33–35] a representative case. On a hanging Hg-drop electrode, it was found to form, very rapidly, a strongly adsorbed film—most likely of monolayer coverage—that was electroactive upon transfer to a protein-free buffer solution. The corresponding adsorbed-film voltammetric response was observed in the potential domain appropriate for the native protein, but it died away upon reduction. Differential capacitance measurements showed, however, that the film was retained upon reduction and, furthermore, that the reduced protein was also adsorbed from solution, in this case to give a film that gave no signals. With cytochrome c_3 in bulk solution, the response was electrochemically reversible and diffusion controlled. These observations were consistent with the ***APEE*** mechanism outlined earlier.

The presence of four tightly bound redox centres evidently enables the protein to form a "conductive" film, even if some inactivation occurs. The polarogram was broad, reflecting the presence of non-interacting redox sites having slightly different potentials. We shall discuss these results further in the next section.

Polarographic studies on mitochondrial cytochrome *c* showed [36–39] that it too adsorbs strongly at Hg, but, unlike cytochrome c_3, the adsorbed layer *inhibits* the electroreduction of protein molecules in bulk solution. Heterogeneous electron-transfer rates were found to be very dependent upon protein concentration. In DC polarography, one criterion for electrochemical reversibility is that the observed half-wave potential $E_{1/2}$ corresponds to the thermodynamic reduction potential for the species of interest. Anderson and co-workers found [36] that the reduction of cytochrome *c(III)* in Tris-cacodylate buffer (pH 6) proceeded essentially reversibly at Hg if the protein concentration was 20 μM or less. However, as the concentration was increased, $E_{1/2}$ underwent a progressive shift to more negative values; the effective overpotential being approx. − 100 mV at a concentration of 105 μM. Nevertheless, it could be demonstrated that reduction of freely diffusing molecules did occur. Analyses of limiting (plateau) currents showed these to be proportional to cytochrome *c(III)* concentration and essentially diffusion-controlled. Cytochrome *c(II)* prepared by bulk electrolysis at a Hg pool was fully active as a reductant for cytochrome *c* oxidase. Electrochemical *reoxidation* of reduced cytochrome *c* was not observable: at the expected potential, this would coincide with oxidation of Hg. The general conclusions from this work and subsequent studies by other groups [37–39] were that the sluggish electrode kinetics displayed in all but the most dilute of solutions were due to a barrier imposed by adsorbed protein molecules. Mechanisms suggested for the electron transfer to bulk molecules included: diffusion (now restricted) of the protein itself through the layer, electron mediation *via* the redox centres of adsorbed protein molecules, and direct electron transfer through flattened denatured adsorbed molecules. Cyclic voltammetry of a pre-adsorbed cytochrome *c(III)* film transferred to buffer solution (10 mM Tris-HCl, pH 7.6) showed a sharp reduction peak at approx. − 340 mV with no return wave [39]. Cytochrome *c(III)* adsorbed at Hg is thus electrochemically active, but it is in a form in which the potential is shifted to more negative values. As mentioned further in Sects 3.2 and 4.2, this pattern of behaviour may be traced to the relatively easy disruption of the Fe(III) active site.

Various studies of the electrochemical behaviour of ferredoxins, small electron-transfer proteins containing Fe-S clusters, have been reported. However, polarograms, and voltammograms using the hanging Hg drop electrode have generally been complicated by the appearance of large signals, attributable to strongly adsorbed species, that do not correspond to the expected potentials for the intact native systems [40–44]. This interference has been acknowledged by several authors [42–44]; it is apparent also in the study of Holm and co-workers who observed [45] that reversible, diffusion-controlled reduction of *Clostridium pasteurianum* ferredoxin occurred at Hg. It is likely that the problem arises, at least in part, from the nature of the metal centres themselves and the well-known affinity of Hg for sulfur ligands. In this class of protein, any chemical degradation of adsorbed molecules releases a number of "new" electroactive agents—$Fe^{3+/2+}$, S^{2-} and exposed cysteine–S^- residues. Most ferredoxins are highly negatively charged proteins, and in recent years it has come to light that those polarographic and voltammetric signals that are attributable to the *native* redox couples can be greatly enhanced by addition of multi-charged cationic species, including polylysine [46, 47]. Some discussion of this follows in Sect. 3.3.

3.2 *Reversible Electrochemistry of Cytochrome c: The Development of Functionalized Electrodes*

Realization that suitably modified or *functionalized* electrode surfaces could interact in a specific and non-degradative manner with proteins, to allow stable and reversible (i.e. "well-behaved") direct electrochemistry that is uncomplicated by artifacts, came about in the late 1970s. The chemical and mechanistic diversity of these so-called "functionalities" is evident from the three first demonstrations of such behaviour.

Eddowes and Hill found [48, 49] that essentially reversible cyclic voltammetry of horse mitochondrial cytochrome *c* could be achieved with a Au electrode onto which was adsorbed, from the same solution, the reagent 4,4′-bipyridyl. The result is shown in Fig. 4. The criteria described by Nicholson and Shain [50] for a one-electron process controlled by linear diffusion of species to a planar electrode surface are met very closely indeed. The value of $E^{0'}$, given by $(E_{pc} + E_{pa})/2$, was 255 mV, in good agreement with values determined by potentiometry. It could be argued that free, reduced 4,4′-bipyridyl played no part in the mechanism, since its reduction potential is much lower than that of cytochrome *c*. It was proposed [7] that the organic adsorbate allowed electron transfer to occur *directly* by providing, at the electrode surface, functionalities with which the protein could interact specifically and reversibly and thereupon donate or accept electrons rapidly. It was thus termed a *promoter* as opposed to a *mediator*, in which the latter is considered to convey electrons in bulk solution.

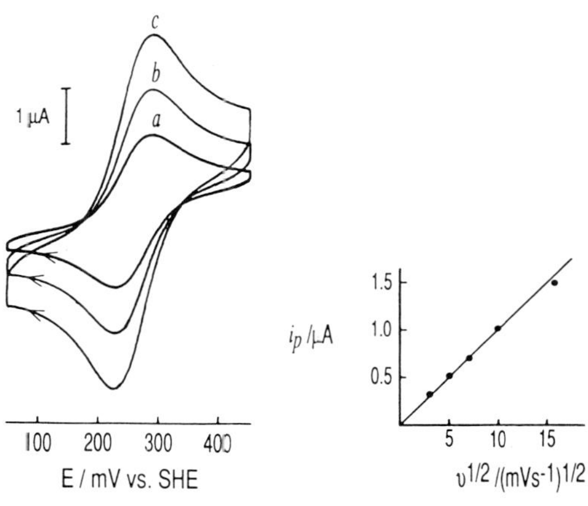

Fig. 4. Cyclic voltammetry of horse cytochrome *c* (approx. 0.4 mM) in $NaClO_4$ (0.1 M), phosphate buffer (0.02 M) at pH 7, in the presence of 0.01 M 4,4′-bipyridyl. Scan rate: (*a*) 20 mVs^{-1}, (*b*) 50 mVs^{-1}, (*c*) 100 mVs^{-1}. Inset shows dependence of peak current upon the square root of the scan rate. From Ref. 49, redrawn with kind permission

Another approach to the problem was reported by Hawkridge and co-workers. They found [51] that spinach ferredoxin gave a cyclic voltammetric response at a Au electrode that had been functionalized with a film of polymerized methyl viologen. A reduction potential in broad agreement with potentiometric data (approx. -430 mV) was estimated from the distorted voltammogram, while spectroelectrochemical experiments with a modified Au "minigrid" electrode demonstrated rapid reduction and reoxidation of the protein in bulk solution at respectably small overpotentials. Further studies [52] carried out with the rotating-disc-electrode technique gave a heterogeneous reduction rate constant of *ca.* 5×10^{-4} cm s^{-1}. The film itself was electroinactive; thus a surface-confined electron mediation effect seemed unlikely.

With a discovery that was free of arguments concerning electron mediation, Yeh and Kuwana showed [53] that cytochrome *c* gave stable electrochemistry (CV and DPV) at a Sn-doped-indium oxide electrode. The cyclic voltammetry at pH 7 was diffusion-controlled and reversible up to a scan rate of 500 mVs^{-1}. In this case we may now regard the electrode surface to be "naturally functionalized" through the presence of the stable oxide groups.

These early discoveries spawned a number of investigations largely devoted to understanding the protein/functionalized electrode interface. Questions to be answered included the following.

1 What role is performed by electrode surface functionalities, either those indigenous to the electrode surface or provided by inclusion of promoters? Do they:
   a) constitute sites that can engage in *reversible* interaction with the protein surface, perhaps analogous to reversible precursor complex formation in physiological processes? (This would be appropriate for the ***RRB*** mechanism.)
   b) prevent electrode fouling caused by strong *non-native* adsorption of protein?
   c) promote strong native adsorption of protein molecules? (This would be required for the ***APEE*** mechanism.)
   d) act by providing facile pathways for electron transfer to and from the protein's redox centre?

2) Can protein selectivity, in other words *macromolecular recognition*, be conferred upon the electrode by variation of the functionality?
3) What information may be derived from analysis of voltammetric data and what assumptions need to be made?

It is useful, first of all, to consider more closely results obtained with "bare" metals like Hg. Other metals, including Pt, Ag or Au have been reported to be generally unsatisfactory as electrodes for metalloproteins. However, there have been some interesting observations that place this view in greater perspective.

Hawkridge and his colleagues searched for reversible electroactivity of proteins at "bare" metals [54]. They carried out experiments using Au and Pt electrode surfaces "cleaned" by various pre-treatments, including soaking in aqua-regia, warm dilute nitric acid or heating in a hydrogen flame. With these freshly prepared, it was possible to observe quasi-reversible cyclic voltammetry of cytochrome *c*. Plots of i_p against $\upsilon^{1/2}$ were linear to $\upsilon = 200$ mVs^{-1} and yielded a respectable diffusion coefficient $D = 1.1 \times 10^{-6}$ cm^2 s^{-1}. However, the response did not persist. (Rapid deterioration of an initially good voltammetric response, or "impersistence", is a central problem in protein electrochemistry. The cause may be either fouling of the surface by irreversible adsorption and denaturation, loss of active electrode surface functionalities, or a combination of both.)

The result here is significant since it provides basic ideas as to the minimum requirements of an electrode for eliciting a response from a protein. In this case a "clean" surface, devoid of strongly adsorbed contaminants such as hydrophobic organics, was evidently active. In later work, Reed and Hawkridge showed [55] that a Ag electrode, polished with alumina then pre-conditioned by repeated cycling between 380 and 100 mV, gave cyclic voltammetry and voltabsorptiometry of purified *un-lyophilized* cytochrome *c* with well-defined peaks for oxidation and reduction. The response was diffusion-controlled at low scan rates, the $E^{0'}$ value was +250 mV and, most importantly, persisted for periods exceeding 12 h. If lyophilized cytochrome *c* was used instead, the response was poor and impersistent. Likewise, if protein contaminant (notably a polymeric form which separates from native protein upon ion-exchange chromatography of lyophilized "pure" material) was added to electro-active cytochrome *c* solutions, the voltammetric response died away.

The Ag experiments are important since it is particularly feasible to detect and characterize adsorbed species by the Surface-Enhanced Raman Scattering (SERS) effect. Studies have indicated [56–58] that strong irreversible adsorption of cytochrome *c* occurs at Ag and that, for the less stable Fe(III) state, this is accompanied by a change in conformation. Direct evidence for this came from the spectroscopically determined reduction potential for the adsorbed protein, which showed a large negative shift, and detection of vibrations associated with non-native forms.

Two further studies on the electrochemistry of cytochrome *c* at Au have produced evidence that a non-native, yet electroactive conformer having a lower reduction potential dominates the voltammetry and inhibits electrochemistry of the native form. Parsons and co-workers reported [59] that pre-adsorbed cytochrome *c* gave a quasi-reversible couple at approx. −250 mV that corresponded to the peak potential obtained in simultaneous potential-modulated electroreflectance and capacitance studies. Bond and colleagues observed [60] a process associated with a reduction peak at approx. −180 mV for which reoxidation was observed only in the presence of I^- ions. As 4,4′-bipyridyl was titrated in, the low-potential reduction wave diminished and reversible waves corresponding to the native couple developed at 260 mV. They proposed that at the unmodified Au surface there might be preferential adsorption of the so-called State IV form [61] of cytochrome *c*(*III*), in which the S ligand from Met-80 has been replaced by NH_2– from a lysine residue (see Sect. 4.2). The transition from native (State III) to State IV conformer occurs normally at higher pH or if alcohols are added. Haladjian and co-workers found [62] that the low-potential response appeared also during experiments performed in the presence of 4,4′-bipyridyl as the solution was made increasingly alkaline.

While these observations point to the formation of an adsorbed film of non-native cytochrome *c* at Au or Ag, with general similarity to the results obtained at Hg electrodes, there is an important factor to consider. In all cases in which SERS or electroreflectance were studied (and for nearly all the voltammetric work) the cytochrome *c* used had not been purified further from its commercial (lyophilized state). In other words, the conditions prescribed by Hawkridge for reversible electrochemistry at clean Ag were not met. It is not difficult to demonstrate the heterogeneity of stored lyophilized cytochrome *c*, including the commercially available material described as "pure". Cation-exchange chromatography [63] on CM cellulose gives several bands attributable to various degradation products.

By contrast with cytochrome *c*, the properties of cytochrome c_3 adsorbed at Ag as determined by SERS or voltammetry are very similar [35] to those of the native state. Voltammograms observed for three types of cytochrome c_3 show resolution of the non-equivalence of redox sites. Also, apparently by contrast to the situation found at Hg electrodes, cyclic voltammetry shows that the reduced form remains electroactive. It is important to note here that the cytochromes c_3 are very stable proteins that survive extended periods at high temperature and extreme pH [64].

What is the nature of the "clean" metal surface that appears active, at least towards pure cytochrome *c*? Hawkridge and his group noted [54] that the active Au surface was very wettable. As projected to cytochrome *c* molecules it must be coated with H_2O or OH^- (and other anions) which comprise the Helmholtz layers. Extensive direct contact between electrode metal atoms and the protein, likely to result in chemical reaction and consequent denaturation, can occur only upon drastic disruption of the hydration layer. However, since Au or Pt have only a relatively low affinity for aquo-adsorbates as compared to hydrophobic organics the surface soon becomes coated with non-native forms of cytochrome *c* or contaminants, protein or otherwise, that are adsorbed more strongly. The native voltammetric response dies away. One view therefore is that the clean hydrated surface is *already* "functionalized" for cytochrome *c* electrochemistry. This may extend to interactions of some specificity since the vicinal H_2O and particularly OH^- may form hydrogen bonds or salt bridges with lysine residues on the protein's natural interaction domain around the exposed heme edge. Whichever rationale is most appropriate, the overriding problem remains that of practicality. We wish to *use* direct electrochemical techniques as a tool for characterization. The "bare" metal surface is a critical entity that appears at best to be easily poisoned by trace amounts of protein impurities and degradation products.

The detailed mechanism by which 4,4′-bipyridyl and (as subsequently found) a large number of other reagents act to promote cytochrome *c* electrochemistry at noble metals has posed a very interesting problem. While studies by Hill's group and others have established a number of points, there still remain elements of controversy. Titration of the reversible electrochemical response (using AC voltammetric peak currents) as a function of the concentration of 4,4′-bipyridyl or 1,2-bis(4-pyridyl)ethene showed [49] Langmuir behaviour indicative of monolayer formation. Respective Langmuir constants were approx. 5 mM and 0.5 mM; thus 1,2-bis-(4-pyridyl)ethene is adsorbed more strongly. The Oxford group has favoured [7] a specific interactive role for the promoter; in this case it was suggested that pyridyl-N atoms projecting from the adsorbate layer formed hydrogen bonds to the lysine-NH_3^+ residues at the protein's interaction domain. As outlined previously, the ***RRB*** mechanism views the protein/functionalized-electrode interaction as being analogous to the physiological protein–protein precursor complex, i.e. the rate of electron transfer depends upon the strength of rapid, but reversible binding (adsorption) as expressed simply in Eq. 2.

A quantitative kinetic investigation of the Au/4,4′-bipyridyl/cytochrome *c* system was carried out [65] with the rotating-ring-disk technique. In terms of the ***RRB*** mechanism, rate constants for adsorption and desorption of cytochrome *c* were, respectively, 3×10^{-2} cm s^{-1} and 50 s^{-1}. The limiting first-order rate constant for electron exchange within the protein–electrode complex was determined to be 50 s^{-1}. This is certainly a reasonable value when compared with recent determinations [12–16] of rate constants for intramolecular electron transfer within protein molecules.

The bipyridyl reagents were also found to promote reversible electrochemistry of cytochrome *c* at Ag [66] and Pt [67] electrodes. As determined by SERS [68, 69] and ellipsometry [70], the molecules are adsorbed in "end-on" i.e. perpendicular fashion, so that one N is bound to surface metal atoms while the other N must be positioned towards the solution.

A practical drawback with these simple bipyridyl promoters was that a significant level of the reagent in solution was necessary in order to achieve the monolayer coverage required for optimal cyclic voltammetry. Promoters with a much higher affinity for adsorption at the metal surface (so all that was required was to dip the clean Au surface in a solution of the reagent, then rinse off the excess) were clearly desirable. The sample solution itself could thus be "promoter-free". If this was attempted [71] with 4,4′-bipyridyl or 1,2-bis(4-pyridyl)ethene, the resulting electrochemistry was relatively poor since the promoter desorbed readily. Taniguchi and his co-workers found [71] that the reagent bis(4-pyridyl)disulfide adsorbed much more strongly, giving stable and reversible cyclic voltammetry of cytochrome *c*. From SERS studies of its adsorption at Ag and Au, they found evidence [69, 72] that the surface contact involved the S atoms. Adsorbed species did not give the –S–S– stretching signal, but an identical SERS spectrum was obtained if 4-mercaptopyridine was adsorbed instead [69]. It was thus proposed that bis(4-pyridyl)disulfide adsorbed with cleavage of the –S–S– bond. Consistent with this, results from ellipsometry studies suggested [70] a kinetically complex adsorption mechanism in which initial adsorption of the reagent was followed by rearrangement. This could correspond to cleavage of the S–S bond in adsorbed molecules following which the two halves then bound to Au in perpendicular fashion.

There are now a large collection of reagents which, like the bipyridyl-types, are electro-inactive in the potential domain of cytochrome *c*, yet promote its direct electrochemistry at metal electrodes [71–81]. A selection of these are shown in Fig. 5. Many of them arise from an extensive survey carried out by Allen *et al* [73]. They described such reagents as "dual-functionality" or "XY" promoters since they have an "X" group that has a high affinity for the metal substrate and a "Y" group that is suitable for interacting with the protein surface. Groups X and Y are separated by a linking structure. The most successful groups X were soft bases like thiols or

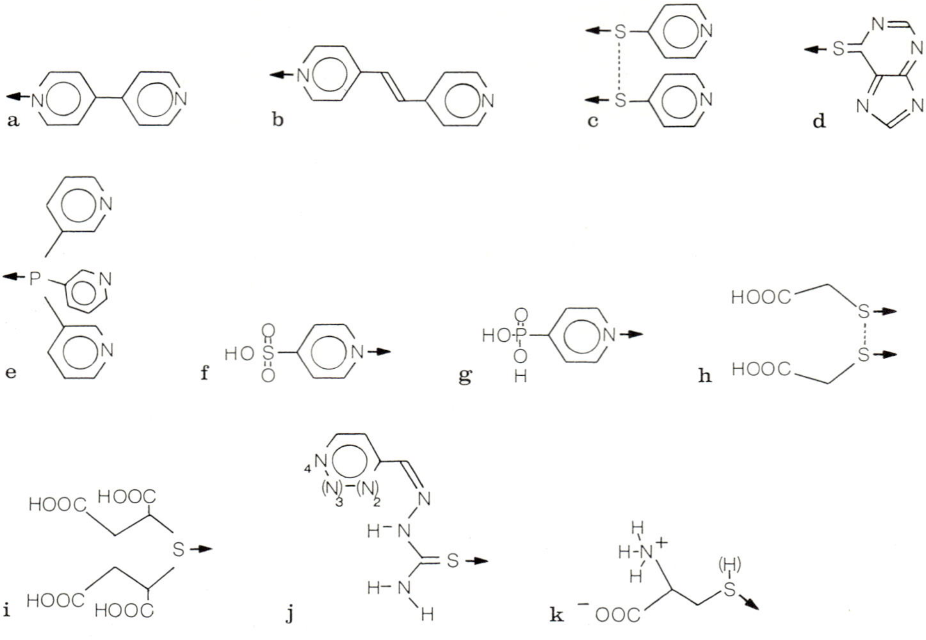

Fig. 5a–k. Structures of a selection of reagents that promote the direct electrochemistry of cytochrome *c* at Au (Ag, Pt) electrodes. Proposed orientations of adsorption are indicated by arrows directed toward the electrode surface. **a**) 4,4′-bipyridyl (Refs. 48, 49, 65): **b**) 1,2-bis(4-pyridyl)ethene (Ref. 49): **c**) bis(4-pyridyl)disulfide (Ref. 71): **d**) 4-mercaptopurine (Ref. 77): **e**) tris(3-pyridyl)phosphine (Ref. 73): **f**) 4-pyridylsulfonic acid (Ref. 73): **g**) 4-pyridylphosphonic acid (Ref. 73): **h**) thiodiethanoic acid (Ref. 73): **i**) 2,2′-thiobis(succinic acid) (Ref. 73): **j**) pyridine-*n*-aldehyde-thiosemicarbazone (PATS-*n*) (Refs. 76, 78, 80): **k**) L-cysteine (Ref. 79)

phosphines, which have a high affinity for metals like Au or Ag. The effective Y groups were either neutral hard bases or negatively charged groups such as $-CO_3^-$ or $-SO_3^-$.

A further clear feature to emerge was that molecules lacking delocalized orbitals in the linking structure could still be good promoters. Possible need for a conducting system had been suggested by the early observation [49] of the inactivity of bis(4-pyridyl)*ethane*. The demonstration that promoters do not function *a priori* by provision of a facile electron-transfer pathway echoes the simple observation of activity at "bare" metals described above. Instead, Allen *et al* proposed [73] that the advantage of a conjugated linking group lay in its ability to create a *rigid* conformation, thus imposing *directionality*, i.e. electrode—X-Y—protein. By preventing both X and Y from binding simultaneously to electrode surface atoms, the availability of X for interaction with the protein would be more certain. According to this hypothesis, bis(4-pyridyl)ethane was ineffective because rotation about the $-CH_2-CH_2-$ linkage allowed *both* pyridyl N atoms to be directed towards the Au. It was found also that for some reagents to act as promoters, particularly those without rigid conformations, "pre-activation" by treatment of the electrode surface with bis(4-pyridyl)disulfide followed by polishing, was required. The "pre-activation" step alone did not give a response with cytochrome *c*, but it was proposed that it could result in partial coverage of Au by S functionalities, thus minimizing simultaneous adsorption through X.

There are further, more subtle items of evidence for the existence not only of these directionality effects, but also for contributions by specific ion binding. For example, Hill et al investigated [78, 80] the promoter activities of reagents termed Pyridine-*n*-Aldehyde-Thio-Semicarbazones ($n = 2, 3, 4$), abbreviated PATS-n. These are included in Fig. 5. They found [80] that, with use of NaCl or cacodylate as the supporting electrolyte, there was a progressive deterioration in the cyclic voltammetric response as the promoter was varied from PATS-4, to PATS-3 and thence to PATS-2. The latter was virtually non-promoting. However, on changing to $NaClO_4$ as the electrolyte, PATS-2 was active [78]. Another interesting effect was observed [78] if PATS-4 was methylated, either at position 4′, in which case much of the activity was retained, or at position 2′, which resulted in loss of activity.

Evidence for the importance of coulombic effects is provided by the success of promoters in which Y is a negatively charged functionality like $-CO_2^-$ or $-SO_3^-$ and the failure [76] of those for which Y is $-NH_3^+$. Moreover Hill's group has extended the XY theme by devising promoters based upon peptides that contain cysteine [79, 81]. The X functionality is now provided by Cys-S while specific combinations of other amino acids provide the appropriate Y groups. For cytochrome *c*, peptides such as $(\text{Cys-Glu})_2$, $(\text{Cys-Glu(OMe)}_2)_2$ and $(\text{Cys-Gly})_2$ have been found to promote electrochemistry [81]. By contrast, $(\text{Lys-Cys})_2$ is ineffective.

Taniguchi and co-workers observed [72] that addition of bis(4-pyridyl)disulfide to a cytochrome *c* solution in contact with the electrode resulted in the complete replacement of the SERS signals due to adsorbed cytochrome *c* molecules by signals associated with adsorbed promoter. They found also [69] that the SERS signals due to adsorbed XY promoters were quite insensitive to the presence of cytochrome *c* in solution. The latter result shows that interaction between the promoter and the protein surface is *weak* and may be taken as evidence for the ***RRB*** mechanism. Certainly, the marked effects of small variations in the structure of the promoter, in particular those relating to *directionality* of hydrogen bonding groups, are reasonably explained in terms of *specific* interactions that orient the protein correctly during a brief encounter at the modified electrode surface.

With a change of emphasis, there is the view that promoters act in a more general manner, by preventing *direct* adsorption that is normally irreversible and may induce denaturation. Some support for this hypothesis stems from experiments [80, 82] carried out with cytochrome c_{551}. This is a smaller, bacterial protein which, although having a similar tertiary structure to that of cytochrome *c*, has only one lysine residue in the vicinity of its exposed heme edge. This domain is instead dominated by more hydrophobic residues. Haladjian and Bianco found [82] that cyclic voltammetry of cytochrome c_{551} could be observed with the presence of 4,4′-bipyridyl in solution. Results were poor; voltammograms showed comparatively rounded peaks, low currents and a large peak separation, but they did yield the appropriate reduction potential. Walton and co-workers found [80] that cytochrome c_{551} gave no cyclic voltammetry, at least in the expected, thermodynamically reversible region, either at bare Au *or* Au modified only by pre-dipping in 1,2-bis(4-pyridyl)ethene. The latter failure was strikingly different from the situation observed with cytochrome *c* where a pre-dipped 1,2-bis(4-pyridyl)ethene-modified Au electrode does elicit a response, albeit impersistent. The observation that both proteins are active at a bis(4-pyridyl)disulfide-modified Au electrode suggested that the difference lay in a greater tendency for cytochrome c_{551} to *displace* the more weakly bound X = N promoter from the electrode surface, thereby adsorbing irreversiby, and presumably with denaturation, in its place. The Y functionality, after all, is common to both pyridyl reagents.

However, prevention of direct adsorption cannot, alone, be regarded as a mode of action of promoters. Walton and co-workers noted [80, 83] that thiophenol (Y = C), although adsorbing strongly, did not act as a promoter either for cytochrome *c* or for the more hydrophobic cytochrome c_{551}.

Another hypothesis on XY promoters, which again does not exclude the basic ideas of reversible binding, is that they prevent denaturation during adsorption. Evidence from IR-reflectance studies has suggested [84] that 4,4′-bipyridyl and cytochrome *c co-adsorb* in an intimate manner at the Au surface, i.e. with direct contact between protein and the electrode surface. Niki and his co-workers proposed [84] that the main role of 4,4′-bipyridyl in this case is to stabilize the native conformation of the *directly-adsorbed* protein. This idea is supported for the action of purine (but not bis(4-pyridyl)disulfide, see above) by SERS studies carried out by Taniguchi and co-workers. They found [72] that for a solution of cytochrome *c* and purine, SERS signals persisted that were due to both species adsorbed *directly and simultaneously* at Ag.

Metallic-oxide electrodes for cytochrome *c* have been studied further, particularly by Hawkridge and co-workers [54, 85–87]. Sn-doped indium oxide and F-doped tin oxide each give well-behaved cyclic voltammetry, although preparation of the electrode surface seems to be a critical factor. With these materials, it was possible to observe, very readily, the effect of protonating electrode-surface functionalities implicated in interactions with the protein. The intrinsic properties of cytochrome *c* are not influenced significantly by changes in pH over the range 7 to 5 or by changes in ionic strength, but the voltammetry at oxide surfaces is very sensitive to these variations. At F-doped tin oxide, lowering the ionic strength from 0.2 to 0.002 resulted [54] in a decrease in peak currents and an increase in ΔE_p. This was attributed either to protein–electrode repulsion or to irreversible, electrostatically induced adsorption of cytochrome *c* rendering the interface "blocked". It was noted also that the response was attenuated if the solution was acidified from pH 7 to pH 6 but was restored upon alkalinization. For RuO_2 electrodes, Harmer and Hill found [88] a pH effect of similar type. With support from published measurements of the point of zero zeta potential (pH 5.1 to 6.1) and the potential of zero charge (235 mV), they proposed that the voltammetric response was linked to deprotonation of electrode-surface functionalities (pK 6). In its protonated state the electrode was inactive, as was to be expected if electron transfer depended upon hydrogen bonding or salt-bridge interactions with lysine-NH_3^+ groups on the protein.

However, Hawkridge and co-workers established also that cytochrome *c* undergoes strong adsorption at Sn-doped indium oxide electrodes with retention of native properties. They observed [54] that cyclic voltammetry could be obtained after transfer of an adsorbed film of cytochrome *c* to buffer solution. The charge passed during scans corresponded to fractional monolayer coverage, and the observed reduction potential was + 245 mV, only a little lower than the accepted thermodynamic value. More extensive adsorption (although weaker *reversible* adsorption would now be contributing) was evident from analysis of current/scan-rate data for the normal bulk-solution voltammetry of cytochrome *c* at the same electrode. Toward higher scan rates the peak current arose increasingly from adsorbed protein molecules [86, 87]. The surface excess thus calculated was high, corresponding roughly to monolayer coverage. This had to be taken into consideration when calculating standard heterogeneous rate constants [87].

Willit and Bowden made further studies of the electrochemistry of adsorbed films of cytochrome *c* on F-doped tin-oxide electrodes [89]. They found adsorption to be favoured at low ionic strength under mildly alkaline conditions (e.g. pH 8.5). One interesting feature of this work was the estimation of first-order rate constants for the electrode-protein electron exchange by measurement of ΔE_p as a function of υ. Values of the first-order rate constant k_{et} were found to be dependent themselves upon υ, thus indicating kinetic complexity that was inconsistent with a simple model for electron exchange within a stable configuration. Measured at a single scan rate, values of k_{et} fell within the range 1.2 to 6.2 s^{-1} and varied somewhat with pH and ionic strength. The results may be compared with the value of $k_{et} = 50\ s^{-1}$ obtained [65] for the 4,4′-bipyridyl-modified Au electrode in which rapid and reversible binding was postulated. Willit and Bowden proposed that the factor limiting k_{et} was a potential-dependent reorientation of adsorbed cytochrome *c* molecules. A further point of interest here that under the conditions for such strong and saturative adsorption to occur, fast electron transfer to free protein molecules would be likely to occur by the ***APEE*** mechanism.

Carbon materials, especially pyrolytic graphite (PG), have been found to be very suitable for the voltammetry of various proteins. From a practical standpoint, the large potential window offered by carbon essentially spans the entire range of biological reduction potentials. Hill and co-workers [90, 91] examined the voltammetric response of cytochrome *c* at various faces and preparations of PG. A piece of PG was mounted as an electrode in one of two ways In the first case, a block of material was oriented with the aromatic layers (the "basal" or "a–b" plane) parallel to the solution interface and the sides were sealed with silicone rubber. This leads to basal plane (PGB) electrodes. The electrode surface was prepared by cleaving with a razor blade to expose a fresh basal layer. In the second case, a block of material was mounted in epoxy resin so that the aromatic layers were perpendicular to the solution interface; this orientation is known as "edge" or "b–c", hence "PGE" electrodes. Here, the electrode surface was prepared by polishing with an aqueous slurry of Al_2O_3 (typically 1.0 or 0.3 μ) or diamond powder, then sonicating to remove adhering particles. Each of the two electrode surfaces gave a very different voltammetric response towards cytochrome *c*. The results are illustrated in Fig. 6. With the polished PGE, a near-reversible cyclic voltammogram was obtained with well-defined "peak-like" reduction and oxidation waves. Peak currents were proportional to $\upsilon^{1/2}$ up to at least 200 mVs^{-1}, thereby demonstrating a diffusion-controlled process. By contrast, the response at the freshly cleaved PGB surface was weak. There were no peaks, only a pair of faint sigmoidal-like waves. If the

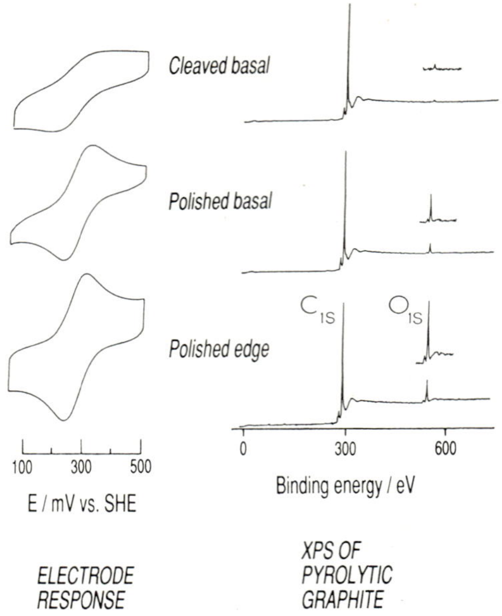

Fig. 6. Cyclic voltammograms (fourth scan, 20 mVs^{-1}) of horse cytochrome *c* at various types of pyrolytic graphite electrode. Protein is 0.15 mM in 5 mM Tricine and 0.10 M NaCl at pH 8. Temperature 20 °C. In each case the corresponding X-ray photo-electron spectrum of the graphite surface is shown. The scale enlargement for the O_{1s} peak is × 3

basal electrode was then polished, a much sharper peak-like voltammogram was generated; it was similar to that obtained with the polished PGE, but having rather broader waves and an apparently larger peak separation.

The chemical distinction between the various PG surfaces became clear from XPS analysis (ESCA). This showed [91] (Fig. 6) that the polished PGE was rich in surface oxides while the cleaved PGB contained only a very small number. This is indeed expected since cleaving between the aromatic layers should not, ideally, generate highly reactive sites for recombination with dioxygen. This is in contrast to any process that ruptures bonds *within* the layers. Polishing, particularly across the basal layers, creates a rich layer of C–O functionalities that include carboxylates, alcohols, ketones and quinones [92–94]. The relationship between surface oxide density and the appearance of the voltammetry provided further evidence for the need to have functionalities on the electrode surface. Another study bearing upon this work was reported by Anderson and co-workers [95] who investigated the electrochemistry of cytochrome *c* at carbon fibre electrodes. They found that voltammetric waves, seeming to arise from adsorbed native cytochrome *c* were obtained only after pre-conditioning the electrode by repeated oxidative and reductive (± 2.5 V) cycling. The effect of this treatment might be either to destroy and release surface contaminants and/or generate oxidized surface functionalities.

The voltammetry of cytochrome *c* at the pyrolytic graphite edge showed a marked pH dependence [91]. Analysis of the reduction current measured at a rotating-disk electrode as a function of pH showed that this was virtually abolished under acid conditions. A plot of $\log(i_{max} - i/i_{max})$ against pH was linear (slope 1.5) between pH 4.2 and 7.0 and yielded an effective pK of 5.6. A similar set of measurements on the reduction of $Fe(CN)_6^{3-}$ under conditions of low ionic strength (to amplify double-layer effects) showed an increase in current as the solution was acidified, and the corresponding analysis also gave pK = 5.6. The results indicated that each of the electrochemical reactions were controlled by a common acid–base equilibrium at an electrode surface functionality. One possibility suggested was a carboxylate with the pK raised by a neighbouring hydrogen-bond acceptor.

The latter results in particular suggested a rather close analogy between the reaction of cytochrome *c* at an electrode and that with a physiological partner for which activity could also be coupled to the protonation state of groups on the proteins' surface. Electrode surface functionalities that are negatively charged or able to act as hydrogen-bond acceptors may be able to bind to cytochrome *c via* the protonated lysine ($-NH_3^+$) residues close by the exposed heme edge. Switch-off occurs if the electrode groups are protonated. Experiments conducted by Taniguchi and co-workers provided further evidence that the interaction between cytochrome *c* and PGE functionalities resembles the physiological situation [96]. They observed that voltammetric waves of cytochrome *c* at PGE were removed very effectively by polylysine at pH 7. Polylysine is an inhibitor of the reduction of cytochrome *c* oxidase where it competes with cytochrome *c* for binding at the enzyme [97]. By contrast, it was found that polylysine was ineffective at inhibiting the electrochemistry of cytochrome *c* at Au electrodes modified by bis(4-pyridyl)disulfide. In the latter case, polylysine neither displaces the promoter from Au nor binds preferentially to the modified surface. Thus the interaction of cytochrome *c* with pyridyl-N groups appears not to resemble, at least in detail, the physiological interaction which is largely controlled by coulombic interactions.

Armstrong and Brown [98] tested the activity of PGE electrodes that had been "silanated". Treatment of electrodes with trimethyl- or triphenyl-silyl reagents generated a hydrophobic surface upon which oxygen-containing functionalities would be derivatised. The resulting cyclic voltammetry of cytochrome *c* consisted of weak sigmoidal-type waves similar to those observed at freshly cleaved PGB electrodes. This type of response was observed also at Au electrodes under conditions in which active promoters like bis(4-pyridyl)-disulfide were partially displaced by the "inert" adsorbate thiophenol [83]. As discussed in Sect. 3.4, subsequent examination of this type of response was to indicate that the attenuation was not due to a decrease in the intrinsic rate of electron transfer, but rather to a change in the conditions assumed for mass transport.

3.3 *Electrode Interfaces for a Wide Variety of Proteins*

Interfacial electrostatics pose a far more critical determinant for the electrochemistry of proteins than is usually the case for simple redox complexes. This is apparent even from the simplest of experiments in which one seeks to examine the electrochemistry of a protein other than cytochrome *c* at one of the electrode interfaces described above. The result is generally poor even with a high concentration of supporting (1:1) electrolyte to screen adverse coulombic interactions. Often, no voltammetric response is observed at all. A major factor in remedying this situation is generation of *attractive* coulombic interactions. Ideally, one needs to consider the nature and distribution of polar and charged groups on the protein surface and then modify the electrode surface or solution interface to provide an environment that *complements* these.

By contrast with cytochrome *c*, the majority of small electron-transfer proteins have an excess of acidic over basic residues. Consequently, they usually bear a significant negative charge which, as with the positive charge of cytochrome *c*, may be localized in what may be identified as an "interaction domain". Examples include the ferredoxins (small Fe–S proteins) and most plastocyanins (photosynthetic "blue" Cu proteins). Reduction potentials for these span a volt or more. Where appropriate, as for cases in which voltammetry or other electrochemical techniques have assisted in their characterization, we shall review structures and properties in the next section. But for the purpose of simplifying our discussion on electrode interactions, we can group them together as proteins that all appear to require positively charged or hydrogen-bond donor functionalities at the electrode interface.

At the PGE surface, these functionalities may be generated simply by protonation of the C–O groups. At pH 4, spinach plastocyanin gives [99] cyclic voltammetry with well-defined peaks, which is diffusion controlled up to 500 mVs^{-1}, whereas above pH 6 only an ill-defined, impersistent response is obtained in 0.1 M KCl and no response is observed at all at low ionic strength (see Sect. 4.3). The origin of this pH effect is less certain than is the case with cytochrome *c* since the isoelectric point (pI) of spinach plastocyanin is around 4. Electroactivity under mild acid conditions could instead be a result of plastocyanin behaving as an uncharged molecule.

However, stable oxide surfaces like PGE can be covalently modified to give positively charged functionalities. It was found [100] that Cr(III) complexes could be generated at graphite electrode surfaces by voltammetrically cycling a solution of $Cr(NH_3)_6^{3+}$ in aqueous ammonia between − 400 and − 1200 mV. Reduction of the complex gave labile Cr(II) species which would become linked to acidic C–O groups upon re-oxidation. The Cr-modified surface was verified by ESCA. Well-defined peak-type cyclic voltammetry of plastocyanin was thus obtained at pH 7.

Another example of electrode modification which places positively charged groups at the surface is a glassy carbon electrode modified by covalent attachment (carbodiimide coupling) of a viologen (DAPV = *N*, *N'*-di(γ-aminopropyl)viologen). This has been reported [101] by van Dijk and co-workers. The DAPV-modified electrode was itself electroactive: the surface-confined species yielding adsorption-type voltammograms with $E^{0'} = -395$ mV. However, when ferredoxins (either the [2Fe–2S] protein from spinach, or the 2[4Fe–4S] protein from *Megasphaera elsdenii*) were added to the solution, they gave excellent quasi-reversible cyclic voltammetry with $E^{0'}$ values of −400 mV and −375 mV respectively, in both cases appropriate for the native species. Although these values were very close to the viologen potential it is important to note that peak currents were *diffusion-controlled*, up to a scan rate of 500 mVs^{-1} for spinach or 200 mVs^{-1} (thereafter curving downwards) for *M. elsdenii*. While in this case a mediatory role for the viologen functionality was certainly feasible, a more direct explanation could be that its approx. 3 + charge acted to promote reversible interaction of the protein at the electrode. A similar rationale might be appropriate for the polymerized methylviologen-Au electrode, reported by Hawkridge and co-workers, which was active toward spinach ferredoxin [51, 52].

Hill and co-workers pursued the theme of XY promoters to investigate compounds in which Y is a basic group [76]. Examples of these, 2-aminoethane and 2,2′-dithiobisethanamine, whose amino groups are protonated at pH 8 and below, gave well-defined and stable cyclic voltammetry of plastocyanin while being inactive toward cytochrome *c*. All three isomers of PATS-n (Fig. 5) were effective, thus demonstrating that one reagent could promote the electrochemistry of two very different proteins. Similar findings were made with a derivative of cytochrome *c*. This had been chemically modified to carry an overall negative charge by attaching carboxydinitrophenyl (CDNP) groups to lysine residues. The CDNP-cytochrome *c* gave a stable and essentially reversible response at Au electrodes modified by adsorption of 2-aminoethane, 2,2′-dithiobisethanamine, and PATS [76], although, like plastocyanin, it was not active at Au electrodes pre-dipped in bis(4-pyridyl)disulfide. The stability of the plastocyanin voltammetry, but not that of modified cytochrome *c*, was dependent upon temperature. Well-defined and persistent electrochemistry was observed provided the solution was kept cold (around 3 °C). At higher temperatures, the response showed impersistence. This problem occurred in studies of plastocyanin at other electrodes, even PGE at which the functionalities are very stable. It implied that plastocyanin might be particularly susceptible to surface-induced denaturation.

Further success with various proteins has been achieved by use of promoters based upon peptides incorporating lysine(s). Hill and his group [81] found that $(Lys\text{-}Cys)_2$ but not $(Cys\text{-}Phe)_2$ or $(Cys\text{-}Tyr)_2$ promote the electrochemistry of plastocyanin at Au. There is considerable scope for developing this idea to provide electrode binding sites that might be specific for one particular protein.

Very effective promotion of electrochemistry of proteins at various electrode surfaces may be achieved by the simple addition to solution of ionic reagents of multiple charge opposite in sign to that carried by the surfaces of protein and electrode. It is well known that a number of electron self-exchange reactions are catalyzed by counter ions (as indicated in Table 1), and such effects may be important in physiological electron-transfer reactions. For example, reduction of the $P700^+$ centre of membrane-bound Photosystem I by plastocyanin, which encounters *coulombic repulsion* at neutral pH, is promoted by Mg^{2+} ions [102]. It is also most likely [88, 91] that the oxide-type electrode surfaces PGE and RuO_2 (and IrO_2) bear a negative charge at pH > 6 over much of the potential range appropriate for aqueous studies.

Promotion of protein electrochemistry by multi-charged counter ions shows several interesting features. A typical case, illustrated in Fig. 7, concerns the 2[4Fe–4S] ferredoxin from *Clostridium pasteurianum*. This is a low-potential electron carrier (MW approx. 6000) that carries a large overall negative charge, possibly as high as − 10 for the fully oxidized state. Structurally, it is very similar to the ferredoxin from *Peptococcus aerogenes* (see Fig. 14). Voltammetry at PGE

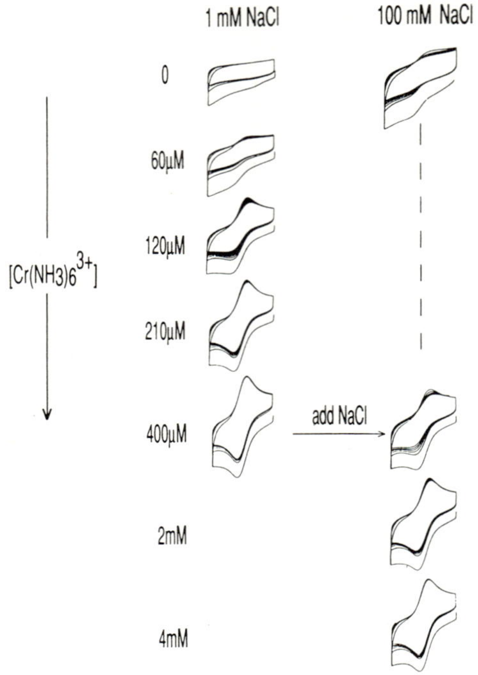

Fig. 7. Cyclic voltammetry of *Clostridium pasteurianum* 2[4Fe–4S] ferredoxin showing the effects of successive additions of $Cr(NH_3)_6^{3+}$ to solutions (buffered with 5 mM Tricine at pH 8) containing 1 mM or 100 mM NaCl as background electrolyte. Temperature = 25 °C. The PGE electrode was freshly polished at each stage. Cyclic voltammograms shown are consecutive cycles (scan rate 20 mVs^{-1}) recorded from the time of initiation at the positive limit

electrodes is promoted [91, 103, 104] by addition of multi-charged cations, typically Mg^{2+} or $Cr(NH_3)_6^{3+}$. It was found [91] that under more conventional conditions of 0.1 M NaCl electrolyte, only a poor and impersistent response was obtained at pH 8 (at which the PGE surface is in the deprotonated state). No response was observed at all if the concentration of supporting electrolyte was at the millimolar level, but as $Cr(NH_3)_6^{3+}$ was titrated in, the single voltammetric response due to both redox centres (which have indistinguishable reduction potentials) grew and became more stable. The electrode was repolished between each addition. The "end point" was judged as being the point at which stable voltammetry with well-defined peak-shaped waves was obtained. An important observation to note here is that, if a large amount of supporting 1:1 electrolyte was then added, the voltammetry reverted back to being impersistent and further relatively large amounts of $Cr(NH_3)_6^{3+}$ were required to restore the well-defined stable voltammogram. At each "end point", peak currents were proportional to $\upsilon^{1/2}$ up to at least $\upsilon = 200\ mVs^{-1}$. If the PGE electrode was replaced by a polished PGB electrode, the response deteriorated again. A freshly cleaved PGB electrode gave only a weak sigmoidal-type response even at a very high concentration of $Cr(NH_3)_6^{3+}$.

The value of $E^{0'}$ obtained from this experiment was typically − 370 mV, which is at the positive side of the range of values obtained for the protein by potentiometric methods. Most importantly, and by contrast with results obtained at Hg electrodes, the voltammetry of an Fe–S protein was demonstrated to be *stable*, essentially *reversible*, and *free of other signals*.

This pattern of behaviour has been encountered [91, 99, 103, 104] for a wide range of proteins bearing negatively charged surface domains with use of a variety of cations. A similar promotion of electrochemistry has also been observed in studies with a RuO_2 electrode [88]. Interestingly, even electrochemistry at Hg electrodes [46] is improved in this way (thus showing that the interactions involved are not limited to surfaces covered by stable oxide functionalities). To qualify this, it should be borne in mind that the Hg surface charge is negative under the conditions of scanning in the low-potential range appropriate for ferredoxins. Bianco and co-workers found [46] that the polarographic wave due to the native redox couple of spinach and *Clostridium thermocellum* ferredoxins was amplified considerably, by comparison with the normally-dominant adjacent wave arising from degraded protein.

Several conclusions stem from the studies of negatively charged proteins at the PGE electrode; these may be valid also for other electrode surfaces.

1) Values of $E^{0'}$ obtained after promotion of optimal cyclic voltammetry are in good agreement with values obtained by potentiometric measurements. Small differences generally manifest themselves as an *increase* in reduction potential.

2) The cations may be termed "promoters" following the preferred definition for electro-inactive adsorbates at metal electrodes. *They do not themselves mediate electrochemistry by carrying electrons.* While it is possible that this could occur given the use of a reagent with redox potentials close to those of the protein's active-site process, the success of such a range of cations that are not redox active under the experimental potential conditions, including Group 2 metal ions and stable organic amines, shows that this is not a primary factor.

3) Promotion by cations is much more marked at PGE than at PGB.

4) There is a close relationship between the development of a well-defined peak-shaped response and the absence of "impersistence".

5) More persistent electrochemistry, with a better-defined peak-shaped response is obtained at low protein concentrations.

6) The primary mechanism of activity involves ion–pair interactions. Promotion of electrochemistry by multi-charged cations is inhibited by a high concentration of background (1:1) electrolyte. The effectiveness of a cation is related to its charge. Monovalent cations are completely ineffective in many cases, even at very high concentrations, whereas cations with a large charge, 4+ or more, can be potent at sub-millimolar levels. Stability is particularly influenced. For example, cyclic voltammetry of plastocyanin as promoted by a small concentration of $Pt(NH_3)_6^{4+}$ is stable at 25 °C, whereas promotion by Mg^{2+} requires temperatures around 5 °C or lower.

7) The potential applied at the PGE electrode has no clear influence on the promotion of electrochemistry, at least within the range spanned by spinach plastocyanin ($E^{0'}$ = + 370 mV) and *Azotobacter chroococcum* ferredoxin ($E^{0'}$ for the [4Fe–4S] cluster is − 645 mV), with which the requirements for promotion are comparable.

8) Proteins show specificity through their promotion requirements. Some, like *Clostridium pasteurianum* ferredoxin, as mentioned above, respond well with Mg^{2+}, and even Na^+ or K^+ give reasonable results at molar concentrations. Others like

spinach plastocyanin give a stable response with Mg^{2+} only at low temperatures and if the protein concentration is low, but good results can be obtained with $Cr(NH_3)_6^{3+}$ under less strict conditions.

The first point 1) is very important since complex formation that is strong enough to cause the protein to interact intimately with the electrode surface may alter intrinsic properties and change the reduction potential. This is particularly relevant since the promoters are free in solution. One example in which two promoters yield different reduction potentials is in the voltammetry [105] of *Azotobacter chroococcum* 7Fe ferredoxin, which is discussed later.

Points 2)–8) together pose the question of what the cation is doing. Their role appears to be (at least) two fold. First, by associating with or between negatively charged areas on the electrode or protein surfaces, they can create an electrostatically favourable situation to stabilize a viable protein–electrode interaction. This is similar to the catalysis of protein–protein electron self-exchange reactions by counterions [106, 107]. An obvious analogy is also to be found with conditions that destabilize colloids. According to the Schultz–Hardy rules [108], coagulation is induced very much more effectively by counter-ions of high charge. Second, one seeks an explanation for the loss of "impersistence". If the interfacial electrostatics are balanced, i.e. all local charge accumulation in the intervening space is minimized by a distribution of charge that is complementary to that of the surface of the native protein, there should be less tendency to unfold. This assumes, as seems likely, that the "impersistence" is equated here with fouling of the electrode surface by denatured and irreversibly adsorbed protein molecules. It was suggested [91] that the situation could be illustrated by treating the cations as if they associated in potential "cavities" created by docking the native protein conformation with an active area of the electrode, i.e. of oxidized functionalities on PGE.

To put this discussion in perspective, there are now several examples of the use of *macromolecules* as promoters of direct electrochemistry. Van dijk and co-workers showed [47] that polarographic responses attributable to native ferredoxins, rubredoxin, and flavodoxin could be generated through the addition of polylysine to the electrolyte solution. Hill and Barker found [109] that cyclic voltammetry of negatively charged cytochrome b_5 at a PGE electrode was promoted by the addition of cytochrome *c*. Equally effective was redox-inactive Zn-substituted cytochrome *c*. These observations are not compatible with the idea of a promoter being "sandwiched" *between* protein and electrode; the promoter is now so large that it would constitute a *barrier* to such an electron-transfer process. A side-by-side or "rolling" interaction is likely.

We thus find ourselves knowing the "ingredients" of the protein–electrode interface, but with little knowledge about their arrangement or dynamics. Possible arrangements are shown in Fig. 8. The direct interaction of protein may be transient leading to the ***RRB*** mechanism, or long-lived to give strong adsorption whereby rapid oxidation and reduction of freely diffusing protein molecules occurs by the ***APEE*** mechanism. Further light upon this has been cast by studies using more complex organic cations, in particular aminoglycosides, which have $-NH_3^+$ groups spaced over a quasi-rigid framework. The structure of *Neomycin B* is shown here. The $-NH_2$ groups become protonated over the pH range 8–9. Such reagents

Neomycin B (base)

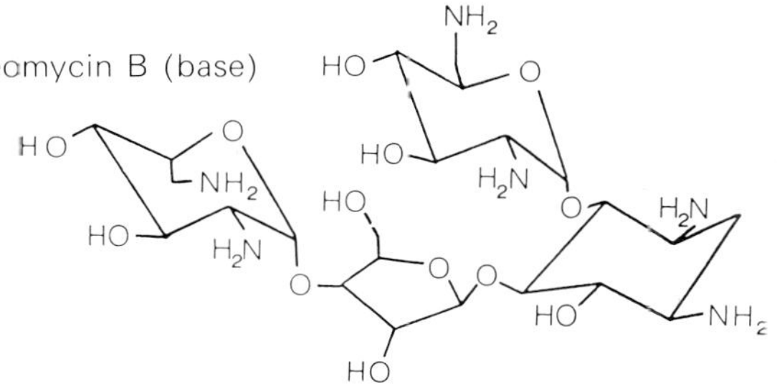

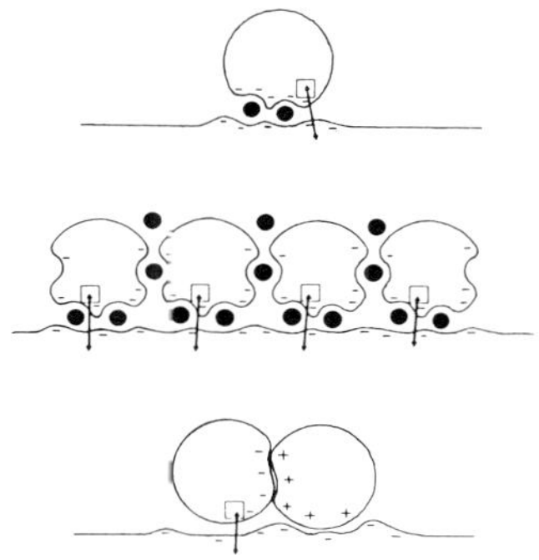

Fig. 8. Some possible arrangements of negatively charged proteins, cations, and the deprotonated PGE electrode surface. *Upper*: Cations bound in potential "cavities" at interface of protein and electrode. *Middle*: Layer stabilized by cations binding at the protein–electrode interface and between adjacent protein molecules. *Lower*: Promotion of protein–electrode interaction by complex formation with a positively charged protein molecule

have been found [105, 110, 111] to be particularly effective promoters of a variety of negatively charged proteins at PGE electrodes, giving rise to well-defined diffusion-controlled cyclic voltammetry at concentrations of around millimolar or lower. Moreover, it has been found [112] that many of these proteins *co-adsorb* strongly with aminoglycosides at PGE electrodes giving transferable films that approach monolayer coverage. Most success so far has been obtained with ferredoxins that contain *two* Fe–S clusters. Further studies are necessary to clarify the importance of multiple sites and rapid intramolecular electron transfer for facilitating an ***APEE*** mechanism. Returning to an earlier point, it is very likely that the reversible bulk solution electrochemistry of cytochrome c_3, which is adsorbed strongly at various electrodes, is a result of the ability of the four metal sites to *relay* electrons through the molecule. Bianco and co-workers have shown recently [113] that a PG electrode modified by pre-adsorption of the positively charged cytochrome c_3 from *Desulfovibrio vulgaris* (Hildenborough) yields voltammetry of ferredoxins in solution without addition of other reagents. Because of rapid electron exchange between redox centres, cytochrome c_3 may be loosely regarded as forming a "conducting" film [114, 115].

How *specific* are the interfacial coulombic interactions which are clearly so important? To make a meaningful assessment will require detailed comparisons of each system based upon electrochemical kinetic data treated according to an appropriate model. This has yet to be done. However, the greater promotional activity shown by multiple-site reagents like aminoglycosides does conform to ideas about coulombic interactions in biological systems as opposed to simple small molecules. While the charge on a simple metal ion or complex is essentially symmetrical, the surface charge on a protein is always irregular, and it is not possible to determine binding constants without considering how this charge is arranged. Tam and Williams [116] have discussed geometry effects in ion pairing and have shown that interactions between complex organic cations and anions cannot be calculated satisfactorily from basic coulombic expressions. Instead, the influences of charge distribution, shape, and rigidity now form the basis for *macromolecular recognition.*

The participation of *hydrophobic* interactions is even less clear. These may be very subtle, involving discreet local contacts within a larger area of polar interactions. Indeed, water molecules, if they intervene, may impede electron transfer so that some dehydration is expected to be advantageous. Examination of the crystal structures of representative electron-transfer proteins; plastocyanin, azurin, ferredoxins, flavodoxins and rubredoxin, show that the surface residues immediately overlying the redox-active site are hydrophobic.

Despite this positive notion, there have been relatively few reports of protein electrochemistry at deliberately prepared hydrophobic electrode surfaces. A PGE electrode at which acidic C–O functionalities were blocked by trimethyl- or triphenyl-silyl groups [98] became almost inactive towards horse-heart cytochrome *c*. However, another electron-transfer protein, azurin, gave cyclic voltammetry with prominent peaks (separation 100–105 mV at 20 mVs^{-1}). This demonstrates that one cannot formulate absolute rules. Azurin is a "blue" Cu protein whose surface bears few resultant charged domains. It may thus have a high propensity for interacting with hydrophobic surfaces, without distortion or disruption of a strong solvation sheath. It has also been found [117] to adsorb strongly with retention of native redox properties at a modified carbon-paste electrode made by mixing together carbon paste and 4,4′-bipyridyl. Stable *adsorbed-film-mode* electrochemistry was observed upon transfer of the protein-coated electrode to a buffer solution. It was proposed that 4,4′-bipyridyl formed a film with the oil component of the paste, into which azurin adsorbed tightly. Some enzymes, as discussed in Sect. 7, display activity if adsorbed on carbon-paste electrodes. It is very likely that the "oily" environment provides a substitute for hydrophobic interactions that are normally operative, for example in stabilizing membrane-bound proteins.

3.4 Summary

While it is interesting to view strongly adsorbed proteins in the role of relaying electrons to bulk species, there is no clear evidence that the deliberate provision of electron-mediating functionalities at the electrode surface is relevant, at least for the small electron-transfer proteins that we have discussed above. Modified electrodes designed to "relay" electrons have been described, and results show that the potential of the surface-confined mediator (as would be the case were these to be free reagents [6]) needs to be matched with that of the protein to be addressed in bulk solution. Wrighton and co-workers modified Pt, Au, and p-type Si with viologen-type reagents. They observed [118] that cytochrome *c* in bulk solution could be reduced, but only at the low potential of the viologen; reoxidation was not observed. Similarly, Elliott and Martin [119] found reduction of cytochrome *c* to occur at low potentials on a viologen polymer film. However, modification of Pt and n-type Si electrodes with a layer of ferrocene molecules retained by a $-(CH_2)-Si-O-$ matrix yielded [120] a surface-confined redox system with $E^{0'} = +280$ mV. This modified electrode showed reversible *catalytic* enhancement upon introduction of cytochrome *c*. Solutions of the protein could be rapidly brought to redox equilibrium and there was no evidence for irreversible adsorption over long periods of time. Another class of systems that have received considerable attention are conducting salts onto which enzymes, including flavocytochrome b_2, can be adsorbed to yield biocatalytic currents upon introduction of substrate [121–123]. It has been debated whether or not these might act by mediating electrons via dissociation from the film. While this is unlikely, since studies have been conducted *within* the potential range of stability of the film, no simple experiments with electron-transfer proteins have been reported.

The drawback with any surface mediatory system is that the protein is not being addressed *directly*. This poses no problem for many applications, including the use of enzymes in biosensors. But for visualizing directly the *chemistry* of an active site through its voltammetry, straightforward interpretation of the current-time-voltage response demands that the flow of electrons is not distorted by the redox properties of the mediator. If electrons are *relayed* via adsorbed protein molecules, the same considerations will apply. distortion of the response and irreversible behaviour may ensue if the reduction potential of the metal centre is changed drastically upon adsorption.

A more general conclusion to be drawn from the studies of the better characterized smaller proteins is that they are *surface-selective*. This is manifested in various ways. Adsorbed proteins require an environment that stabilizes their structure while allowing fast electron transfer. To achieve this, it is necessary to provide stable and compatible electrode-surface functionalities Additional "tailoring", for example by ion complexation, may be needed to provide favourable electrostatics. Freely diffusing molecules need to interact rapidly and reversibly with "electron distribution sites". These are suitable patches located either on the electrode surface, for example as generated by promoters (***RRB*** mechanism), or on adsorbed protein molecules that are capable of relaying electrons (***APEE*** mechanism).

A number of questions remain concerning the dynamics of the protein–electrode interaction. Do experimental data give any idea about the rates, relative or otherwise, of the electron-transfer step? The clearest result so far has been the determination of k_{et} for horse heart cytochrome *c* at 4,4′-bipyridyl-modified Au with the rotating ring-disk technique as mentioned above [65]. There have been a number of determinations of compound heterogeneous rate constants for protein electrochemistry, mostly using Nicholson's method [124] for their estimation from CV peak separations. All calculations have assumed that the mass transport can be treated in terms of linear diffusion to a uniform planar electrode surface. Bond and co-workers have pointed out [125, 126] that in many instances this is unlikely to be the case. The central theme of the *microscopic model* concerns a protein's selectivity for suitable sites on the electrode. Electron transfer occurs only at these and it is fast. Electroactivity at other sites is negligible. The situation is similar to that of an electrode under conditions of variable partial blockage [127–129]. There are two limiting cases. On the one hand, the active sites may be so dilute and separated that mass transport to them is radial. On the other hand, the site density may be so high as to constitute a macroscopic planar surface to which mass transport from the bulk will be linear. These are illustrated in Fig. 9.

The latter gives rise to the familiar peak shaped cyclic voltammetry that is termed "well-behaved" and with which one may use Nicholson's analyses [124] for estimating kinetic constants. If the kinetics are sufficiently fast, peak currents obey the *Randles–Sevcik* equation—Eq. (5).

The former case is intriguing since the discrete active electrode site is a "microelectrode". The radial diffusion gives rise to steady-state sigmoidal voltammetric waves, but since mass transport is very efficient, the current that is obtained from an array of sites need not be greatly lower than that obtained for a uniformly active surface. For a reversible electrode reaction the potential $E_{1/2}$ at which half-maximal current is obtained is equivalent to the formal reduction potential $E^{0'}$.

An intermediate situation occurs if the predominance of radial diffusion is destroyed by the electrode active sites becoming larger, or if their packing becomes dense enough so that diffusion layers overlap. The size of the diffusion layers depend, of course, upon the time domain of the experiment. The resulting cyclic voltammogram comprises a mixture of linear- and microscopic radial-diffusion terms. Peaks appear more flattened and rounded.

Thus the use of peak separation for kinetic analysis is valid only if the electrode surface is uniformly active, i.e. homogeneous. With this view, it was possible to

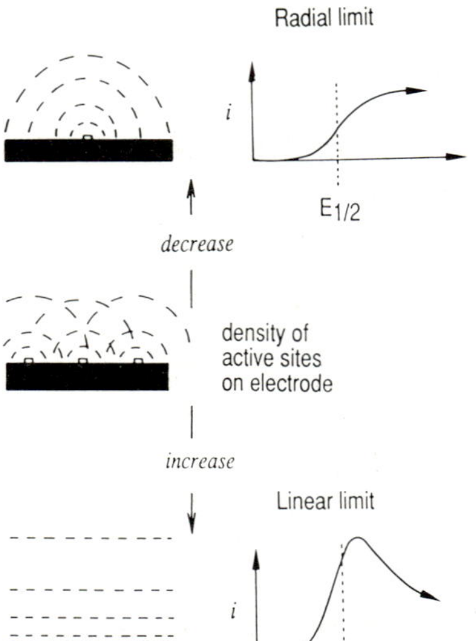

Fig. 9. Schematic representation of the transformation between *radial* and *linear* diffusion modes, and the corresponding voltammetric waveforms expected for limiting cases

rationalize the shapes of voltammograms that arise with the use of graphite electrodes. The response of cytochrome *c* at a freshly cleaved PGB electrode surface (see Fig. 6) or that of ferredoxin at a PGE electrode under conditions of low $Cr(NH_3)_6^{3+}$ concentration (Fig. 7) may be interpreted in terms of *reversible* electrochemistry occurring at so few electrode sites as to constitute a microelectrode array subject to radial diffusion. Thus while reduction and oxidation waves appear flattened and broad, their $E_{1/2}$ values virtually coincide. The same rationale could be applied to electrode surfaces "aged" through "impersistence". The deterioration of electrochemical response, which probably arises from denaturation of protein molecules, is not due to a decrease in the heterogeneous rate constant, but rather due to destruction of sites and their resulting dilution. A similar result is found for voltammetry at modified Au electrodes if the coverage of promoter adsorbate is much less than monolayer.

The implication here is that proteins, once they bind to their respective electrode sites with (presumably) the correct orientation, may actually transfer electrons very rapidly indeed. The microscopic model [125, 126] is in accordance with the view, expressed at the beginning of this article—that intrinsically the reactivity of electron-transfer proteins is high, but is tempered by specificity.

4 Studies of Thermodynamic and Kinetic Properties

4.1 Advantages Offered by Voltammetric Techniques

The most obvious application of voltammetric techniques in studies of metalloproteins might be seen as lying in the measurement of reduction potentials; however, as I hope to show here and in the following sections, their scope extends throughout *and beyond* simple redox equilibria. One needs to ask, "For what type of problem does voltammetry offer any advantage over potentiometry?" We have, after all, just discussed how a voltammetric response depends *critically* upon the behaviour of the protein at the electrode–solution interface. By contrast, potentiometry [5, 6] represents a tried and tested methodology with wide applicability.

There is, for example, an extensive databank of thermodynamic data, including half-cell enthalpies and entropies of reduction, that has been built up from investigations that use small mediators to carry electrons between protein and electrode. In these potentiometric studies, one measures the equilibrium concentrations of components in oxidized and reduced states at various values of the electrode potential. There are a number of variations on this theme. For example, a determination may also be carried out without using an electrode, by equilibrating the couple of interest with a titrant whose reduction potential is known accurately. Most importantly, it is necessary that the component of interest (or the titrant) exhibits some difference, in a readily measurable property, between oxidized and reduced forms. Light adsorption is the most convenient parameter since it may be monitored conveniently *in situ*. An excellent method, which has now gained wide acceptance, involves the use of a thin-layer cell incorporating optically transparent oxide or Au "minigrid" electrodes. An "OTTLE" [130, 131] (optically transparent thin-layer electrode) cell enables precise measurements to be made with rapid equilibration times and economy with sample size. The "ideal" subjects for this type of study have been cytochromes and "blue" Cu proteins, for which large and distinct spectral changes occur between oxidized and reduced forms, and which are usually stable enough to survive a titration in both directions.

On a wider note, a number of considerations determine the suitability of potentiometry and the various modes of monitoring redox status, for the purpose of measuring reduction potentials. A potentiometric study conducted with an OTTLE can yield information with ease and reliability for "ideal" systems. But in certain cases, as discussed now, the approach is fraught with various complications.

1) Absorption spectra of the redox states may be weak and indistinctive so that other characteristics must be monitored. In suitable cases, NMR offers opportunities to examine thermodynamic and structural aspects of protein redox equilibria. The most widely exploited alternative is EPR, but since room-temperature spectra are not generally observed for metal centres, electrochemical and spectroscopic measurements are made under very different conditions. After establishing equilibrium under ambient conditions, a sample is withdrawn and frozen for examination at cryogenic temperatures. Molybdenum, "Type 2" Cu, and—for the most part—Fe–S centres, do not constitute strong or distinctive chromophores and low-temperature EPR is the method of choice.

Alternatively, the extent of oxidation or reduction of protein-active sites at various applied potentials may be determined by *coulometric* measurements in the presence of a suitable mediator [132]. Here, the extent of reduction or oxidation of the group is determined from the charge that is passed in response to a potential perturbation. The need for spectral monitoring is thus removed. Heineman and coworkers have reviewed [133] the use of two techniques, thin-layer pulse coulometry and thin-layer staircase coulometry, each of which are suitable for studies on biological molecules.

2) There may be several redox centres present having similar spectra. In this case the result obtained by deconvolution of titration curves, whether using optical absorption, EPR, or other modes of monitoring, may not alone be definitive.

3) The protein may be unstable in one or more of its oxidation states and may not survive a titration cycle without significant degradation. Correction of data for this loss of material is an undesirable necessity.

4) Making measurements of systems that exhibit extremes of reduction potential can be particularly difficult. For active sites with very low potentials, long-term anaerobicity is essential, and this requires working with a sealed cell or within a glove box. Furthermore, there are few mediators or titrants that are suitable for equilibration at potential extremes, i.e. close to or beyond the thermodynamic limits of aqueous solvent/electrolyte stability.

Aside from these points, specific oxidation states may be active with regard to ligand exchange or other chemical processes that are directly relevant to the biological function of the protein. In a potentiometric investigation these interesting reactions present a disturbance that is difficult to define. By contrast, voltammetry in such cases can yield a coherent picture. The voltammogram itself may be regarded as a spectrum. It displays the *quantitative time-resolved redox chemistry* of the active sites. In other words we acquire the means to *visualize* and *integrate* complex redox behaviour. Great economy with sample size is possible. This is particularly apparent in cases where it is possible to study electroactive *thin films* of strongly adsorbed native protein. But even for macroscopic bulk-solution experiments, the electrochemical perturbation can be restricted to the diffusion layer. Consequently, each reaction, including a coupled irreversible process, may be examined afresh after replenishing this relatively small fraction of the total sample by brief stirring.

4.2 Cytochromes

Studies of cytochromes have provided excellent demonstrations of the ability of voltammetry to obtain acceptable quantitative data on protein redox equilibria. Reduction potentials of cytochromes are influenced by a number of factors [134]. These include the nature of the axial ligands to Fe [135], the degree to which the heme group is exposed to solvent [136], electrostatic interaction between the Fe and propionate side chains of the porphyrin [137], and specific binding of anions [138].

Cytochrome *c* itself shows pH-dependent variations in conformation that are manifested as changes in the axial ligation to Fe [61]. These are depicted in Scheme 1. For cytochrome *c(III)*, State III is the physiologically important form, with low-spin Fe coordinated as shown in Fig. 1. State IV is also low-spin, but Met-80(S) is replaced by another strong donor, established to be NH_2— from a lysine residue. By contrast the Fe(II) form retains a stable conformation (termed here as State II_R) with both axial ligands intact between pH 4 and 12. The transition of cytochrome *c(III)* from State III to State IV, which can be resolved upon rapid alkalization or rapid oxidation of the reduced protein under mild alkaline conditions, has a relaxation time in the order of hundreds of milliseconds [139]. State IV is not reducible by ascorbate and stronger reductants such as dithionite are required. This is reflected in potentiometric measurements which

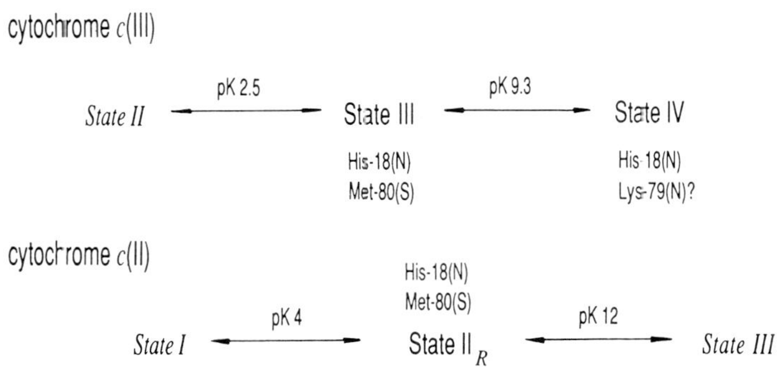

Scheme 1

show a steady decrease in reduction potential above pH 8 [140]. Direct electrochemistry provides an informative, dynamic picture; the situation being an example of multiple redox-active conformers.

Haladjian and co-workers [62] investigated the cyclic voltammetry of horse cytochrome *c* under alkaline conditions using a Au electrode modified by 4,4′-bipyridyl. With a solution of cytochrome *c(III)* at pH 9.3, they observed two electrode reactions at very different potentials, as shown in Fig. 10. At high potential, in the region expected for cytochrome *c* electrochemistry under conditions of neutral pH, there is a large oxidation wave (1a) with a very small reduction counterpart. By contrast, at low-potential there is only a large reduction wave (2c). Within a narrow range of low scan rates, peak currents of both waves were found to be diffusion-controlled. Qualitatively, at least, this behaviour is consistent with kinetic studies [139] and is not inconsistent with the equilibrium data [140]. The reactions may be rationalized in terms of the cycle depicted in Scheme 2. If the Fe(III) form exists as State IV, its reduction does not occur until a much lower electrode potential is applied. Since the Fe(II) product associated with the low-potential process reverts spontaneously to the native form State II_R (resembling State III of the oxidized protein) i.e. with re-coordination of Met-80(S),

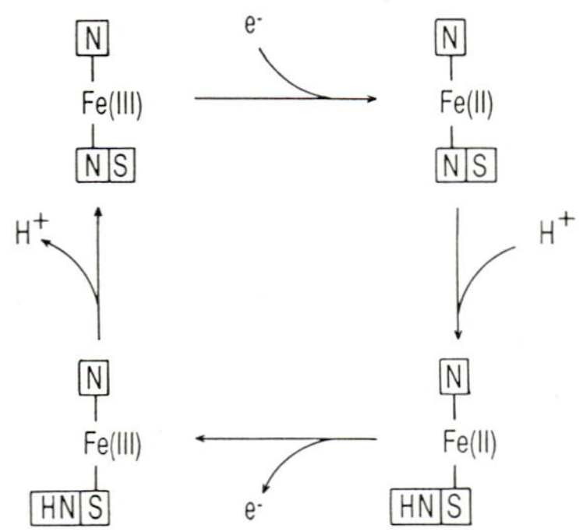

Scheme 2

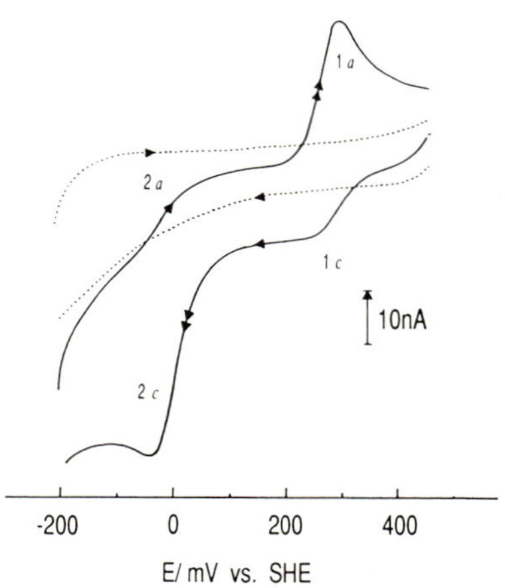

Fig. 10. Cyclic voltammogram of horse cytochrome *c* at pH 9.30. Solution contained 0.35 mM oxidized cytochrome *c* in a medium consisting of 0.10 M sodium perchlorate, 0.02 M sodium borate and 0.01 M 4,4′-bipyridyl as promoter. An Au electrode, area 0.0079 cm^2 was used. Scan rate 5 mVs^{-1}, temperature 25 °C. Two chemically non-reversible redox processes are observed, Wave 2*c* is associated with reduction of the State IV conformer which prevails at this pH. Note the virtual absence of wave 1*c*, which would be observed for reduction of the State III conformer. The corresponding return wave 2*a* is not observed because the Fe(II) product reverts immediately to the State II_R conformer resembling State III of the Fe(III) form. Instead, re-oxidation (wave 1*a*) is observed at a potential appropriate for the native State III system. From Ref. 62, redrawn with kind permission of the authors

reoxidation occurs in the normal potential regime. The immediate product (State III) then reverts to State IV, and the cycle of events continues.

The existence of interconverting conformers having different reduction potentials means that the pH profile given by potentiometric measurements is a complex summation of several terms. On the basis of peak heights from differential pulse voltammetry, it was estimated that the effective pK for the State III to State IV equilibrium was 8.1, a value which is, however, much lower than that determined optically [61]. Haladjian and co-workers found that the wave corresponding to reduction of State IV was only well defined at low scan rates, and they attributed this to slow electrode kinetics. It might otherwise have been possible to measure the rates of interconversion among conformers according to the methods described by Nicholson and Shain [50].

The groups of Taniguchi and of Hawkridge have each undertaken detailed studies of the influence of temperature, electrolyte composition and pH upon the reduction potential of horse cytochrome *c*. Reduction potentials were measured by cyclic voltammetry with a non-isothermal cell. With this configuration, in which the reference temperature is kept constant while the sample temperature is varied, data lead directly to the reaction centre entropy change ΔS^0_{rc} as given in Eq. (6).

$$\Delta S^0_{rc} = nF(dE^{0'}/dT) = S^0_{red} - S^0_{ox} \tag{6}$$

Taniguchi and coworkers used a Au electrode "pre-dip modified" with bis(4-pyridyl)disulfide. They examined the temperature dependence of $E^{0'}$ over the range 0 to 55 °C at pH 6, 7 and 8 ($I_{phosphate} = 0.10$) with addition of 0.10 M $NaClO_4$ [141] or NaCl [142]. Over this range of conditions the structure of the promoter interface was not observed to change significantly as monitored by SERS. Values of $E^{0'}$ were obtained from voltammograms recorded at various scan rates 20 to 200 mVs^{-1} over which range the electrode reaction appears diffusion-controlled.

Koller and Hawkridge used a tin-doped indium-oxide electrode and made measurements over the temperature range 5 to 75 °C at pH 7 using phosphate (I = 0.20 M) or Tris/cacodylate (I = 0.20 M) buffer media [86]. In further work they extended the pH range to 5.3 and 8.0 [87].

The result of each group are displayed for comparison in Table 2 together with data from potentiometric studies made with redox mediators. First, we see that the general agreement is excellent; ΔS^0_{rc} values lie within a range of $\pm$ 10 $JK^{-1} mol^{-1}$. The negative entropy change is consistent with cytochrome *c(II)* having a more ordered structure. Second, there are small differences due to choice of electrolyte. Taniguchi and coworkers found [142] that $E^{0'}$ values were typically 10 mV higher if 0.10 M NaCl was present instead of $NaClO_4$. Koller and Hawkridge observed that $E^{0'}$ values measured in the presence of Tris/cacodylate were approx. 8 mV higher than those measured in phosphate [86]. Thus, small differences in ion-binding affinities could be detected easily; for example the ratio of specific anion binding constants ($K_{c(II)}/K_{c(III)}$) is greater for Cl^- than for ClO_4^-. Third, each group noted sharp downward breaks in the $E^{0'}$ vs T plots at increased temperature. Both groups found a transition at approx. 40 °C that occurred only at pH 8. Previous potentiometric studies [143] had also established this type of behaviour although it was observed at pH 7 and only in the presence of

Table 2. Thermodynamic parameters measured for horse cytochrome *c* by potentiometric and voltammetric methods

pH	Method	Electrolyte	$E^{0'}$ (25°)/ mV vs SHE	ΔS_{rc} (temp. range)/ $JK^{-1} mol^{-1}$	Ref.
5.3	V[a]	Tris/cacodylate I = 0.20	268	– 52.7 (5–75°)	[87]
6.1	V[b]	phosphate/ClO_4^- I = 0.20	261	– 53.1 (3–52°)	[141]
6.1	V[b]	phosphate/Cl^- I = 0.20	272	– 55.6 (1–56°)	[142]
7.0	V[a]	Tris/cacodylate I = 0.20	264	– 56.1 (5–65°)	[86]
7.0	V[a]	phosphate I = 0.20	256	– 54.0 (5–55°)	[86]
7.0	V[b]	phosphate/ClO_4^- I = 0.20	259	– 49.4 (3–53°)	[141]
7.0	V[b]	phosphate/Cl^- I = 0.20	268	– 51.9 (2–55°)	[142]
7.0	P[c]	phosphate/Cl^- I > 0.20	262	– 42.7 (25–42°) – 123 (> 42°)	[143]
7.0	P[d]	phosphate I = 0.10	262	– 54.0 (9–39°)	[e]
7.9	V[b]	phosphate/ClO_4^- I = 0.20	258	– 43.1 (< 42°) – 172 (> 42°)	[141]
7.9	V[b]	phosphate/Cl^- I = 0.20	266	– 51.5 (< 40°) – 132 (> 42°)	[142]
8.0	V[a]	Tris/cacodylate I = 0.20	264	– 53.1 (5–40°) – 100 (45–60°)	[86]
8.0	V[b]	phosphate I = 0.20	256	– 59.4 (5–40°) – 135 (45–55°)	[86]

[a] cyclic voltammetry at Sn-doped indium oxide electrode;
[b] cyclic voltammetry at Au electrode modified by bis-(4-pyridyl)disulfide;
[c] OTTLE with dichlorophenol-indophenol as mediator;
[d] OTTLE with $Ru(NH_3)_6^{3+}$ as mediator;
[e] Taniguchi VT, Ellis, WR Jr, Cammarata V, Webb J, Anson FC, Gray HB (1982) In: Kadish KM (ed) Electrochemical and spectrochemical studies of biological redox components, ACS Adv. Chem. Ser. 201: 51

Cl^- ions. This observation had led to the conclusion that the discontinuity was essentially extrinsic, and it was suggested [143] that it might coincide with a phase transition in the bulk water structure due to changes in Cl^--ion hydration. The extensive voltammetric studies showed the discontinuity to occur in various electrolyte media and thus indicated that its origin was probably intrinsic. Taniguchi et al found [141] spectral evidence for some limited disruption of the Fe(III)–Met-80(S) bond occurring at 40 °C under conditions of pH 8 (phosphate) with 0.1 M $NaClO_4$. The absorption band at 695 nm (which is assigned to Met-80(S)-to-Fe(III) charge transfer) was diminished in intensity although there was no major change in the resonance Raman spectra. Thus, a subtle effect was operative, in other words a conformational change yielding a species that resembles more closely the low-potential State IV (alkaline) form. It was suggested that the new high-temperature form corresponded to an intermediate state (IIIb) as proposed by Myer et al [144].

Taniguchi and co-workers used a Au electrode "pre-dip modified" with 6-mercaptopurine to examine the redox chemistry of horse cytochrome *c* under acidic conditions [77]. This promoter gives well-defined voltammograms down to pH 2.8. It was easy to see that $E^{0'}$ values decreased sharply below pH 3–4, with the effective pK being somewhat dependent upon the supporting electrolyte. By analogy with the transition that occurs at high temperatures in mild alkali, a correlation could be made with changes in the corresponding absorption spectra of cytochrome *c(III)*, for which acidification produced a decrease in the intensity of the band at 695 nm. The acid species was suggested to be the so-called State III*a* [144] in which the Fe–S ligation is again weakened by a conformational change (although the Fe(III) remains in the low-spin state).

Four further reports show how voltammetric techniques may be applied conveniently to determine factors that influence reduction potentials of cytochrome *c*.

A modified (horse) cytochrome *c*, in which Met-80 is carboxymethylated and therefore unable to coordinate to Fe, was studied by Di Marino et al [145]. At pH 7 the Fe(II) form, by analogy with deoxymyoglobin, is able to bind O_2 and CO. Thus the sixth coordination position is vacant or is occupied by weakly bound H_2O. With a Au electrode, either as wire or coated onto carbon, but without further modification, cyclic voltammetry was observed with peaks due to reduction and oxidation. There was clear evidence for specific adsorption of oxidized reactant, and the authors did not analyze their results further. The reduction potential was determined to be -218 mV from equilibrium measurements.

Santucci and co-workers [146] studied "microperoxidase", an Fe-porphyrin undecapeptide (MW 1900), pI = 4.7) obtained by hydrolysis of cytochrome *c*. Cyclic voltammograms, showing diffusion-controlled reduction and oxidation waves, were obtained with a PGE electrode at pH 7. At 25°, $E^{0'}$ was -160 mV, i.e. considerably lower than for the intact native protein. This shift was attributed to the far greater solvent exposure for the microperoxidase heme group. By contrast with most electron-transfer proteins, and with behaviour now much more reminiscent of small redox molecules, well-defined electrochemistry was not critically dependent upon electrolyte conditions. There was merely some enhancement of the electrochemical rate (decrease in ΔE_p) upon changing from 0.10 M $NaClO_4$ to 0.025 M $Mg(ClO_4)_2$.

Hill and Whitford [147] examined a number of derivatives of horse cytochrome *c* in which specific lysine residues had been modified with 4-chloro-3,5-dinitrophenol (CDNP). The primary effect of this is to replace each 1+ charge by 1−. Using a PGE, or a Au electrode modified with bis(4-pyridyl)disulfide, they found only small variations in $E^{0'}$ and electrochemical activity (ΔE_p) among cytochromes singly modified at Lys-13, -72, or -60, and only a small decrease in $E^{0'}$ (with greater ΔE_p) for di-substituted species. However, cytochrome *c* that had been extensively modified (6-7 CDNP per molecule) gave a much higher reduction potential (+450 mV) and required multi-charged cations to promote its electrochemistry at PGE electrodes [76] (see Sect. 3.3).

Voltammetric characterization of precious samples in limited supply was demonstrated in an interesting report by Sorrell and co-workers [148]. They used genetic engineering methods to prepare a specific mutant form of yeast iso-2 cytochrome *c* in which His-18 (which normally ligates the Fe and is conserved in all species) had been replaced by Arginine. The redox properties of the mutant protein, termed C2-18R and obtained in small yield, were compared with those of the wild type by cyclic voltammetry, using a tin-doped indium-oxide electrode. The wild-type protein showed the expected reversible response with $E^{0'} = 268$ mV. Under the same conditions, 0.1 M phosphate (pH 7.8) and 0.8 M NaCl with a scan rate of 2 mVs^{-1}, C2-18R showed a much larger peak separation, 200 mV, indicative of a rather slower electron-transfer rate, yet $E^{0'}$ was *unchanged*. This was a surprising result. Yeast strains lacking the ability to produce cytochrome *c* grew only if they were transformed by plasmids containing mutant genes coding for His-18 or Arg-18, although growth of colonies producing C2-18R was much slower. Several other mutant cytochrome *c* genes, which would give Tyr-18, Thr-18, Cys-18, Gln-18, Ser-18, or Leu-18 did not give viable proteins. The specificity suggested that Arg-18 must be coordinated to Fe. If so, the electrochemical findings show that the direct effect of this major alteration to the active site, substitution of imidazole by guanidyl, is to alter the *kinetic activity* but not, remarkably, the reduction potential.

The cytochromes c_3 from *Desulfovibrio* give a good demonstration of the investigation of complex proteins containing several metal centres with similar reduction potentials. It was mentioned in Sect. 3 that these proteins gave reversible diffusion-controlled electrochemistry at Hg and other electrode surfaces without any requirement for special modification of the interface [33–35]. They contain four covalently bound heme groups, each of which is more exposed to solvent than that of mitochondrial cytochrome *c*. Two histidines comprise the axial ligands to Fe. The reduction potentials are very negative as is appropriate for electron-transfer partners of hydrogenases. Depending upon the source of protein, the cyclic voltammetry consists of a single broad wave [34] (*D. vulgaris* Strain Miyazaki) or two distinct waves [149, 150] (*D. baculatus* Strain Norway), in each case revealing the presence of sites that differ (to an increasing extent respectively) in their reduction potentials. Differential pulse voltammetry is a more suitable technique for resolving individual electrode reactions with small differences in potential [151]. Values obtained are *macroscopic* reduction potentials since they refer directly not to the intrinsic properties of each individual active site, but to equilibrium distributions of protein molecules with 0, 1, 2, 3, and 4 electrons added. Different approaches to deconvolution of voltammograms have been taken by various groups. On the one hand, Bianco and co-workers, conducting experiments with *D. baculatus* cytochrome c_3, have treated [149, 150] the voltammogram as being the sum of contributions from four different species each undergoing a reversible one-electron process. Alternatively, it has been argued [152, 153] that the appropriate analysis should consider a fully oxidized molecule diffusing to the electrode surface from bulk solution and undergoing four consecutive one-electron reductions. Figure 11 shows the DP polarogram obtained for *D. vulgaris* (Miyazaki) cytochrome c_3 with the simulated curve based upon a model incorporating four consecutive reversible reduction processes. As an illustration of the validity of voltammetric data in this type of problem, a comparison between reduction potentials obtained by various methods is presented in Table 3.

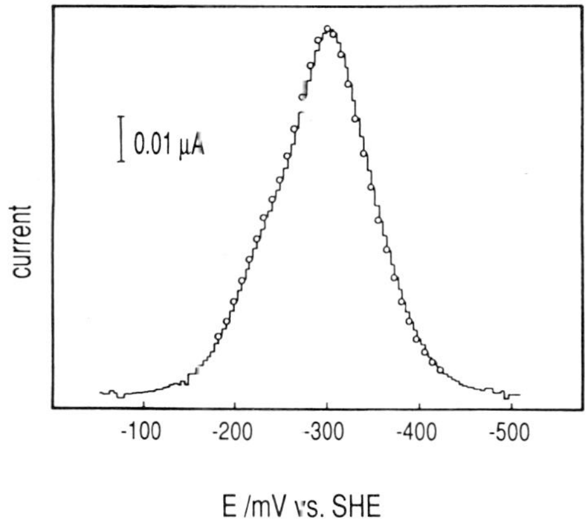

Fig. 11. Experimental (——) and simulated (ooo) differential pulse polarograms of cytochrome c_3 from *Desulfovibrio vulgaris*, strain Miyazaki, in 0.03 M phosphate buffer pH 7.0, temperature 25 °C. Drop time 2 s, pulse amplitude 10 mV. Simulation is for four reversible consecutive one-electron processes having macroscopic formal potentials of − 240, − 297, − 315 and − 357 mV. Original plot kindly supplied by Dr Katsumi Niki (Yokohama National University)

Table 3. Comparison of macroscopic reduction potentials for cytochromes c_3 measured by electrochemical or potentiometric (EPR) methods

Source	Method	Conditions	E_1	E_2 /mV vs SHE	E_3	E_4	Ref.
D. baculatus	EPR	pH 8.1, 18° 0.1 M Tris	− 150	− 270	− 325	− 355	[a]
	DPV[b]	pH 7.6, 25° 0.01 M Tris	− 160	− 300	− 330	− 380	[150]
	DPP[c]	pH 7.6, 25° 0.01 M Tris	− 165	− 305	− 365	− 400	[149]
	DPP[c]	pH 8.5, 25° 0.01 M borate	− 160	− 290	− 360	− 390	[154]
	DPP[c]	pH 7.0, 25° 0.025 M Tris	− 126	− 246	− 290	− 339	[35]
D. vulgaris (Miyazaki)	EPR	pH 8.1, 21° 0.1 M Tris	− 227	− 287	− 320	− 366	[d]
	DPP[c]	pH 7.0, 25° 0.03 M phosphate	− 240	− 297	− 315	− 357	[35, 153]
D. vulgaris (Hildenborough)	DPP[c]	pH 7.4, 25° 0.03 M phosphate	− 263	− 321	− 329	− 381	[35, 153]

[a] Gayda, J-P, Bertrand P, More C, Guerlesquin F, Bruschi M (1985) Biochim. Biophys. Acta 829: 262;
[b] DPV = differential pulse voltammetry at glassy carbon electrode;
[c] DPP = differential pulse polarography;
[d] Gayda J-P, Yagi T, Benosman H, Bertrand P (1987) FEBS Lett. 217: 57

Attempts have been made to assign individual potential values obtained from voltammetry to specific heme groups. In one example, the bulk-solution voltammetry of cytochromes c_3 was compared to voltammetry of the proteins *adsorbed* at silver electrodes and the corresponding SERS "potentiometric" data [35]. Since the adsorbed-film voltammetry gave potentials very close to the bulk solution values, it was concluded that the proteins were adsorbed essentially in their native forms. The SERS potentials each corresponded more closely to the electrochemical process of highest potential. Since SERS is sensitive only to groups located very close to the electrode surface, i.e. within 5 Å, it was argued that the SERS-derived reduction potential should be assigned to the heme whose environment best allowed close contact with the electrode. The group termed heme-1 [32] was suggested to be the best candidate since in each case the environment comprised a significant excess of amino acids with N-donor side groups (several lysines and a glutamine) that would make the preferred contacts with the Ag surface. On the basis of crystallographic data, however, this group is not the most buried.

In another example, Dolla and co-workers examined [154] a chemical derivative of *D. baculatus* cytochrome c_3 in which the solitary arginine group, Arg-73, had been modified with cyclohexane 1,2-dione in the presence of borate. The DP polarogram was significantly altered as compared to the native protein; in particular the normally well-separated high-potential couple was shifted considerably, by approx. − 50 mV. This result was considered in view of the structural evidence that Arg-73 is situated closest to heme-1 (heme-4 in the authors' system of numbering) and therefore this redox group would be expected to incur the greatest alteration to its environment upon modification.

4.3 Blue Copper Proteins

The "blue" Cu proteins are so named because they contain a Cu active site that is the origin of an intense blue colour exhibited in the oxidized state [155]. "Blue" or "Type 1" Cu centres as they are variously termed have been characterized through extensive spectroscopic efforts and by X-ray diffraction studies on several proteins. Of these, the photosynthetic electron carrier *plastocyanin* has been examined in most detail. The role of plastocyanin is to convey electrons between two membrane-bound components of the plant chloroplast electron-transport chain [156]. It is a soluble protein of molecular weight approx. 10500 that is located in the intrathylakoid space. Here it is reduced by cytochrome *f* and reoxidized by the $P700^+$ centre associated with Photosystem I. The crystal structure [157] of poplar plastocyanin shows that the Cu is coordinated by four ligands; two imidazole-N's from His-37 and His-87, a thiolate-S from Cys-84, and a thioether-S from Met-92; the latter constituting what may be regarded as an exceptionally long bond. Active-site structures and interatomic distances (Å) appropriate for various states (see below) are depicted in Scheme 3.

The ligand set is a compromise between the requirements of Cu(I) (soft or class "a") and Cu(II) (intermediate). The geometries of both Cu(II) and Cu(I) forms at neutral pH may be described as rather distorted tetrahedra; again the imposed geometry gives little preference to either oxidation state. The active site is located close to the protein surface in an area that is dominated by hydrophobic residues. Plastocyanins from higher plants bear a significant excess of acidic amino acids, which results in an isoelectric point (pI) of around 4 or even lower. The negative charge tends to be localized into an "acidic patch" some distance away [158]. The overall charge on spinach plastocyanin at pH 7, as estimated from its composition, is − 8.

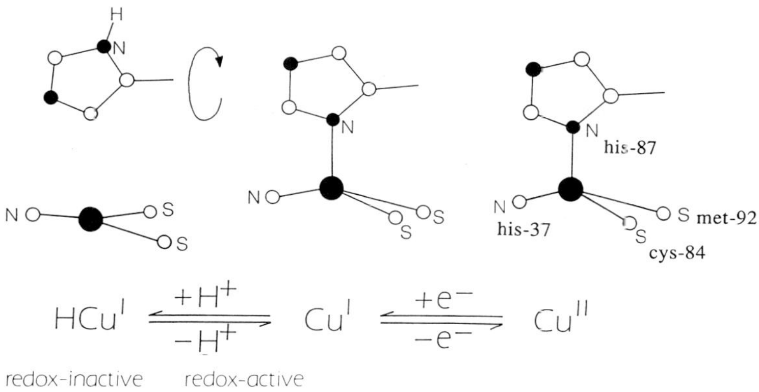

Scheme 3

What factors influence or modulate the activity of plastocyanin? Different lines of research reveal two effects. On the one hand, reduction of $P700^+$ by plastocyanin [102] is optimized under conditions of raised Mg^{2+} levels or a pH of around 4.7. This may be understood broadly in terms of electrostatics. The pI of the isolated P700-chlorophyll *a* complex is also low, at around 5. Thus at higher pH, formation of a precursory electron-transfer complex with plastocyanin involves an unfavourable (overall) coulombic interaction. Binding is promoted by protonation of surface residues or association of divalent cations. On the other hand, detailed kinetic studies on the oxidation of plastocyanin by small inorganic reagents have shown [159, 160] that protonation of a group on the protein (pK = 4.9 and 5.5, respectively, for spinach and parsley plastocyanin) yields a redox-inactive form. The reaction (in which the oxidant is typically $Fe(CN_6)^{3-}$) is described in Eqs. (7)–(9).

$$\mathrm{PCu(I)\text{-}H^+} \rightleftharpoons \mathrm{PCu(I)} + \mathrm{H^+} \qquad K_H \tag{7}$$

$$\mathrm{PCu(I)} + \mathrm{OX} \longrightarrow \mathrm{PCu(II)} + \mathrm{RED} \qquad k \tag{8}$$

$$\mathrm{PCu(I)\text{-}H^+} + \mathrm{OX} \longrightarrow \mathrm{PCu(II)} + \mathrm{RED} + \mathrm{H^+} \qquad k_H \tag{9}$$

If k_H is small, i.e. if the protonated form is essentially inactive, the rate is given by Eq. (10).

$$k_{obs} = \frac{k[\mathrm{OX}]}{1 + K_H[\mathrm{H^+}]} \tag{10}$$

Freeman and his group conducted a crystallographic-pH titration to determine structural changes that are coupled to pH equilibria [157]. They found that while the active-site geometry for the oxidized protein is essentially retained over the pH range 4 to 8, the reduced form undergoes major alteration (Scheme 3). As the pH is lowered, the Cu(I) moves away from His-87 to become effectively three-coordinate. Furthermore the imidazole ring of His-87 rotates by 180° so that the original Cu–N bond is replaced by a Cu–H–C (His-87) van der Waal's contact. The kinetic inactivation at low pH as discovered by Segal and Sykes [159] is thus identified as a major structural change at the active site which does not comply with the Franck–Condon requirement for minimal reorganization.

Cyclic voltammetry of spinach plastocyanin portrays an interesting view of how these factors affect its ability to transfer electrons. It was observed [99] that the conditions required to promote electrochemistry at a PGE electrode were broadly similar to those pertaining to the photosynthetic electron-transport system. For a solution of the protein (oxidized or reduced) at low ionic strength, well-defined diffusion-controlled voltammetric waves were observed upon addition of Mg^{2+} or by acidification to pH 4. The peak-current response as a function of these variables is shown in Fig. 12. At pH 7 (3 °C), $E^{0'}$ was found to be 375 mV, in good agreement with the potentiometric value reported by Katoh and co-workers [156]. The electrode reaction was found to be essentially reversible at pH 4. Whilst this appears at first to be in conflict with the evidence for plastocyanin being *inactive* at this pH, closer consideration shows that this is a consistent result. The corresponding electrode reaction may be written as in Eqs. (11)–(12)

$$\mathrm{PCu(II)} + e^- \rightleftharpoons \mathrm{PCu(I)} \tag{11}$$

$$\mathrm{PCu(I)} + H^+ \rightleftharpoons \mathrm{PCu(I)\text{-}H^+} \tag{12}$$

with Eq. (12) being described by $k_f[H^+]$ ($= k_f'$), k_b, and $K[H^+]$ ($= K'$). The point about the coupled equilibrium, Eqs. (7) or (12), which refers to protonation and reorganization of the Cu(I) geometry as shown in Scheme 3, is that it is established *very rapidly*. At the electrode, the consequence of Eq. (12) is that $E^{0'}$ increases by the increment $(RT/F)\ln(1 + K')$ as the pH is lowered. At pH 4, $E^{0'}$ is 430 mV. With rapid equilibrium between redox-active and redox-inactive forms, the result is the same as may be obtained by potentiometry. The original studies showed [161] that

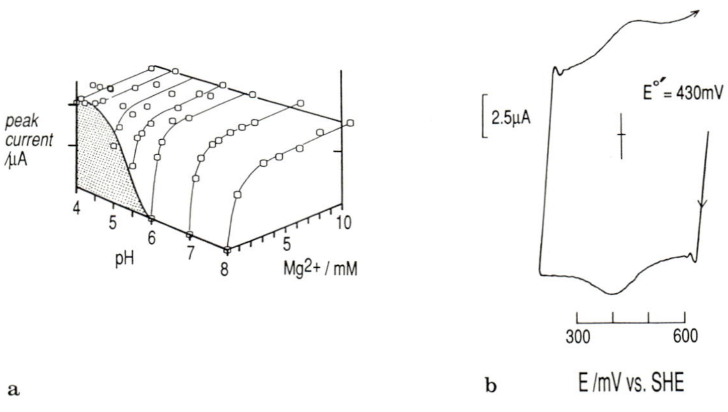

Fig. 12a, b. Electrochemistry of spinach plastocyanin at a PGE electrode. **a)** 3-D representation of the effects of pH and Mg^{2+} concentration upon observed peak currents (initial scan at 20 mVs^{-1} normalized with respect to electrode surface area. Plastocyanin 25 μM in 5 mM buffer (acetate, MES, HEPES, Tris) with 1 mM KCl at 3 °C. **b)** Initial-scan cyclic voltammetry at 500 mVs^{-1} obtained for oxidized plastocyanin (28 μM) at pH 4.0 (5 mM acetate, 1 mM KCl, 10 mM $MgCl_2$). Temperature = 3 °C

the reduction potential increases by 2.3 RT/F volts per pH unit below the pK of approx. 5. In terms of the electrochemical kinetics, a greater overpotential now has to be applied in order to oxidize PCu(I), but correspondingly less is required to reduce PCu(II).

More interesting are the actual values of k_f and k_b, since in the physiological reduction of $P700^+$ under conditions of lowered pH, deprotonation-linked reorganization of the Cu coordination site could become a rate-determining factor in turnover. In principle, rate constants for reactions coupled to an electrochemical process may be obtained by analysis of the voltammetric waveforms according to procedures described by Nicholson and Shain [50]. At pH 4, any limitation on electron-transfer reactivity resulting from an inability to reorganize *rapidly* would be observed as a progressive change in the shape of the cyclic voltammogram as the scan rate is increased. Sampling a solution of PCu(II) and recording the first cycle, the cathodic (reduction) peak should move towards more negative potentials and the anodic (reoxidation) wave should be attenuated. As shown in Fig. 12, the pair of cyclic voltammetric waves remain essentially symmetrical and of comparable amplitude at least up to $\upsilon = 500\ mVs^{-1}$. Estimation of k_b based upon a limiting ratio $i_{pa}/i_{pc} > 0.75$, showed [99] that this must be greater than $640\ s^{-1}$. A similar result was obtained with a Au electrode modified by adsorption of 2,2′-dithiobisethanamine [76]. This lower limit on the reorganizational rate agrees broadly with results of NMR studies which indicated [162] that proton relaxation on N^δ of His-37 occurs with τ of the order of 0.1 ms, i.e. k_b must be around $1000\ s^{-1}$. It is likely that a precise evaluation of the control of electron transfer by conformational dynamics may be achieved by employing voltammetric techniques with superior kinetic resolution.

Other "blue" Cu proteins have been studied by voltammetry. As mentioned earlier, Azurin, an electron-transport protein from *Pseudomonas*, is interesting since it appears to interact reversibly at hydrophobic electrode surfaces such as PGB [91] and silanated PGE [98]. It also exhibits activity at a carbon-paste electrode in the presence of 4,4′-bipyridyl, a reaction that appeared to involve formation of an adsorbed film of native protein [117]. In each of these cases, $E^{0'}$ values were obtained that were in agreement with results obtained by potentiometry. Another protein, rusticyanin, has been investigated [163] by Lappin and co-workers. This is an electron-transfer protein produced by *Thiobacillus ferrooxidans*, an organism whose energy is derived from the oxidation of Fe(II) by O_2 at pH 2. Rusticyanin is characterized by a high reduction potential and may be closely involved in Fe(II) oxidation. Like azurin, it was found to adsorb at a 4,4′-bipyridyl-modified carbon-paste electrode to give a transferable electro-active film. This provided an easy means of determining the pH dependence of the reduction potential since the protein-modified electrode could be immersed in various solutions. Cyclic voltammograms were obtained at different pH values between 1 and 3 over which range $E^{0'}$ was found to be invariant at $+670$ mV. This high reduction potential is certainly suited thermodynamically for serving in an electron-transport chain that involves the FeIII/II couple at low pH. Rusticyanin is not, however, unusual in this respect since the "blue" Cu centres in fungal laccase also have high reduction potentials (see Sect. 7.2).

4.4. Iron–Sulfur Proteins

The techniques of direct electrochemistry are put to their best use in the study and manipulation of proteins for which the redox chemistry is not addressed effectively by other methods. Subjects to benefit particularly are proteins containing metal centres that may be intrinsically unstable or have redox chemistry at potentials beyond the stability threshold of the solvent system. Many proteins containing Fe–S clusters fall into this category. These centres are widely distributed in biological systems [164] where their most widely accepted role is as electron-transfer agents.

Four structural classes of Fe–S centre have been identified to date; these are depicted in Fig. 13. All of them feature high-spin tetrahedral Fe(II) or Fe(III) coordinated typically by four sulfur donors. Apart from the monomeric centre found in proteins known as rubredoxins, they are all clusters that contain both protein donors and "inorganic" bridging (μ) sulfido ligands. Most of our knowledge stems from studies made on the small electron-transport proteins known as ferredoxins (Fd's) and from work on "model" compounds. Figure 14 shows the structure of a ferredoxin isolated from the anaerobe *Peptococcus aerogenes* [165].

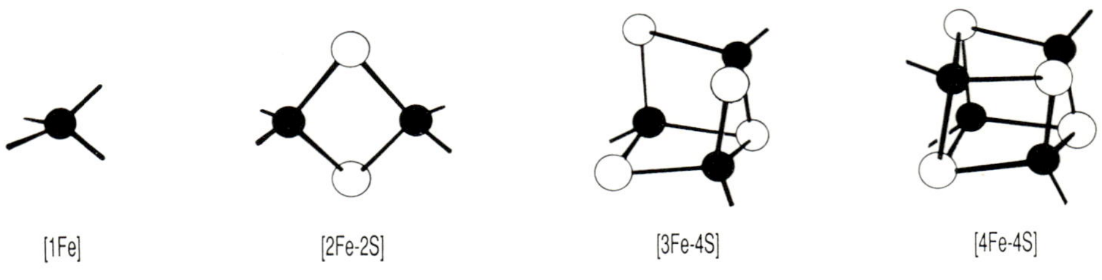

Fig. 13. Structures of the four currently established classes of Fe–S centre

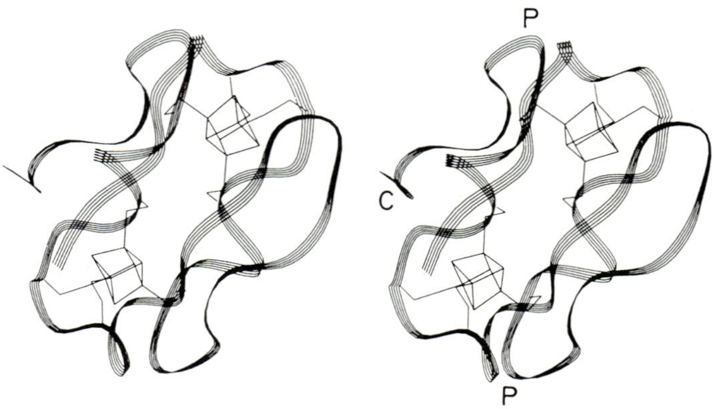

Fig. 14. Stereo "ribbon" structure of *Peptococcus aerogenes* ferredoxin showing the two [4Fe–4S] clusters. The "*C*" terminal and positions of the two prolines (*P*) are also indicated. This illustration was kindly provided by Larry Sieker (University of Washington, Seattle)

Two [4Fe–4S] clusters are coordinated within a polypeptide fold that is remarkably conserved [166] among bacterial ferredoxins of differing size and composition (see also Sect. 5.2). The complex redox chemistry of these systems is best described by referring to the overall charge on the "core"; that is, the active-site structure minus protein ligands. As evident from Scheme 4, the clusters function in one-electron reactions, shuttling between species that may be categorized according to whether their total d-electron count is odd (in which case the centre is usually detectable by EPR) or even (for which EPR is generally not very useful).

Redox Relationships among Fe-S Clusters

Bold type indicates that the oxidation level is common
Light type indicates that the oxidation level is rare
Italics indicate that the oxidation level has never been detected (but see Sect. 5.2).

			Fe^{3+} —	**Fe^{2+}**
		$[2Fe\text{-}2S]^{2+}$ —	**$[2Fe\text{-}2S]^{1+}$** ---	*$[2Fe\text{-}2S]^{0}$*
$[4Fe\text{-}4S]^{4+}$ ---	$[4Fe\text{-}4S]^{3+}$ —	**$[4Fe\text{-}4S]^{2+}$** —	**$[4Fe\text{-}4S]^{1+}$** —	$[4Fe\text{-}4S]^{0}$
	$[3Fe\text{-}4S]^{1+}$ —	**$[3Fe\text{-}4S]^{0}$** ---	*$[3Fe\text{-}4S]^{1-}$* ---	*$[3Fe\text{-}4S]^{2-}$*
d-electron count				
EVEN	*ODD*	*EVEN*	*ODD*	*EVEN*

Scheme 4

As a rule the protein environment appears capable of accommodating only single changes in the oxidation "level", thus even [4Fe–4S] displays only *one* of the several redox couples that are, in principle, available. Generally this is the 2 + /1 + couple as found also for [2Fe–2S]. In either case the reduced (1 +) oxidation level is a resultant spin doublet and gives an EPR spectrum characterized by $g_{av} < 2$ (but see Sect. 5.2). This level is strongly reducing and usually very air-sensitive. In certain cases the protein environment does not permit the existence of $[4Fe\text{–}4S]^{1+}$, but instead favours oxidation of $[4Fe\text{–}4S]^{2+}$ to the 3+ level. This, again, is identified by EPR ($S = 1/2$, but in this case $g_{av} > 2$). The proteins in which this occurs are termed "High-potential Iron Proteins" or HiPIPs. In all cases so far studied, the 2− oxidation level has a diamagnetic ground state.

Tuning of cluster redox potentials by the protein environment is very fine. From studies with synthetic analogues [167], it is known that greater stabilization of the 1+ level is afforded by aromatic thiol ligands as compared to the more electron-releasing groups such as aliphatic thiols, which can provide relative stabilization of the 3+ level. Solvation is also important [168]. With thiolate ligation the centres carry a negative charge that increases as the oxidation level decreases. The 1+ oxidation level is thus stabilized by exposure to a high dielectric solvent such as water. Further stabilization of the more electron-rich reduced levels is provided by hydrogen bonding between μ-sulfido subsites and hydrogen atoms from solvent or polypeptide [169].

The [3Fe–4S] clusters may be regarded as being derived from [4Fe–4S] with one Fe subsite vacant [170–172]. Only two oxidation levels have so far been identified. These are the 1+ level, which gives a distinctive EPR spectrum ($S = 1/2$ and $g_{av} > 2$), and the 0 level ($S = 2$). In some instances, for example aconitase, interconversion between [3Fe–4S] and [4Fe–4S] clusters—associated with the inactive and active enzyme, respectively—is known to occur readily [173–175].

Studies on Fe–S clusters are generally hindered by the lack of distinctive optical-absorption features and by low reduction potentials and O_2 sensitivity. Techniques established for their characterization do not, furthermore, allow the investigator to view and monitor rapid *changes* in cluster structure *as* they may occur or be induced *during* an experiment. Consequently, in what may turn out to be a large number of cases, the rich chemistry that stems from the framework of Scheme 4 has been difficult to define or control. As discussed now and in Sect. 5, voltammetric methods are providing new possibilities and insight.

The main experimental problem to be overcome is the provision of an electrode interface that is stable at very low potentials (−1 V or lower at neutral pH) whilst accommodating the specific interaction requirements of the protein as discussed in Sect. 3. The PGE electrode is well suited since it is electrochemically quite stable and displays a high overpotential for H_2 evolution. Ferredoxins usually feature a large excess of negatively charged amino-acid residues, and electrochemistry is promoted by multi-charged cations, particularly aminoglycosides. In addition to being stable towards reduction, the latter reagents are colourless and diamagnetic; thus they do not interfere in the spectroscopy of species generated electrolytically.

The 7-Fe ferredoxin isolated from *Azotobacter* species provides an example of how direct electrochemistry can reveal redox chemistry and generate species that are not accessible through chemical reductants like sodium dithionite. The crystal structure of *Azotobacter vinelandii* Fd I reveals [170, 171] two clusters, one [4Fe–4S] and one [3Fe–4S], incorporated into the protein, MW approx. 12000, within a folding pattern that is related [166] to that of *Peptococcus aerogenes* Fd shown in Fig. 14. There had long been some controversy not only concerning the structure of the 3Fe cluster, but also with regard to the redox properties of the 4Fe site. Previous potentiometric studies showed [176] that the 3Fe centre had a reduction potential of −420 mV at pH 7, but the [4Fe–4S] cluster was reported as being of the 3+/2+ type (see Scheme 4). The latter conclusion was drawn on the basis that it was not reducible by dithionite, but gave instead an EPR signal with $g_{av} > 2$ upon treatment with $Fe(CN_6)^{3-}$. Later, Stephens and co-workers suggested [177] that this cluster was probably of the "normal" type, i.e. reducible to the 1+ level, but with an unusually low reduction potential that made this state inaccessible by dithionite at neutral pH.

Ferredoxin I from *Azotobacter chroococcum* [178], which is spectroscopically indistinguishable from the *A.v.* protein, was investigated [105] by direct electrochemical methods. Cyclic voltammograms and the pH dependence of reduction potentials are shown in Fig. 15. A PGE electrode was used with three different

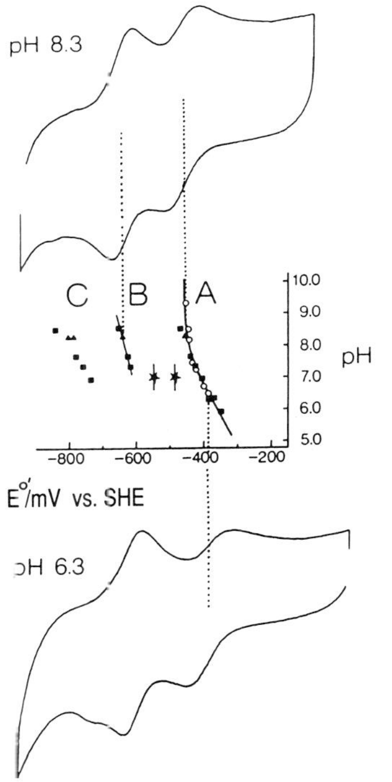

Fig. 15. Cyclic voltammetry of *Azotobacter chroococcum* ferredoxin I, in 0.10 M NaCl at pH 8.3 (20 mM TAPS) and pH 6.3 (20 mM PIPES). Temperature 3 °C, scan rate 10 mVs^{-1}. The electrochemistry was promoted by 1–2 mM levels of aminoglycosides neomycin or tobramycin. Shown at the centre is the pH dependence of $E^{0'}$ values, for which solid lines correspond to: (**A**), $E^{0'}$ (alkaline) = − 460 mV, $\Delta E^{0'}/\Delta pH$ = − 55 mV, pK_{red} = 7.8; (**B**), $E^{0'}$ (pH 8.3) = − 645 mV, $\Delta E^{0'}/\Delta pH$ = − 25 mV. Promoters used were: (▲) neomycin 1–2 mM; (■) tobramycin 1–3 mM; (○) $Cr(NH_3)_6^{3+}$ 8 mM. Average estimated values for the third redox couple (**C**) are also given. Limits for the effective reduction potential of dithionite at pH 7 are indicated by ⫳ ⫳. Lower and upper limits indicate the midpoint potentials appropriate for dithionite in solutions initially 10 μM and 1 mM in dithionite respectively. For a discussion of the reduction potentials of dithionite solutions, see Mayhew SG (1978) Eur. J. Biochem. *85*: 535

promoters in order to make measurements over a wide range of pH and potential. The range of promoters also allowed some assessment of the variation in reduction potentials that might arise from specific binding of promoters to the protein. In this case, both of the aminoglycosides used had an upper pH limit due to deprotonation of $-NH_3^+$ groups while neomycin, but not tobramycin, was found to cause precipitation at pH values < 6.5. The use of $Cr(NH_3)_6^{3+}$, on the other hand, allowed measurements to be made under more alkaline conditions, but the potential range was restricted by its redox activity below − 700 mV.

The voltammetry showed several interesting features. First, the waves (A) could be assigned to the dithionite-reducible [3Fe–4S] cluster. The pH dependence of $E^{0'}$ was analyzed in terms of electron transfer that is coupled to a rapidly established acid–base equilibrium [50] with pK = 7.8. The limiting low-pH slope was − 59 mV per pH unit while the limiting $E^{0'}$ value under alkaline conditions was − 460 mV. Values obtained with $Cr(NH_3)_6^{3+}$ as promoter were in good agreement. The processes involved can be written as Eqs. (13) and (14).

$$[3Fe\text{–}4S]^{1+} + e^- \rightleftharpoons [3Fe\text{–}4S]^0 \tag{13}$$

$$[3Fe\text{–}4S]^0 + H^+ \rightleftharpoons [3Fe\text{–}4S]^0\text{–}H^+ \tag{14}$$

What is particularly interesting about this system is that the acid and alkaline forms of the reduced cluster differ greatly in their magnetic-circular-dichroism (MCD) spectra. While that of $[3Fe\text{–}4S]^0$ is similar to those found for other 3Fe systems, the acid form gives a very different spectrum that indicates a substantial change in the electronic structure [178, 179]. This could be a result of H^+ binding directly to the cluster or inducing a rapid conformational change.

The second process (B) is the reduction of $[4Fe–4S]^{2+}$ to the 1+ level. This was established by carrying out bulk reduction of the protein at a potential of −850 mV vs SHE in a stirred anaerobic cell. The product, now reduced by 1.9 electron equivalents, gave an EPR spectrum typical of $[4Fe–4S]^{1+}$ with evidence for spin-coupling to the nearby paramagnetic species $[3Fe–4S]^0$. Double integration gave 0.9 spins per molecule. Analysis of the voltammetric waves showed that the system conformed well to "ideal" criteria for a diffusion-controlled one-electron process. Plots of i_p vs $\upsilon^{1/2}$ were linear up to at least 1 Vs^{-1} with ΔE_p remaining at around 60 mV. The reduction potential was found to be mildly dependent upon pH, but there was no discontinuity. At pH 8.3 and 3 °C, $E^{0'}$ was determined to be −645 mV, a value considerably more negative than as yet found for any biological Fe–S cluster under conditions of neutral pH. Using $Cr(NH_3)_6^{3+}$, a somewhat higher value (−600 mV) was obtained that was independent of pH. The shift indicated binding of $Cr(NH_3)_6^{3+}$ to a site close to the [4Fe–4S] cluster.

No redox couple could be observed in the potential range −300 to +600 mV, thus showing that the oxidation process with $Fe(CN)_6^{3-}$ previously reported [176] could not be a reversible one-electron reaction, i.e. the 4Fe cluster was not of the 3+/2+ type. However, at more negative potentials, a third pair of waves (couple C) was observed, whose amplitude and $E^{0'}$ value were found to be pH dependent. Such an additional redox couple has been observed for other proteins that contain a [3Fe–4S] cluster. For example, ferredoxin III from *Desulfovibrio africanus* [110, 111] and other 7-Fe ferrodoxins, from the thermophiles *Sulfolobus acidocaldarius*, *Thermus aquaticus*, and *Thermoplasma acidophilum* (Armstrong FA, Butt JN, Cammack R, George SJ, Thomson AJ, unpublished results) each show three pairs of waves, although there are just two clusters. For *Desulfovibrio africanus* Fd III, as with *Azotobacter chroococcum* Fd I, the two couples having the higher reduction potentials were assigned [110] to $[3Fe–4S]^{1+/0}$ and $[4Fe–4S]^{2+/1+}$, through preparation of spectroscopic samples by bulk electrolysis. The third couple, displaying pH-dependent $E^{0'}$ values appeared to be largely confined to adsorbed species. *Desulfovibrio africanus* Fd III has proved to be a most interesting and unusual protein, and I have delayed discussion of the results pertaining to the additional redox couple until Sect. 5.2. At this juncture, it may be mentioned that it proved possible to link the low-potential process unambiguously with the presence of the [3Fe–4S] cluster, and to establish that it was a *two*-electron reaction accompanied by net uptake and release of H^+ [112]. Thus, with voltammetry, it is possible to probe unexpected (and unprecedented) *multiple* electron-transfer activity—in this case, the chemically reversible generation of a state corresponding with or equivalent to $[3Fe–4S]^{2-}$ (see Scheme 4).

Since the discovery of [3Fe–4S] clusters, it has frequently been argued [173] that they must, in many cases, be artifacts produced by degradation of [4Fe–4S] during isolation and exposure to air. Oxidation is probably a key factor since *Clostridium pasteurianum* 2[4Fe–4S] ferredoxin (whose structure and properties are analogous to *Peptococcus aerogenes* Fd) is converted [173, 180], upon treatment with $Fe(CN)_6^{3-}$, to a form containing a [3Fe–4S] cluster. With the exception of HiPIPs, simple one-electron oxidation to produce $[4Fe–4S]^{3+}$ is not known and might be expected to result in decomposition. With voltammetric techniques, it was possible to study the oxidation of *Clostridium pasteurianum* 2[4Fe–4S] Fd at high potentials [181]. Using DC cyclic voltammetry, two broad and chemically irreversible processes were observed. With differential pulse voltammetry, these were defined quite clearly, as shown in Fig. 16. Corrected for pulse height, the effective potential values were +793 and +1120 mV at 5 °C (pH 7). Their identities were suggested by results of an analogous experiment with the 2[4Fe–4Se] analogue. This gave values of +797 and +1090 mV under the same conditions. From the magnitude of perturbation of each of the two peak positions, the lower potential process was assigned to a cluster redox couple (presumably 3+/2+) while the greater-influenced higher-potential process was more likely to arise from secondary oxidation associated with S^{2-} or Se^{2-}. The reaction sequence thus proposed is described by Eqs. (15) and (16) where X is S or Se.

$$[4Fe–4X]^{2+} \rightleftharpoons [4Fe–4X]^{3+} + e^- \tag{15}$$

$$[4Fe–4X]^{3+} \xrightarrow{fast} \text{products} \tag{16}$$

Analysis of high-frequency square-wave voltammograms showed that the putative electrode product $[4Fe–4S]^{3+}$ did not survive for any measurable length of time. Failure to detect the reverse of Eq. (15) even at a frequency of 200 Hz meant that its half-life was certainly less than 1.6 ms at 5 °C. With the coupled, irreversible reaction Eq. (16) taken into account, a lower limit of 360 mV was assigned for the thermodynamic reduction potential of the $[4Fe–4S]^{3+/2+}$ couple. This is much higher than the reduction potential, $E^{o'} = 350$ mV, that is established for *Chromatium vinosum* HiPIP [182].

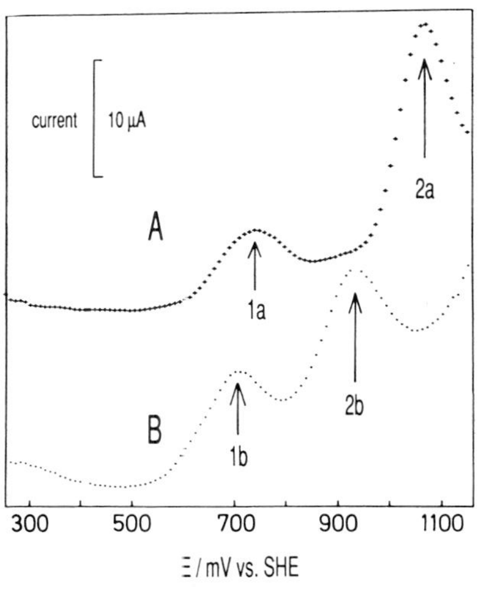

Fig. 16. Differential pulse voltammograms of *Clostridium pasteurianum* ferredoxin at high potentials: (**A**), native protein containing [4Fe–4S] clusters; (**B**) derivative containing [4Fe–4Se] clusters. Protein 0.2 mM in 20 mM phosphate, 0.40 M NaCl, pH 7.0, temperature 5 °C. Pulse amplitude + 100 mV, scan rate 14 mV s^{-1}. Signals 1*a* and 1*b* are assigned to $[4Fe–4X]^{3+/2+}$ whereas 2*a* and 2*b* are, most likely, due to further oxidation of S or Se species

4.5 *Modified Proteins for Studies of Long-range Intramolecular Electron Transfer*

A current strategy in investigations of long-range electron transfer in biological systems is the study of intramolecular processes in specifically-modified multi-centred redox proteins [12–16]. In one class of experiment [13, 183–187], the system for study is prepared by covalent attachment of redox-active complexes to specific amino-acid residues on a suitable protein. The distance between the "native" and the "synthetic" site and the nature of the intervening medium may be deduced from crystallographic data. In typical investigations, surface-accessible His-imidazole groups have been derivatized with the $Ru(NH_3)_5^{3+}$-functionality by treatment of the protein with $Ru(NH_3)_5H_2O^{2+}$ followed by oxidation. The product is then purified and characterized. The position of the Ru label may be confirmed by enzymatic digestion and examination of fragments. A typical kinetic experiment involves pulsed one-electron reduction of protein molecules in which, initially, both native and synthetic sites are in the oxidized state. This reduction may be performed by pulse radiolysis or laser photolysis, but the key requirement is that the kinetics of the initial second-order reaction are such as to generate a significant amount of the state in which the low-potential centre is preferentially reduced. This differs from the final equilibrium state by a free energy corresponding to $F\Delta E'$. As long as interaction between the two sites is negligible, $\Delta E'$ corresponds to ΔE, the difference between the reduction potentials of native and synthetic redox sites that may be derived from equilibrium measurements. The relaxation process, corresponding to intramolecular electron transfer between the two sites, is now monitored.

An example of such a system is horse cytochrome *c* to which the $Ru(NH_3)_5^{3+}$-entity is coordinated to imidazole at His-33 [183, 184]. The distance between the Ru site and the heme is estimated to be about 15 Å based upon the structure for tuna cytochrome *c*. Rate constants reported for the intramolecular electron transfer from Ru(II) to Fe(III) vary between 30 and 52 s^{-1} at room temperature.

In order to enlarge the database, investigators have sought to vary parameters within a single protein host. With regard to distance and medium, the major restriction is the number of protein sites that are suitable and available for modification. With regard to driving force, the nature of the synthetic site can be varied, for example (with Ru) by substitution of different ligands for NH_3. Use of the $Ru(NH_3)_4$ (isonicotinamide) functionality gives a high-potential synthetic site [185], such that with cytochrome *c*, the direction of spontaneous intramolecular electron transfer is reversed, becoming Fe(II)-to-Ru(III).

Voltammetric methods are very suitable for characterizing these modified electron-transfer proteins. With regard to determining ΔE, the Ru chromophore is relatively weak and, for the case of cytochromes, is hidden by the intense heme Soret absorptions. Consequently, potentiometric determination of the reduction potential for the Ru site using, for example, an OTTLE is not viable. In several studies, investigators have adopted the reduction potentials of the solution analogues, for example $[Ru(NH_3)_5(His)]^{3+/2+}$ for which $E^{0'} = 80$ mV. However, since the thermodynamic redox properties of complexes will certainly vary in a protein environment, direct measurement is desirable. The voltammogram itself also provides a useful "fingerprint" by which the integrity of modified proteins may be ascertained.

Horse cytochrome *c* modified at His-33 by attachment of $Ru(NH_3)_5^{3+}$ has been studied independently by the groups of Gray [183] and Isied [184, 185]. They used a Au electrode in the presence of 4,4′-bipyridyl to determine the reduction potentials of each site. In either case, differential pulse voltammetry showed two overlapping peaks. Values of ΔE thus measured showed a temperature dependence (arising mainly from the Ru site) as well as some variability between groups, most likely due to differing experimental conditions. As observed at 25 °C the two signals in each case had very different amplitudes, the Ru peak being smaller and broader (although Gray and co-workers noted [183] that upon lowering the temperature to 5 °C the waves became much more equivalent). Isied and co-workers also prepared and characterized cytochrome *c* modified at His-33 by attachment of the $Ru(NH_3)_4(isn)^{3+}$- entity [185]. Differential pulse voltammograms for both derivatives are shown in Fig. 17.

Detailed electrochemical studies have been carried out on other modified proteins. Two specific examples, plastocyanin from *Scenedesmus obliquis* modified by $Ru(NH_3)_5$- at His-59 [186] and HIPIP ($[4Fe\text{–}4S]^{3+/2+}$) from *Chromatium vinosum* modified by $Ru(NH_3)_5$- at His-42 [187], were studied by cyclic voltammetry [188]. Results obtained with a PGE electrode and comparisons with voltammograms of the native proteins are shown in Fig. 18. By contrast with the native proteins, multi-charged cations were not required to promote a stable and well-defined response at pH 7. Modified plastocyanin gave excellent electrochemistry, with i_p for both redox sites being of equal amplitude and diffusion-controlled up to 50 mVs^{-1}. In view of the fact that this protein still retains an overall negative charge, this observation gives an interesting illustration of the sublety of coulombic interactions with the electrode surface and support for *localized* effects. For modified HiPIP (which is virtually neutral at pH 7) the situation was less clear-cut.

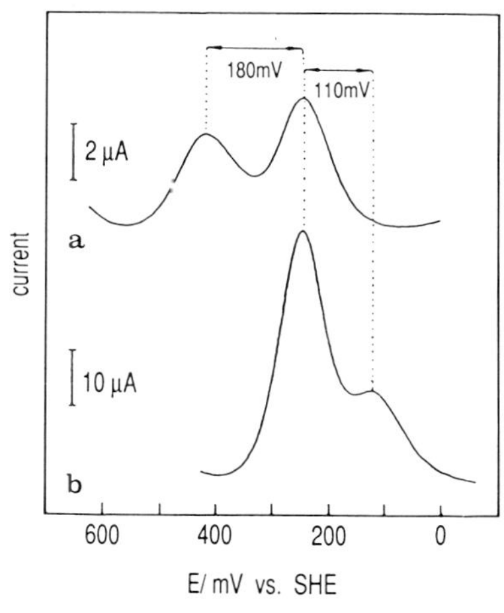

Fig. 17. Differential pulse voltammograms of two derivatives of horse cytochrome *c* in which a Ru complex has been covalently attached at His-33. (*a*), [$Ru(NH_3)_4$(isn)-cyt. *c*]; (*b*) [$Ru(NH_3)_5$-cyt. *c*]. Solutions were in 0.1 M $NaClO_4$, 0.08 M phosphate (pH 7) with 0.01 M 4,4′-bipyridyl to promote electrochemistry at the Au working electrode. Pulse amplitude 25 mV, scan rate 2 mVs^{-1}. Redrawn from Ref. 185, with kind permission of the authors

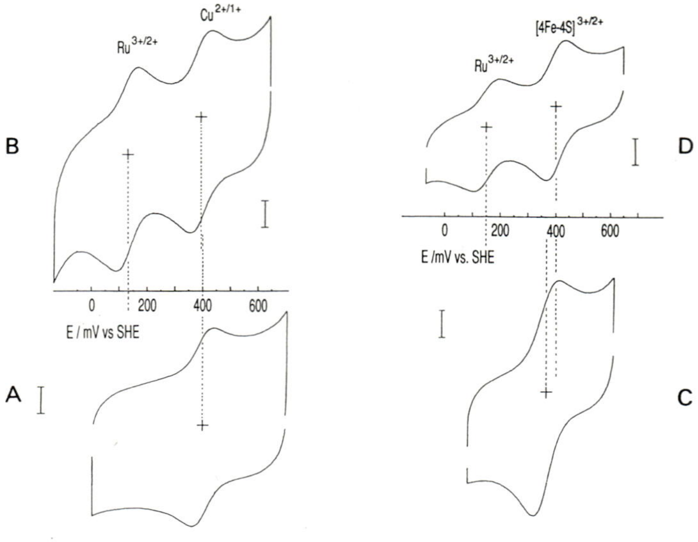

Fig. 18. Cyclic voltammetry of plastocyanin (*Scenedesmus obliquus*) and HiPIP (*Chromatiun vinosum*) and their $Ru(NH_3)^{3+}$-derivatives. (**A**). Plastocyanin, 0.10 mM in 0.1 M NaCl, 20 mM AMHT*, 0.4 mM neomycin, pH 7.0, temperature 2 °C. Scan rate 100 mVs^{-1}, current scale 1.0 μA. (**B**). $Ru(NH_3)_5^{3+}$-His(59) plastocyanin, 0.10 mM in 0.1 M NaCl, 20 mM AMHT*, pH 7.0, temperature 2 °C. Scan rate 20 mVs^{-1}, current scale 0.2 μA. (**C**). HiPIP, 0.10 mM in 0.1 M NaCl, 20 mM HEPES, 0.4 mM neomycin, pH 7.0, temperature 4 °C. Scan rate 10 mVs^{-1}, current scale 0.1 μA. (**D**). $Ru(NH_3)_5^{3+}$-His(42) HiPIP, 0.13 mM in 0.1 M NaCl, 20 mM AMHT*, pH 6.9, temperature 1 °C. Scan rate 10 mVs^{-1}, current scale 0.2 μA. * 20 mM AMHT is a mixed buffer consisting of 5 mM each in acetate, MES, HEPES and TAPS

While the waves for $[4Fe\text{–}4S]^{3+/2+}$ couple were well-defined and diffusion-controlled up to 100 mVs^{-1}, those for the Ru site were somewhat broader.

Taken together, the results on each of these modified proteins show that reduction potentials may *deviate* significantly from values predicted from isolated systems. In the examples shown in Figs. 17 and 18, $E^{0'}$ for the Ru site is shifted considerably from that of the $[Ru(NH_3)_5(imidazole)]^{3+/2+}$ couple in aqueous solution (+ 109 mV at 25 °C). Interaction between the redox sites may also be gauged. In the case of HiPIP, $E^{0'}$ for the $[4Fe\text{–}4S]^{3+/2+}$ couple is shifted + 30 mV upon introduction of the Ru centre so close by. By contrast, reduction potentials of the native sites in modified cytochrome *c* or plastocyanin, in which the Ru centre is further away, appear essentially unaffected.

In an earlier discussion (Sect. 3.4) it was mentioned that the electrode reactions of proteins could be governed by radial-type diffusion, even at a macroscopically planar electrode surface, because of specificity for electroactive sites having microscopic dimensions. Deviation from linear diffusion behaviour depends upon the size and density of these sites [125, 126]. Thus while the differences in waveshape between native and Ru couples that was noted in these studies may have a purely chemical-kinetic origin, the interesting possibility also arises that there exists different effective mass transport behaviour for the electrode reactions of each redox centre contained in a multi-centred macromolecule. Further evidence for this suggestion is presented in Sect. 5.2.

5 Studies of Metal-Ion Speciation

5.1 Voltammetric Signals as Analytical "Signatures"

Once electroactivity has been achieved for a metalloprotein, the voltammetric waves become a valuable *signature* of the active site(s). Voltammetry can therefore be used to define states and to monitor changes in structure and composition that may result from appropriate chemical or electrochemical perturbations of the system. This has interesting and important consequences since an electrochemical experiment provides conditions of strict potential control within a small and often microscopic sample size. The applications are, at present, in their infancy, but I shall endeavour to illustrate the opportunities for studies of metal site speciation by reference to one topical area, that of metal-ion uptake and loss among Fe–S clusters.

5.2 Studies of Fe–S Cluster Transformations

It is now widely appreciated that the old, established view of Fe–S clusters as a class of integral structural units whose activity is limited to electron transfer is in error. We have become aware of other roles. For example, in the Fe-hydrogenases, it is certain that a cluster serves as the catalytic site with which the H_2/H^+ substrates interact [164]. Moreover, following the discovery by Beinert's group that aconitase is an Fe–S protein [175], there are now further examples of clusters occurring in enzymes that do not catalyze, at least obviously, a redox reaction. With aconitase itself, it is established that substrates bind to one Fe subsite of a [4Fe–4S] cluster without major disturbance of the core structure [189]. Such findings demonstrate that "real chemistry" is available, reactivity that may be exploited once tuned by the protein environment.

Furthermore, studies are showing that certain proteins may permit variations in the composition and nuclearity of the core. So far this has been illustrated by several systems in which the core types [3Fe–4S] and [4Fe–4S] (see Fig. 13) are interconvertible [173–175]. Of these, aconitase itself is the best example. Its isolation under aerobic conditions yields an inactive form containing a [3Fe–4S] cluster [174]. Activation involves addition of Fe^{2+} under reducing conditions. Beinert and his co-workers were able to establish that this Fe entered the vacant subsite on [3Fe–4S] and that this rather labile component was the site for substrate binding [189]. The aconitase example showed that specific subsites of a cluster could be endowed with unusual activity.

The smaller proteins, i.e. ferredoxins, which show [173] interconvertibility between [3Fe–4S] and [4Fe–4S] core types are valuable since representative 3-D structures are available and there are extensive sequence data. It thus becomes possible to investigate the features that determine which structure is more stable and how rapidly they may interconvert. In most cases there is a clear preference for one core type over the other, and interconversion requires prolonged treatment with excess reagents followed by chromatography to remove unreacted protein and side products. The cluster status is then verified, typically by EPR, with MCD or Mössbauer as further probes. In general, conversion of [4Fe–4S] to [3Fe–4S] is achieved by reaction with an oxidant such as $Fe(CN)_6^{3-}$ (or O_2), while incorporation of Fe^{2+} into [3Fe–4S] requires a reductant such as ascorbate or a thiol [165–167]. This is described by Eq. (17).

$$[3Fe\text{–}4S] + Fe^{2+} \underset{\text{oxidant}}{\overset{\text{reductant}}{\rightleftharpoons}} [4Fe\text{–}4S] \tag{17}$$

It has been demonstrated [190, 191] that metals other than Fe may be incorporated into clusters in a corresponding manner. Particular interest in this conversion stems from the possibility that *heterometal* clusters may be active sites of enzymes, as indeed is the case in nitrogenase [164].

The application of direct electrochemical techniques in this area is proving to be very rewarding. Just why this approach should be so suitable may be understood by considering the problems associated with conventional techniques.
1) As discussed earlier, *in situ* monitoring of cluster status is difficult. Absorption spectra are generally broad and do not differ appreciably from [3Fe–4S] to [4Fe–4S]. The problem is readily appreciated if only one out of two or more clusters in a protein is reactive towards transformation and if other coloured reagents e.g. $Fe(CN)_6^{3-}$ are present. On the other hand, cluster interconversion may be accompanied by a sizeable change in reduction potential; thus the reaction may be followed by loss and appearance of well-defined voltammetric responses. These responses now become a sensitive and discriminating "handle". Significantly, in this respect, differential pulse polarography was the technique of choice for studying the kinetics of ligand exchange at small [4Fe–4S] analogue complexes [192].
2) Transformations in each direction are associated with oxidation or reduction. Thus the obvious way to supply redox equivalents is electrochemically. The charge passed can be monitored; furthermore the oxidation level (and hence reactivity) of each type of cluster is controlled instrumentally.

A most unusual ferredoxin that has recently been characterized provides us with a valuable "model" system. It has importance not only for understanding cluster interconversions, but also for defining the effects of non-cysteine ligation upon the spectroscopic and chemical properties of Fe–S clusters. As isolated, ferredoxin III from the sulfate-reducing bacterium *Desulfovibrio africanus* contains 7Fe atoms in a molecule of molecular mass approx. 6000 [193]. Electrochemical and spectroscopic investigations showed [110] that there are two clusters, one [3Fe–4S] and one [4Fe–4S] while the amino acid sequence revealed [193] that there is considerable homology with 2[4Fe–4S] ferredoxins as typified by those from *Clostridium pasteurianum* and *Peptococcus aerogenes* (see Fig. 14). A comparison of the primary structures of a number of ferredoxins containing [3Fe–4S] and [4Fe–4S] clusters is shown in Fig. 19. The 2[4Fe–4S] ferredoxins have two cluster-binding domains each with the sequence –Cys–X–X–Cys–X–X–Cys– - - - - - - - -Cys–Pro–. In the case of *D.a.* Fd III and some other 7Fe ferredoxins, one of these domains is modified. In *D.a.* FdIII, there are two notable changes to the N-terminal domain; the middle –Cys– is replaced by –Asp(D)–, and the –Pro– that

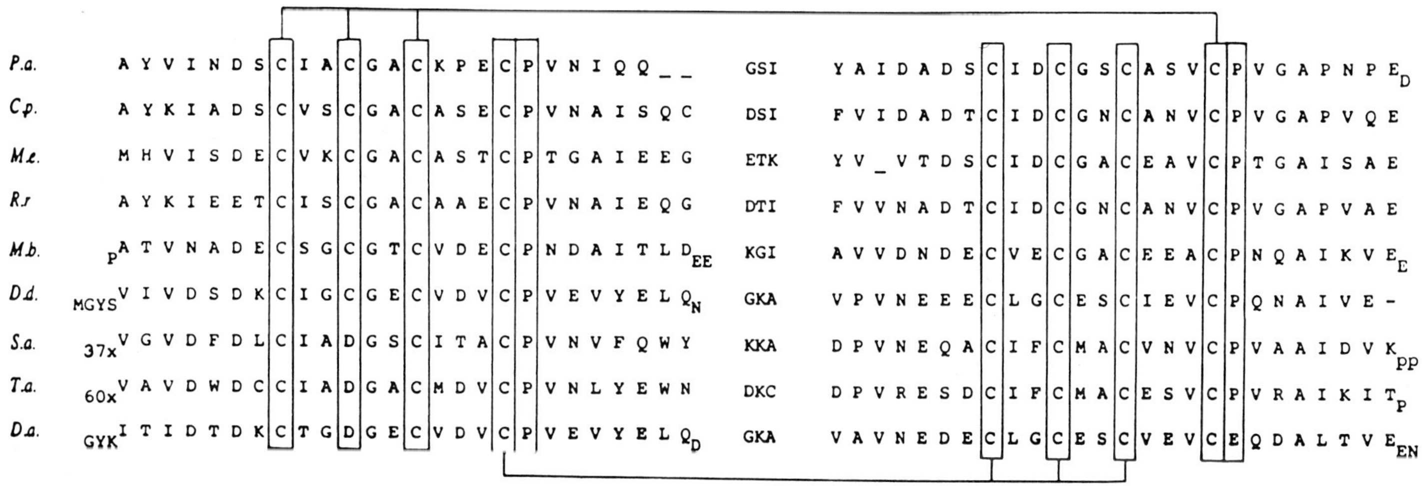

P.a. *Peptococcus aerogenes*

C.p. *Clostridium pasteurianum*

M.e. *Megasphaera elsdenii*

R.r. *Rhodospirillum rubrum*

M.b. *Methanosarcina barkeri*

D.d. *Desulfovibrio desulfuricans II*

S.a. *Sulfolobus acidocaldarius*

T.a. *Thermoplasma acidophilum*

D.a. *Desulfovibrio africanus III*

Fig. 19. Sequence homology among ferredoxins containing two [4Fe–4S] or one [4Fe–4S] and one [3Fe–4S] cluster. All data taken from: Otaka E, Ooi T (1987) *J. Mol. Evol. 26*: 257, except sequence of *D.a.* ferredoxin III (Ref. 193)

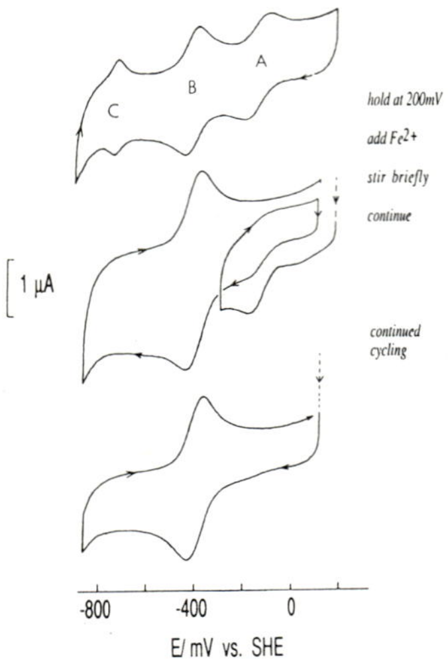

Fig. 20. Cyclic voltammetry of *Desulfovibrio africanus* ferredoxin III, showing how cluster transformation may be controlled and monitored. Scan rate 16 mVs^{-1}, temperature 2 °C. Protein 0.1 mM in solution containing 0.1 M NaCl, 20 mM HEPES, 0.1 mM EGTA and 1.1 mM neomycin at pH 7.4. The EGTA sequesters Fe that is released during slow degradation of the protein during handling.
Top. Prior to addition of Fe^{2+}
Middle. After holding at + 200 mV while Fe^{2+} is added with stirring (to total of 0.2 mM)
Lower. Continuation of scan with no further stirring. Waves A associated with $[3Fe\text{–}4S]^{1+/0}$ have disappeared completely. Waves C have also vanished.

succeeds the remote –Cys– has been replaced by –Glu (E). This domain is thus expected to accommodate the [3Fe–4S] cluster.

Direct electrochemistry of *D.a.* Fd III was achieved [110] with a PGE electrode in the presence of low levels of aminoglycoside promoters such as neomycin. As shown in Fig. 20 (top), three sets of voltammetric waves were observed. Spectroscopic examinations of the oxidized protein and of the products of bulk electrolyses carried out at − 320 mV and − 610 mV identified couples A and B as $[3Fe\text{–}4S]^{1+/0}$ ($E^{0'}$ = − 140 mV) and $[4Fe\text{–}4S]^{2+/1+}$ ($E^{0'}$ = − 410 mV), respectively. In each case, reduction potentials were essentially invariant with pH over the range pH 6 to 8. Early in the investigations it was noted that the relative amplitudes of waves A and B varied erratically from one sample to another. But following extensive aerobic dialysis against EGTA (or EDTA) and with the complexing agent in solution and provided a low scan rate ($<$ 16 mVs^{-1}) was used, voltammograms (as shown) were obtained that displayed A and B at equal intensity. Without this treatment, or at scan rates $>$ 40 mVs^{-1}, waves A were always smaller than B. (The choice of EGTA over EDTA lay in the fact that the reduction potential of Fe(EGTA) is sufficiently positive that its own electrode reaction does not interfere with the voltammetry of the protein.)

This finding was an important indication of what was to happen when Fe^{2+} is added to the solution. As shown in Fig. 20 (middle), additions of Fe^{2+} in stoichiometric or excess amounts (after allowance for EGTA complexation) led to dramatic changes in the voltammetry *after passage through* the $[3Fe\text{–}4S]^{1+}$ reduction wave [111]. The corresponding return (oxidation) wave was lost, and there was a marked (almost two-fold) increase in the amplitude of the original $[4Fe\text{–}4S]^{2+/1+}$ waves (Fig. 20, bottom). An appealing aspect of this experiment is that the irreversible reaction thus initiated is restricted to the diffusion layer; thus the experiment could be repeated a number of times after brief stirring at a poten-

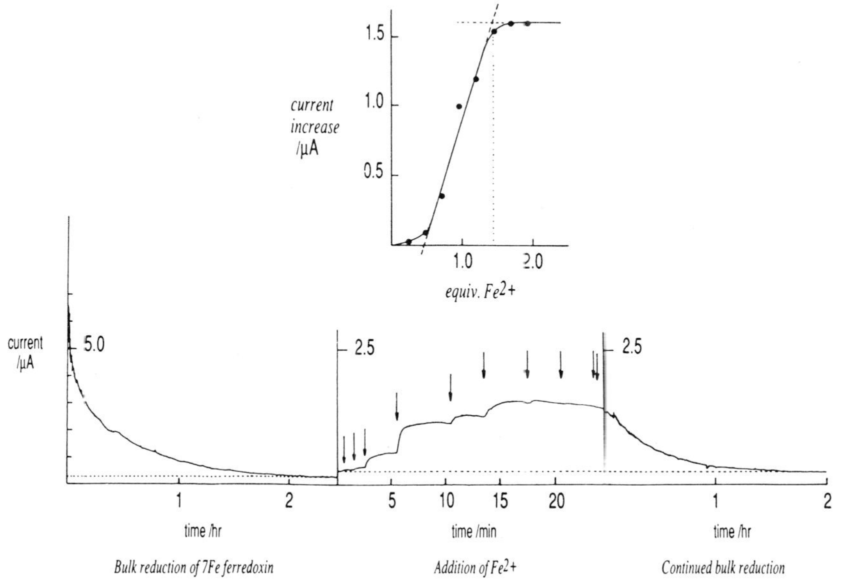

Fig. 21. A quantitative analysis of cluster transformation in *Desulfovibrio africanus* ferredoxin III by direct electrochemistry. The experiment starts with anaerobic bulk reduction, at -610 mV, of the oxidized protein (400 µL of a 0.1 mM solution including 20 mM HEPES, 0.1 M NaCl, 0.1 mM EGTA and 1.5 mM neomycin) contained in a sealed cell equipped with a PGE working electrode. Temperature = 3 °C. At exhaustion, aliquots of Fe^{2+} were added (indicated by *arrows*) whilst holding the potential at -610 mV. The graph above shows the increase in current observed as a function of the number of equivalents of Fe^{2+} added. The line drawn through the points corresponds to uptake of 1.0 Fe^{2+} per protein molecule. The lag corresponds to uptake of Fe^{2+} by remaining free EGTA. Finally, the second-phase bulk reduction was continued to exhaustion. This phase consumed (in total) one further electron equivalent

tial positive with respect to couple A, i.e. in a region in which $[3Fe–4S]^{1+}$ is not reduced.

The product of this electrochemically induced reaction was generated for spectroscopic characterization by carrying out controlled electrolysis. This was done in a way that allowed the reaction stoichiometry to be determined. As shown in Fig. 21, initial reduction of the 7Fe protein at -610 mV, in the presence of EGTA, required approx. 2.0 electrons. Then addition of aliquots of Fe^{2+} were made, still maintaining the potential at -610 mV, until they induced no further rises in current. These current increases occurred in each case with a half-life of a few seconds. Exhaustive reduction then consumed a further charge of approx. 0.9 electrons. The amount of added Fe^{2+} needed to effect the total change, after subtracting the lag phase due to complexation by free EGTA, was 1.0 atom equivalents. Detailed spectroscopic examination revealed that the [3Fe–4S] cluster had been replaced by a [4Fe–4S] cluster of particular novelty in that the reduced 1+ form had the ground state $S = 3/2$ instead of 1/2. The transformed cluster gave an EPR spectrum characterized by a resonance at $g = 5.2$ instead of the typical "$g = 1.94$" signal. The MCD spectrum of the transformed $[4Fe–4S]^{1+}$ cluster was similar to that of reduced *Clostridium pasteurianum* 2[4Fe–4S] ferredoxin except that it was considerably more intense.

The amino-acid sequence demanded that one of the [4Fe–4S] clusters, certainly the new one, must have one Fe subsite coordinated to the protein by a non-cysteine ligand. As evident from Fig. 19, this is likely to be the carboxylate-O atom(s) from aspartate. The reduction potential value of the new cluster is very close to that of the original one and is, again, not dependent upon pH, so that only one resultant couple (of greater amplitude but shifted + 10 mV) is observed. The situation which results is essentially analogous to that found for *Clostridium pasteurianum* 2[4Fe–4S] ferredoxin; this shows (Fig. 7) simple voltammetry due to two similar non-interacting sites. Thus the only property easily distinguishing the new type of center is the magnetic ground state. A few other examples of $[4Fe–4S]^{1+}$ are known to exhibit $S = 3/2$ ground states, for example the Fe protein of nitrogenase, but the protein structures are less well defined.

The manner by which Fe uptake occurred spontaneously upon reduction allowed unequivocal assignment of the route indicated in Eqs. (18)–(19) in which generation of $[3Fe–4S]^0$ at the electrode is followed by spontaneous incorporation of Fe^{2+}. The product, $[4Fe–4S]^{2+}$, is capable of further reduction—Eq. (20).

$$[3Fe–4S]^{1+} + e \rightleftharpoons [3Fe–4S]^0 \tag{18}$$

$$[3Fe–4S]^0 + Fe^{2+} \rightleftharpoons [4Fe–4S]^{2+} \tag{19}$$

$$[4Fe–4S]^{2+} + e^- \rightleftharpoons [4Fe–4S]^{1+} \tag{20}$$

This is a good demonstration of how the electrochemical approach can permit the quantitative study of a system that otherwise exhibits complex and confusing behaviour. Each reaction of this ferredoxin could be *monitored* and *controlled*. It is an unstable protein that degrades quite rapidly on standing in solution. As a result of trace Fe continuously released by cluster decomposition, conversion to the 8Fe form occurs spontaneously in the presence of reducing equivalents unless the Fe is sequestered by agents such as EGTA.

In a similar manner it has proved easy to demonstrate and study the stoichiometric incorporation of Zn^{2+}, Co^{2+} and Cd^{2+} to produce clusters with characteristic electrochemical and spectroscopic properties (Armstrong FA, Butt JN, George SJ, Hatchikian EC, Thomson AJ, unpublished results). Reduction potentials decrease in the order [Co–3Fe–4S] (approx. -300 mV) > [4Fe–4S] (approx. -390 mV) > [Zn–3Fe–4S] (approx. -480 mV) > [Cd–3Fe–4S] (approx. -580 mV). In each of these cases, metal-ion uptake into $[3Fe–4S]^0$ was very rapid. Incorporation into the oxidized (1+) cluster was *not* observed.

The investigations outlined above demonstrate the utility of electrochemical techniques in probing cluster reactions. The titrimetric bulk electrolysis procedure allows new cluster types to be prepared easily, with quantitative information on stoichiometry. The cyclic voltammetry approach permitted a systematic study of other ferredoxins to determine the factors that allow rapid reactions of this type to occur. It could be shown, for example, that the presence of Asp in place of Cys in the sequence was not sufficient to confer this striking ambivalence between cluster types. The 7Fe ferredoxin from *Sulfolobus acidocaldarius*, which also has the

Cys-X-X-Asp-X-X-Cys sequence (Fig. 19) shows stable cyclic voltammetry in the presence of 0.1 mM Fe^{2+} even at 60 °C (Armstrong FA, Butt JN, Cammack R, George SJ, Thomson AJ, unpublished results). Clearly in this case, the ability of $[3Fe–4S]^0$ to accept the fourth Fe is greatly diminished. Factors that may be influential include active-site exposure (*Sulfolobus* Fd is much larger) and, perhaps, the presence, in *D.a.* FdIII of a Glu instead of the usual Pro residue adjacent to the remote Cys coordinating the transformable cluster. Proline is well known as a creator of kinks or bends in the polypeptide chain.

These reactions could be studied with adsorbed films of protein that self-assemble spontaneously on PGE in the presence of aminoglycosides [112]. This provides a means to study, quite extensively, the properties of systems such as this which are in limited supply. As shown in Fig. 22a, an adsorbed film of *D.a.* Fd III gives a voltammogram showing three waves; A′, B′, and C′. These waves are sharp and well-defined, and, for A′ and B′, ΔE_p is small even at a scan rate of 500 mVs^{-1}, thus showing that electron transfer is fast. Couple C′ is more complex; the reduction wave was observed to broaden at fast scan rates and higher pH values while the oxidation wave remained sharp. Values of $E^{0'}$ for each couple corresponded closely with those measured for the protein studied in solution under diffusion-controlled conditions. When the electrode modified by adsorption of *D.a.* Fd III was transferred to a cell solution containing Fe^{2+}, rapid cycling (Fig. 22b) showed simultaneous disappearance of A′ and C′ with growth of B′. Together,

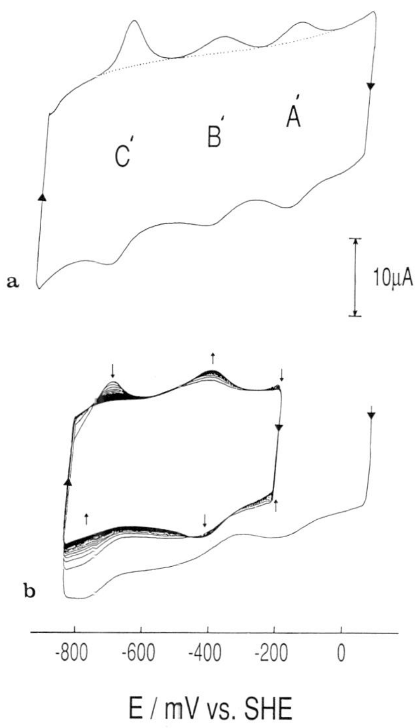

Fig. 22a, b. Cyclic voltammogram of an adsorbed film of *Desulfovibrio africanus* ferredoxin III. The protein was adsorbed onto a PGE electrode from a solution containing 100 µM ferredoxin in 0.1 M NaCl, 5 mM each in TAPS, HEPES, MES and acetate, 0.1 mM EGTA, and 2 mM neomycin at pH 7.0. The voltammetry was measured in the same medium at 0 °C, adjusted to pH 6.25, with a scan rate of 500 mVs^{-1}. A background trace, obtained without adsorbed protein, is indicated (·····) for the oxidative scan. **b**. Effect of cycling the PGE electrode with adsorbed ferredoxin in a solution as above, but adjusted to pH 7.0 and containing 100 µM Fe^{2+} and no EGTA. Scan rate is 500 mVs^{-1} thus each cycle, following *initiation* of the [3Fe–4S]-to-[4Fe–4S] transformation by the reduction of [3Fe–4S] (A′), corresponds to *ca.* 1.3 s.

these observations indicated that the native properties of the clusters and polypeptide environment are retained in the aminoglycoside film. Indeed, the adsorbed film voltammetry was thus demonstrated to be a viable method for studying the kinetics of cluster interconversion in proteins.

As outlined previously in Sect. 4.4, and as indicated above, couples A′(A) (i.e. $[3Fe\text{–}4S]^{1+/0}$) and C′(C) appear to be intimately linked. Since voltammograms of thin films (and particularly of adsorbed species) give waves of essentially finite width within which *all* the redox capacity is contained, it was easy to make a direct comparison of the number of electrons transferred for each couple by integration. In this way, it was determined [112] that the charge passed for re-oxidation of C′ was *twice* that passed for re-oxidation of A′. This ratio was observed regardless of pH (range 6.25 to 7.8) or scan rate (10–1000 mVs^{-1}). It was concluded that couple C′ represents further chemically-reversible *two*-electron reduction of $[3Fe\text{–}4S]^{0}$, to a form corresponding to $[3Fe\text{–}4S]^{2-}$. Formally, this would be equivalent to an all-Fe(II) cluster. The dependence of $E^{0'}$ on pH suggested that two H^+ were bound or released respectively upon reduction and re-oxidation.

Other observations made with this system are important for understanding the mechanism of protein electrochemistry. The absolute charge passed for the couples in Fig. 22a approached or equalled that expected for *monolayer* coverage. This was significant when considering the mechanism of electron transfer in bulk solution electrochemistry. Since solution conditions (temperature, electrolyte, promoter concentration in solution) were the same in each case, the bulk solution electrochemistry described above and shown in Fig. 20 must proceed in the *presence* of this layer of protein. Consequently, assuming a uniform layer, the ***APEE*** mechanism would play a major role in this electrode reaction. Fast scans with the protein in solution actually revealed [110] the adsorption wave A′ for the $[3Fe\text{–}4S]^{1+/0}$ couple since the normal diffusive-type waves broadened and virtually disappeared above $\upsilon = 100\ mVs^{-1}$. The latter observation provided further evidence for differing dynamic behaviour of two centres on *one* protein molecule (see Sect. 4.5).

5.3 The Wider Opportunities for Voltammetric Methods

The use of voltammetry in studies of metal ion speciation is unlikely to be restricted to Fe–S clusters. For example, the kinetics and thermodynamics of Fe distribution in biological systems, including the storage protein *ferritin*, offer a very interesting challenge [194]. Watt and co-workers [195], in an indirect coulommetric study with methyl viologen as mediator, have already demonstrated the general advantage of electrochemical methods in the quantitative study of ferritin, which accommodates up to 4500 Fe atoms in its core. Direct voltammetry may provide a means to distinguish, simultaneously, Fe in various environments, and to monitor its movement under conditions of strict potential control. Metal-ion uptake by metallothionein is another example where applications are obvious. One such study has recently been reported [196]. Voltammetric methods, ideally with the components contained within a thin-layer configuration, are particularly well-suited.

6 Protein Electrochemistry Coupled to Biological Electron-Transport Systems

6.1 What Information is Sought?

By far the greater proportion of protein direct electrochemistry so far reported has involved relatively small molecules whose role is to transport electrons between biological sites of catalysis or energy transduction at which they are produced or required. As the natural mediators for such systems, their own electrochemistry can now be extended to afford interesting opportunities for study. There are several ways in which coupling of the electrochemistry of an electron carrier protein to reactions with its natural partners may be useful.

1) Rate constants for coupled homogeneous electron-transfer reactions may be measured by analysis of voltammograms.

2) Redox titrations or assays of active sites in complex enzymes may be carried out without use of *artificial* mediators. The advantage here of course is that the electron-carrier protein, as the natural mediator, is more likely to be *site specific*. Side reactions, such as non-enzymatic oxidation by dioxygen or peroxide (a frequent problem when using certain small-molecule mediators) are less likely to interfere.

3) Since no small mediators are involved, it is possible to determine the specific accessibility to the electroactive electron-carrier protein, of various enzymes compartmentalized in intact membrane-bound environments.

6.2 Reactions of Electrochemically Transformed Cytochrome c

Most investigations of coupled electron transfer have explored systems that extend the now well-established direct electrochemistry of cytochrome *c*. Some illustrative reaction pathways [18] are shown in Scheme 5. Since cytochrome *c* is a specific electron carrier in a number of biological redox systems, the approach has considerable scope. One of the first studies in this area, however, examined a non-physiological reaction—the coupling of cytochrome *c* electrochemistry to the respiratory chain of *Pseudomonas aeruginosa*. In this respiratory chain the terminal oxidase, cytochrome cd_1 (also known as nitrite reductase since it catalyzes reduction of NO_2^- to NO as well as O_2 to H_2O) is normally reduced not by cytochrome *c*, but by either one of two other electron-carrier proteins, cytochrome c_{551} or azurin. This work, by Hill and Walton [197, 198], established direct electrochemistry as a viable yet previously unexploited way of determining protein–protein electron-transfer rate constants and some detail of the approach will be given here.

It was found [198] that none of the three *Pseudomonas aeruginosa* components, the oxidase or either of the natural reductants, gave a cyclic voltammetric response at a Au electrode modified by 4,4′-bipyridyl. (Haladjian and Bianco later observed [82] that cytochrome c_{551} did respond, albeit poorly, at this electrode.) Since this

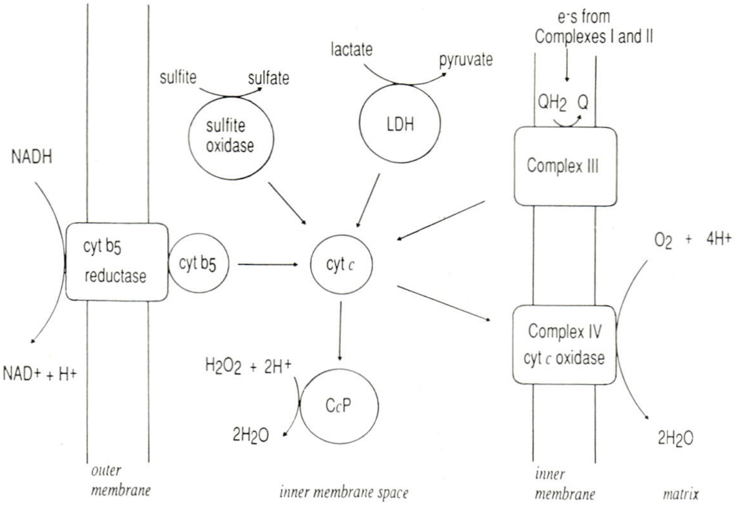

Scheme 5

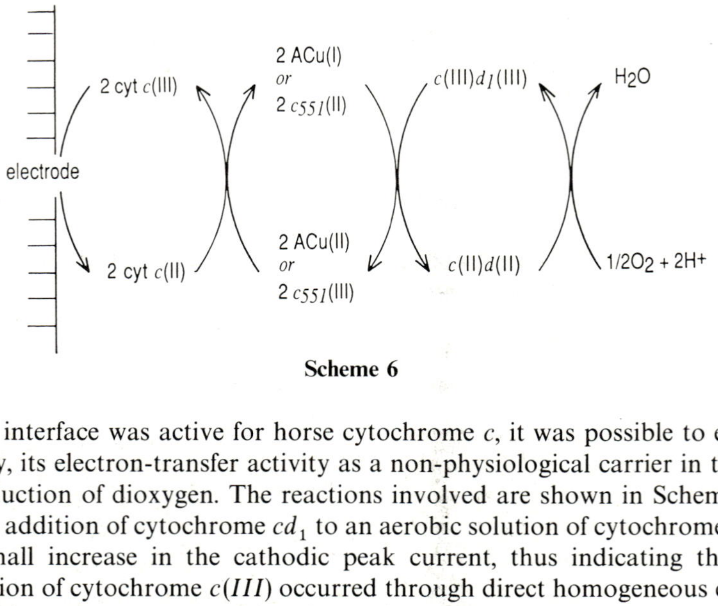

Scheme 6

electrode interface was active for horse cytochrome *c*, it was possible to examine, selectively, its electron-transfer activity as a non-physiological carrier in the enzymatic reduction of dioxygen. The reactions involved are shown in Scheme 6.

Upon addition of cytochrome cd_1 to an aerobic solution of cytochrome *c*, there was a small increase in the cathodic peak current, thus indicating that some regeneration of cytochrome *c(III)* occurred through direct homogeneous coupling to the oxidase. However, as evident from Fig. 23, a drastic change in the voltammetry occurred when cytochrome c_{551} or azurin was added. The sigmoidal wave shape showed attainment of a steady state, with the rate of heterogeneous reduction of cytochrome *c* being balanced by its homogeneous reoxidation, ultimately by O_2. Four-electron reduction of O_2 to $2H_2O$ was demonstrated [197] by integration of the current-time profile for bulk reduction in a sealed cell.

The result showed that cytochrome *c* itself is kinetically incompetent with regard to reduction of cytochrome cd_1, but it is able to transfer electrons to

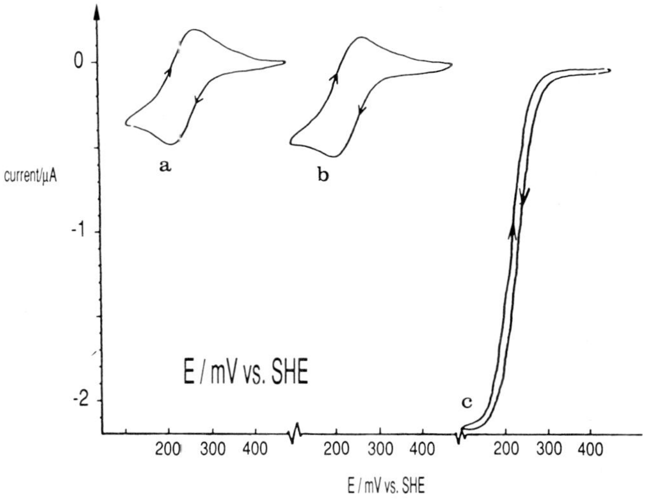

Fig. 23. Coupling the cyclic voltammetry of horse cytochrome *c* (as promoted at a Au electrode in the presence of 4,4′-bipyridyl) to reduction of O_2 by *Pseudomonas aeruginosa* cytochrome cd_1 via a sequence of protein–protein electron-transfer reactions. Aerobic solutions contained 0.1 M $NaClO_4$, 0.02 M phosphate, pH 7.0. Scan rate 1 mVs^{-1}. *a*) horse cytochrome *c* (0.44 mM) alone. *b*) after an addition, of cytochrome cd_1 to 6 μM. *c*) after a further addition, of azurin to 0.25 μM. Redrawn from Ref. 198, with kind permission

cytochrome c_{551} or azurin, each of which have much higher specific activities with the oxidase. Hill and Walton used the voltammetric theory for coupled homogeneous reactions derived by Nicholson and Shain [50] to investigate the kinetics of the protein–protein electron-transfer reactions [198]. Catalytic currents were proportional to cd_1 concentrations so that the rate-determining step appeared to be oxidation of cytochrome c_{551} or azurin. In such a situation, the pseudo first-order rate constant may be obtained by plotting the term $i_k/\upsilon^{1/2}$ against $\log \upsilon$ (i_k is the catalytic current) and then comparing with working curves. Performing this analysis with each concentration of cytochrome c_{551} or azurin yields the corresponding pseudo first-order rate constants. However, experimental conditions limited the concentrations of these proteins so that pseudo-first-order conditions were not achieved. Hill and Walton overcame this difficulty by determining, via an extrapolative procedure, rate constants that would be obtained at infinite scan rates, i.e. if no reagent was actually transformed. By plotting these corrected pseudo first-order rate constants against concentration of cytochrome c_{551} or azurin, the second-order rate constants for electron transfer to cytochrome cd_1 were obtained. These were, respectively, $2 \times 10^4\ M^{-1}s^{-1}$ and $1 \times 10^4\ M^{-1}s^{-1}$, at pH 7 and ionic strength 0.135, in good agreement with data available from independent studies.

Various aspects of the coupling of cytochrome *c* electrochemistry to the reaction with mitochondrial cytochrome *c* oxidase have been studied. The mammalian enzyme comprises 12 subunits, two of which (I and II) contain the four redox-active sites [199]. The centres termed *cytochrome* a_3 and Cu_B constitute a magnetically coupled binuclear site that binds O_2 and executes its reduction to H_2O. The other two, cytochrome *a* and Cu_A, are sites for electron mediation and storage. Cytochrome *c* oxidase spans the mitochondrial inner membrane and it is widely accepted that intramolecular electron transfer is coupled to vectorial movement of H^+. Thus the enzyme constitutes one of the sites at which energy is conserved through the maintenance of an H^+ gradient to drive ATP formation. The active sites and their intrinsic redox properties are more readily investigated after solubilization of the enzyme by suitable detergents. By contrast, the topographical and energy conservation aspects can only be probed if the enzyme is retained in vesicular membranes. The redox chemistries of both soluble and membrane-bound forms of cytochrome *c* oxidase have been coupled to cytochrome *c* electrochemistry.

For the soluble form, free of mass-transport restrictions, rate constants for the cytochrome *c*–cytochrome *c* oxidase electron-transfer reaction have been obtained by analysis of the catalytic current obtained when enzyme and O_2 are added to cytochrome *c* [200, 201]. This is a fast reaction, $k = 10^6$–10^7 $M^{-1}s^{-1}$, that is very sensitive to various factors, including the method of enzyme preparation and such experimental parameters as the ionic strength or the nature of the detergent. This sensitivity being taken into consideration, it may be stated that the direct electrochemistry approach can yield rate-constant values that are very similar to those obtained by more conventional methods, typically the stopped-flow technique. Another application lies with titrations of the cytochrome *c* oxidase redox sites without the involvement of small nonphysiological mediators that may perturb any relevant coupling between sites that is specific to the reaction with cytochrome *c*. Using a tin-doped indium oxide OTTLE, Hawkridge and co-workers [202] demonstrated a spectroelectrochemical coulometric method for the measurement of reduction potentials of cytochromes *a* and a_3. They later extended the coupled system to investigate the thermal denaturation characteristics of cytochrome *c* and cytochrome *c* oxidase [203]. For this investigation they monitored electrochemical turnover (in the form of the catalytic current) at increasing temperature. The onset of retardation could be correlated with data obtained from differential calorimetry.

With membrane-bound cytochrome *c* oxidase, an interesting route into mechanism and organization in *intact* redox systems is displayed. Hill and co-workers coupled [204] the electrochemistry of cytochrome *c* (as achieved at Au electrodes modifed by adsorption of bis(4-pyridyl)disulfide) to electron transport in mitochondria. A demonstration of the manner in which one may thus "tap into" the respiratory chain itself without using non-specific, small-molecule mediators is given in Fig. 24. Upon applying a potential of + 95 mV to an aerobic solution of cytochrome *c(III)* enclosed within a stirred reaction chamber, the electrolytic reduction current decreased exponentially to zero and there was no consumption of O_2 as measured with a Clark-type electrode. If, instead, a suspension of rat-liver mitochondria was added, the reduction current still decreased, but then settled at a steady-state level. There was simultaneous consumption of O_2, the rate of which correspondingly increased to a steady value. If the electrode potential was increased to + 395 mV, the reduction of O_2 ceased abruptly. Upon returning the electrode potential to + 95 mV, the steady-state reduction current, with concomitant con-

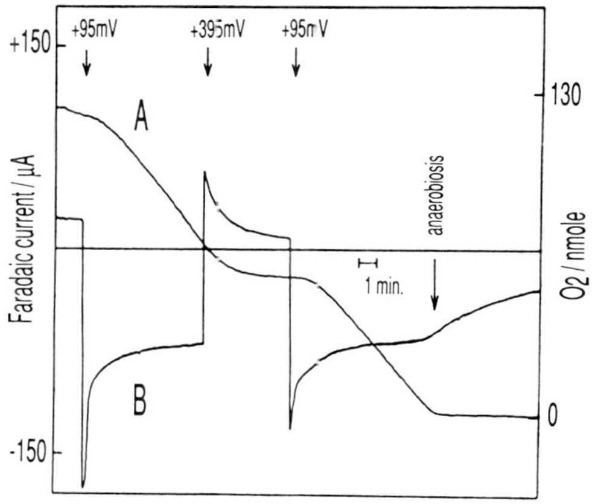

Fig. 24. Coupling the electrochemistry of cytochrome *c* to respiration of mitochondria in a sealed cell. The consumption of O_2 (**A**) and Faradaic current (**B**) as functions of time for 0.7 mM horse cytochrome *c* and rat liver mitochondria (2.0 mg protein/ml). The Au foil electrode was modified by pre-dipping in a solution of bis(4-pyridyl)disulfide. Redrawn from Ref. 204, with kind permission

sumption of O_2, was rapidly resumed. Finally, at anaerobiosis, the reduction current decayed to zero.

There are other variations on this type of experiment. Protoplasts from *Paracoccus denitrificans* showed a similar type of response [204] except that there was a higher level of endogeneous cytochrome *c* reduction. Respiration was "tapped" oxidatively at + 395 mV under conditions of which anaerobiosis was marked by a sharp increase in current. Further increase was obtained by additions of succinate, which is a reductant for cytochrome *c* via the sequence of membrane-bound enzymes succinate dehydrogenase (Complex II) and ubiquinone-cytochrome *c* reductase (Complex III). By contrast, respiratory coupling could not be observed with protoplasts of *Escherichia coli*, an organism in which the terminal oxidase system is not reduced via cytochrome *c*.

Possibilities for investigating the *photosynthetic* electron-transport chain were brought to light by experiments [205] in which the reduction of $P700^+$ was coupled to cytochrome *c* electrochemistry. Although plastocyanin, not cytochrome *c*, is the physiological reductant for $P700^+$, the results demonstrated, as with mitochondria, the ability of a protein (as opposed to a small-molecule mediator) to function in electron transport.

Another experiment carried out by Hill's group was an attempt [206] to detect ejection of H^+ from the mitochondrial matrix, as associated with the cytochrome *c* oxidase "proton pump". With a Au OTTLE modified by pre-adsorption of bis(4-pyridyl)disulfide, to reduce cytochrome *c*, "H^+ pumping" was expected to be observed as a transient change in the light absorption of phenol red—in an essentially unbuffered suspension of rat-liver mitochondria—as the electrode potential was "jumped" to a reducing level. However, the experiment succeeded only in detecting the alkalinization expected from chemical depletion of H^+ as O_2 was reduced.

Whilst lying outside the scope of this article it is worth mentioning that many enzymes with which electron-transfer proteins are natural partners catalyze reactions of considerable analytical interest [10]. Examples of these include cytochrome *c* peroxidase and lactate dehydrogenase. In each of these cases, coupling to cytochrome *c* electrochemistry has been achieved [207, 208], to yield a device that is sensitive to low levels of substrate. Another system thus far demonstrated involves the enzyme *p*-cresol methylhydroxylase, whose natural protein electron donor is the blue Cu protein azurin. In this case the electrochemistry of azurin was coupled [209] to a useful stereospecific synthesis. As *applications* of coupled reactions, however, a definite operational advantage of using the natural protein redox partner in preference to a small mediator must be demonstrated. Since the latter invariably win in terms of cost, stability, and, usually, electrochemistry (since protein direct electrochemistry is so sensitive to the nature of the electrode surface) it is to be expected that mediated amperometric enzyme electrode devices are less likely to make extensive use of electron-transport proteins.

7 Direct Electrochemistry of Metalloenzymes

7.1 Fast Interfacial Electron Exchange with Active Sites of Enzymes

An ability to address the catalytic redox chemistry of an enzyme *via* fast interfacial electron transfer at an electrode without mediators, natural or otherwise, has interesting implications. The action of many redox enzymes may be thought of as the transduction of a "simple" electrical current into a bond-making or -breaking reaction of remarkable specificity. To achieve this, several centres may be required, each playing its own important role in the system. The electrode surface is the ideal "platform" on which to study this chemistry. But enzymes are larger and more complex than electron-transport proteins, and the strategy and rationale behind providing a suitable interface is less well defined. While many are soluble aqueous systems, others are closely associated with membranes and have highly lipophilic surfaces. The area thus presents a challenge with regard to its successful execution and understanding of the interactions involved. The goal is to obtain a well-defined voltammetric response that can be identified as a clean manifestation of the active-site catalytic chemistry without the need for an overpotential that exceeds the requirement in the physiological system. Electron mediation by other reagents must be discountable.

Although there have been a number of reports of direct electrochemistry of enzymes, cases in which these criteria have been met clearly are few in number. Since they lie outside the context of this article, examples in which the sole enzyme active site is an organic group, for example glucose oxidase, have been omitted.

To predict systems most likely to yield meaningful and useful direct electrochemistry, redox enzymes may be divided into two categories. The first comprises those for which one of the redox processes in the catalytic cycle is a discrete outer-sphere (and probably long-range) electron-transfer reaction. Examples include the "blue" Cu oxidases, ferredoxin-linked reductases, and cytochrome *c* oxidase and other enzymes of electron-transport chains. In such systems there is exchange of electrons with an *extrinsic* agent, i.e. one that does not form chemical bonds with the catalytic active site. The agent may be a small molecule or an electron-transport protein. Such enzymes may be referred to as *extrinsic* redox enzymes [11]. Often, long-range *intra*-molecular electron transfer occurs between several redox sites within the enzyme.

The second category comprises enzymes for which *all* electron transfer occurs within a highly localized assembly of the metal centre and redox substrates. These may be termed *intrinsic* redox enzymes.

Since their electron-transfer activity is so confined and may be contained entirely within the coordination sphere of the active site, *intrinsic* redox enzymes may be silent towards non-physiological reaction partners particularly, of course, electrodes. The *extrinsic* enzymes, on the other hand, have generally evolved facile routes for electron transfer between the active site and specific areas of the protein's surface (the interaction site) at which the redox partner binds. It follows that electron exchange with a suitable electrode should be feasible. Where the natural redox partner is an aromatic amine or alcohol, local *hydrophobic* electrode/enzyme contacts might be envisioned to be important. This could be particularly relevant in the case of integral membrane proteins for which part of the structure is buried in a hydrophobic matrix and with which one of the physiological redox partners is the membrane-confined quinone/hydroquinone system. Alternatively, viewed in the light of what we have learnt so far from electron-transport proteins (Sect. 3), we might expect that the most generally amenable of all systems may be those *extrinsic* enzymes whose physiological redox partners are *proteins*. This expectation arises because the need for intimate interaction with another macromolecule suggests that the interaction domain should be open and unobscured. Indeed, in earlier sections it has been suggested often that the "ideal" protein–electrode interaction, permitting fast electron transfer between the active-site metal centre and the electrode surface, is likely to resemble a protein–protein interaction appropriate to the physiological process. The hypothesis should extend naturally to such *extrinsic* enzymes.

7.2 Oxidases and Peroxidases

Among the first accounts of direct electrochemistry of an enzyme that met some of the criteria mentioned above were studies [210, 211] of laccase, a soluble "blue" Cu oxidase which catalyzes the rapid four-electron reduction of dioxygen to water. Laccase is isolated from various plants and fungi from which it is secreted to carry out oxidation of various extracellular aromatic alcohols and amines. The enzymes from the Japanese lacquer tree *Rhus vernicifera* and from the fungus *Polyporus versicolor* have been particularly well characterized [212]. They each contain four Cu centres. One is a "blue" (or Type 1) Cu having similar spectral properties to plastocyanin, one is "non-blue" (or Type 2) Cu, and the remaining two constitute a magnetically coupled pair (the Type 3 centre) which is the site directly responsible for binding and reducing O_2. The reduction potentials of the Cu sites are known from potentiometric studies [213]. Those for fungal laccase are particularly high: as measured at pH = 5.5, the "blue" Cu site has a potential of + 785 mV while that of the type 3 centre is at + 782 mV [213]. Since the four-electron reduction potential of O_2 at pH 7 is 815 mV, very little free energy is conceded during the oxidation of reduced enzyme.

An efficient electrocatalysis of O_2 reduction by fungal laccase was first described by Tarasevich and co-workers [211]. Their most successful electrode surfaces were of carbon materials; pyrolytic graphite, glassy carbon and CO_2-treated carbon black cemented with Teflon varnish. They observed that laccase adsorbed at these materials to yield biocatalytic electrochemistry. Whereas reduction of O_2 is normally slow at unmodified electrodes and generally proceeds, under mild conditions, only as far as hydrogen peroxide, they found that the *four-electron* reduction of O_2 at "laccase-modified" electrodes proceeded easily at potentials close to that for the reversible couple. At pH 5, as used in the experiments, this is + 934 mV. This is rather higher than the reduction potentials of the Cu centres since reduction of O_2 is fast enough to maintain the oxidized sites at a high steady-state level. Electrocatalysis was suppressed by addition of N_3^- or F^-, each of which are specific inhibitors of laccase. Activity was also decreased by addition of H_2O_2, another inhibitor, and this aspect was reflected, interestingly, in the suppression of reduction currents below a potential of approx. + 300 mV at which the uncatalyzed two-electron reduction of O_2 (and thus formation of H_2O_2) commences at carbon-black electrodes. Taken together, the observations showed that O_2 reduction must be occurring *via* rapid and direct electron transfer from the electrode to the Cu sites of the enzyme.

Further studies on the electrochemistry of fungal laccase were reported by Anson and co-workers [214]. In this case the enzyme was adsorbed at a PGE electrode. Reduction of O_2 was studied by DC cyclic voltammetry and by the rotating-disc-electrode technique. Cyclic voltammograms measured for O_2-saturated solutions at pH 3.1 showed a nearly reversible sigmoidal (i.e. steady-state) waveform with $E_{1/2}$ close to + 700 mV. The rotating disc experiments gave curved Levich plots characteristic of a limiting chemical step. The slope and intercept of the resulting Koutecky-Levich (double-reciprocal) plot yielded the number of electrons transferred (four) and the limiting rate constant ($1.5 \times 10^4\ M^{-1}s^{-1}$) corresponding to the reaction of laccase with O_2. Four-electron reduction was further substantiated by the failure to detect H_2O_2 in the reacted solution provided the electrode potential had not been held low enough to generate H_2O_2 non-catalytically. Qualitatively, there was good agreement with the results obtained by Tarasevich and co-workers. They also reported a direct voltammetric response from the adsorbed enzyme in the absence of O_2. To observe this, they needed to add the reagents 2,9-dimethylphenanthroline or 4,4′-bipyridyl. The voltammograms yielded $E^{0'}$ values of approx. 600 mV and 700 mV respectively, which showed no marked variation between pH 3.7 and 5.6. These are significantly lower than the potentiometric values for Types 1 and 3 Cu centres. The requirement for these agents must be regarded as puzzling since the catalytically enhanced electroactivity was clearly visible without them.

Peroxidases catalyze the two-electron reduction of H_2O_2 (or organic peroxides) to H_2O (or alcohols) by electron-transfer proteins or small organic reagents. A general catalytic cycle for the heme-containing enzymes is shown in Scheme 7.

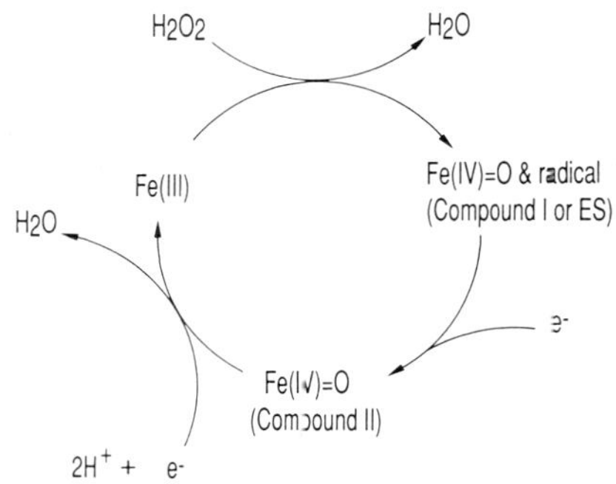

$H_2O_2 + 2e^- + 2H^+$	$\rightleftharpoons$	$2H_2O$	$E^{\circ\prime}_{(pH7)}$ = *ca.* 1.3V[a]
Compound I + e^-	$\rightleftharpoons$	Compound II	$E^{\circ\prime}_{(pH7)}$ = *ca.* 900mV (HRP)[b]
Compound II + e^-	$\rightleftharpoons$	Fe(III)	$E^{\circ\prime}_{(pH7)}$ = *ca.* 930mV (HRP)[b]
Fe(III) + e^-	$\rightleftharpoons$	Fe(II)	$E^{\circ\prime}_{(pH7)}$ = -270mV (HRP)[c] -190mV (CcP)[d]

[a] Hoare JP (1974) in "Encyclopaedia of Electrochemistry of the Elements" (Bard AJ, ed.) Vol. II, p 194;
[b] Hayashi Y, Yamazaki I (1979) J. Biol. Chem. 254: 9101;
[c] Harbury HA (1957) J. Biol. Chem. 225: 1009;
[d] Conroy CW, Tyma PH, Erman JE (1978) Biochim. Biophys. Acta 537: 62.

Scheme 7

Cytochrome *c* peroxidase (C*c*P) [215] isolated from yeast mitochondria uses cytochrome *c(II)* as electron donor. Its crystal structure has been determined [216]. It is a monomer of molecular mass 34000 containing a single *b*-type heme group that is largely buried within the molecule. The Fe is coordinated on one side by a histidine residue while the sixth position is occupied by a weakly bound H_2O in a channel suitable for entry and exit of small substrates.

The Fe(III) centre reacts rapidly with peroxide to yield a two-electron oxidized species termed ES or Compound I. This contains Fe(IV) as ferryl (Fe=O), and a cation radical species that is located either on the porphyrin ring system (plant peroxidases) or on a nearby amino-acid residue(s) (cytochrome *c* peroxidase). The Fe(III) form is regenerated by reduction *via* another species, Compound II, which contains Fe(IV) but no radical.

For cytochrome *c* peroxidase, the reduction involves two protein–protein one-electron transfers. There have been extensive studies of the physiologically relevant reaction that provide much support for very rapid formation of a precursory 1:1 electron-transfer complex [217]. Electrostatic interactions are a major factor in this process. Examination of the structure of C*c*P reveals a ring of aspartate residues that are arranged so as to be complementary with the distribution of highly conserved lysines that surround the exposed heme edge region of the cytochrome *c* molecule (see Fig. 1). A hypothetical complex that optimizes the polar interactions between surface groups has been proposed [21]. The resultant heme-edge–to–heme-edge distance is 16.5 Å.

The C*c*P redox reactions encompass a wide thermodynamic range. The relevant half-cell reactions and their reduction potentials are indicated also in Scheme 7. With $E^{0'}$ for cytochrome *c* being + 260 mV, the driving force for reduction of Compounds I and II is of the order of 0.75 V. Energy is not conserved by coupling. Instead, the large driving force renders the reduction of H_2O_2 irreversible and accelerates the rate of long-range protein–protein electron transfer. Estimates of the rate constant for this reaction [218] are of the order of 10^3 s^{-1}.

Direct electrochemical transformations of C*c*P were achieved by Assefa and Bowden who showed [219], using optical monitoring, that the Fe(III) form could be reduced to Fe(II) and then reoxidized at a fluorine-doped tin oxide OTTLE. The reactions were electrochemically irreversible, and large overpotentials (– 490 mV for reduction and + 610 mV for oxidation) were used to achieve rapid reaction.

Since the non-physiological redox couple could be addressed in this way, the question remained, "What is observed if the electrode potential is modulated within the range of activity for the catalytic cycle?" To address this question would involve application of electrode potentials greater than 1 V, and problems of stability might be expected. Armstrong and Lannon searched instead for the biocatalytic reduction of H_2O_2 that must result if direct electrochemistry is able to address the catalytically active states [220]. Overpotentials of at least 1 V are normally required for reduction of H_2O_2 at electrodes. This was indeed found to be the case at polished PGE electrodes. However, in the presence of small (μM) levels of C*c*P in solution, large reduction currents were obtained at potentials that, under some conditions, were higher than + 700 mV. No catalysis occurred if the free prosthetic group Fe protoporphyrin IX was used instead of enzyme. It was thus concluded that the facile reduction of substrate could only be proceeding *via* direct electrochemistry of the enzyme. The high potentials at which this occurred indicated that the interfacial electron-transfer reaction was, in terms of kinetic facility, at least comparable to the physiological reduction by cytochrome *c*(*II*). The voltammetry is shown in Fig. 25. There are a number of points of interest.

1) The catalytic activity is associated with *adsorbed* C*c*P. At low enzyme concentrations, i.e. < 1 μM, time-dependent changes in wave form and current could be monitored conveniently. The initial sigmoidal-like voltammetric response developed into a sharp peak whose potential position was dependent upon scan rate and H_2O_2 concentration, and, to a lesser extent, C*c*P concentration and temperature. It was shown, by parallel monitoring of the changes in ellipsometric parameters for the reflection of polarized light from the electrode surface [221], that this time dependence corresponded to the time course for adsorption of C*c*P molecules, most likely resulting in monolayer coverage. The development of a peak-like voltammetric waveform occurred within the first and major change in ellipsometric parameters and with a rate that was broadly proportional to the solution concentration of C*c*P. The change from sigmoidal- to peak-type response could be interpreted in terms of a "microscopic model" [125], mentioned in section 3.4, in which growing clusters of adsorbed enzyme molecules behave as expanding *micro-electrodes*. Progressive adsorption of enzyme results in a transformation

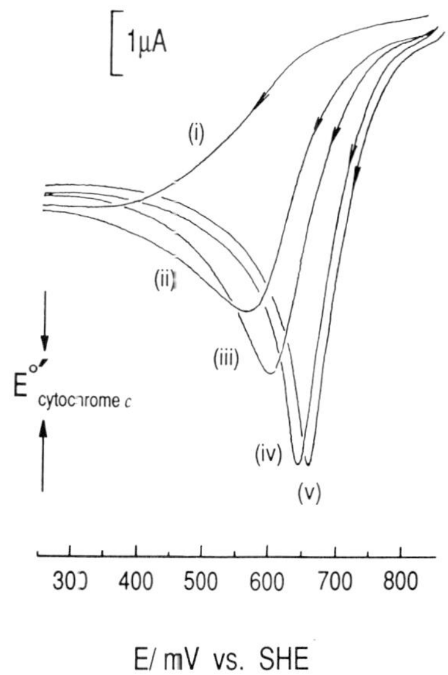

Fig. 25. Linear-sweep voltammograms of H_2O_2 (56 μM) in 0.1 M KCl, 5 mM neomycin, 5 mM HEPES, pH 7.0, initiated at various times following addition of cytochrome *c* peroxidase to a concentration of 0.2 μM. Scan rate 8 mVs^{-1}, temperature = 0 °C: *i)* 30 s; *ii)* 178 s; *iii)* 325 s; *iv)* 750 s; *v)* 1080 s. The position of $E^{0'}$ for cytochrome C, the natural reductant of CcP is indicated.

from microelectrode behaviour (radial diffusion of substrate to isolated sites) to that of a uniform planar macro-electrode (linear diffusion to a planar surface of essentially packed sites). At the limit of zero time such a microelectrode would have the dimensions of the enzyme molecule itself.

2) The peak currents obtained were proportional to the concentration of H_2O_2. The dependence on the square root of the scan rate curved downwards above $\upsilon = 5\ mVs^{-1}$ although the initial slope was consistent with a diffusion coefficient of $1 \times 10^5\ cm^2\ s^{-1}$, a value appropriate for H_2O_2.

3) The voltammetric response appeared to require the presence of neomycin or gentamycin, aminoglycosides observed [105, 110–112] to be active promoters of the electrochemistry of other proteins with negatively charged surfaces. Here, it was proposed that the spatial distribution of $-NH_3^+$ groups presented by these sugars at the electrode interface might resemble, broadly, that presented to CcP by the lysine residues on cytochrome *c*. Thus the adsorption process might by guided by the same recognition factors as are responsible for assembly of the physiological 1 : 1 complex. Subsequent studies indicated [221] that this might not be the case since it has proved possible to obtain a biocatalytic response of CcP at PGE electrodes in the absence of these promoters. This finding is indeed described independently in a recent paper by Paddock and Bowden [236].

4) From the appropriate reduction potentials for H_2O_2 and CcP Compounds I (ES) and II, it is clear that the biocatalytic response due to CcP adsorbed at PGE electrodes still occurs at a significant overpotential (approx. 650 mV as related to the H_2O_2/H_2O couple at approx. + 1.3 V). However, as indicated in Fig. 25, the driving force is actually considerably smaller than that which would be operative based upon the reduction potential of cytochrome *c*. Thus the experiment demonstrated the ability of an electrode to function at least as effectively as a physiological (protein) electron-transfer partner.

Plant (horseradish) peroxidase (HRP) has also been studied [222–225]. Like C*c*P, it can be reduced to a non-physiological Fe(II) state. Razumas and co-workers investigated this reaction at Au electrodes modified by electrochemically induced polymerization of viologens [222, 223]. The clearest evidence for direct electron transfer was obtained in the case of the methylviologen-modified electrode, which gave no voltammetric signals in buffer solution alone. Upon introduction of HRP, a reduction peak appeared at -470 mV while a poorly defined oxidation wave was observed between -160 mV and -360 mV [222]. The reduction current was proportional to the concentration of HRP over the range 7 to 27 μM. Reduction of enzyme molecules in bulk solution was demonstrated by changes in the absorption spectrum. The product thus detected was not the Fe(II) form, but the oxygenated Fe(II)-O_2 derivative. This oxygenated product occurred because of the presence of O_2 in the sample solution since the Fe(II) form, like deoxymyoglobin, rapidly forms an O_2 adduct. A kinetic study of the reduction of HRP at these electrodes was carried out [223] with the rotating-disc technique. Heterogeneous rate constants for the methyl- and benzyl-viologen derivatives were, respectively, 1.3×10^{-5} and 8.4×10^{-4} cm s^{-1} at pH 7. Razumas and co-workers argued that mediated electron transfer was also unlikely in the case of the benzylviologen/Au electrode system, despite its electroactivity in buffer solution alone, since the benzylviologen cation radical is a relatively poor mediator to HRP in solution. The much higher activity of the benzylviologen/Au electrode was attributed to the presence of the aromatic rings, which could aid in specific adsorption of HRP or assist electron transfer itself by penetrating the protein interior.

Biocatalytic responses have been reported [224, 225] for HRP adsorbed at various carbon electrodes. Where clearly evident [225], the overpotentials required to effect rapid reduction of H_2O_2 appear rather higher (approx. 850 mV related to the H_2O_2/H_2O couple) than values appropriate for C*c*P discussed above.

7.3 Dehydrogenases

There is a wealth of intriguing and useful chemistry available within those enzymes that catalyze specific oxido-reduction or oxygen-atom insertion (monooxygenation) reactions with organic substrates. Clear demonstrations of direct electrochemistry have been made for a few cases, and the last discussions in this article are concerned with these.

Ikeda and co-workers have studied [226] the voltammetric properties of D-gluconate dehydrogenase (GADH) from *Pseudomonas fluorescens*. This is a membrane-bound enzyme, MW 130 000, that catalyzes the oxidation of D-gluconate to 2-keto-D-gluconate (Eq. 21) and conveys the two electrons to ubiquinone in the lipid layer, thereby linking up with the main respiratory chain. The enzyme consists of three non-identical subunits which contain FAD (the likely site of D-gluconate oxidation) a *c*-type heme, and an Fe-S cluster. Either or both of the latter one-electron-active groups probably receive electrons from FAD and act as immediate electron donors to ubiquinone [227].

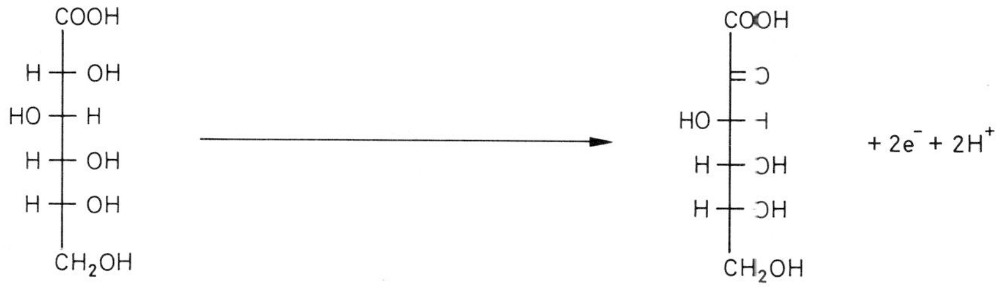

(21)

Carbon-paste electrodes were modified [226] with GADH by placing small aliquots of a solution of the detergent-solubilized enzyme onto the surface, allowing the solvent to evaporate, and then rinsing away weakly adsorbed GADH and low-MW solutes. Cyclic voltammetry of the modified electrode in buffer (0.1 M acetate, pH 5) showed no waves in the range − 80 to + 380 mV. However, when D-gluconate was added to the solution, there was a large catalytic enhancement of the oxidation current above a potential of + 100 mV with a steady-state plateau ($\upsilon = 5\ \mathrm{mV\,s^{-1}}$) being reached at approx. + 320 mV The result demonstrated achievement of a specific organic oxidation process, Eq. (21), by an enzyme adsorbed at an electrode. Even without observation of a reversible redox couple in the absence of substrate (this is presumably a problem of sensitivity) the evidence points strongly, as with laccase and peroxidases, towards a direct transfer of electrons between electrode and the enzyme centres. The onset of catalytic activity occurred at a potential close to that determined (+ 120 mV) for the *c*-type heme.

Another dehydrogenase, *p*-cresolmethylhydroxylase (PCMH), has recently been shown [228] to display high catalytic activity at PGE electrodes. Here, the interfacial dynamics are more distinguished since there is a marked and discriminatory requirement for and among aminoglycoside promoters. The enzyme is isolated from *Pseudomonas putida* in which it catalyzes [229] the specific oxygenation of *p*-cresol to *p*-hydroxybenzaldehyde, Eq. (22):

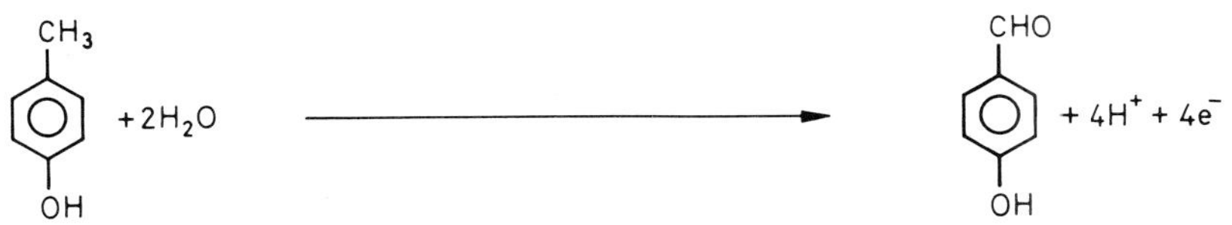

(22)

It is an $\alpha_2\beta_2$ tetramer of MW 115000. The larger of the two subunit types (MW 49000) each contain a covalently bound FAD that is the site of substrate oxygenation. The other subunits, *c*-type cytochromes of MW 8500, serve to relay electrons to an accepter molecule, which is probably azurin (reduction potential + 320 mV). In their electrochemical study, Hill and his group effectively provided an electrode surface as a substitute for this protein partner.

As shown in Fig. 26, direct electrochemistry could be observed both through the electrocatalytic oxidation of *p*-cresol by low (sub-micromolar) concentrations of

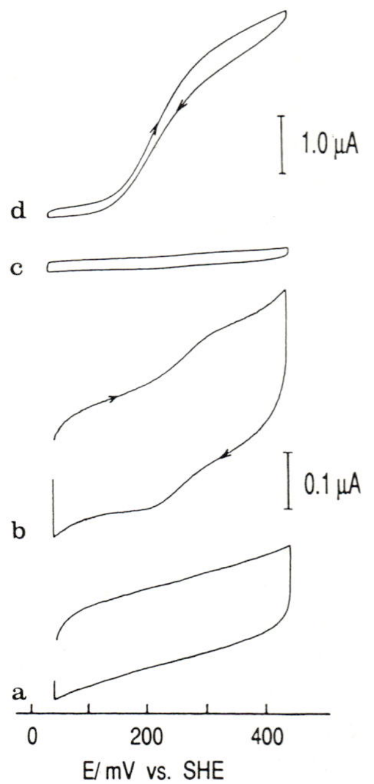

Fig. 26a–d. Cyclic voltammetry of *p*-cresolmethylhydroxylase (PCMH) at a PGE electrode in the presence of spermine as promoter. Electrolyte comprised 10 mM HEPES, 10 mM KCl (pH 7.4), scan rate 5 mVs^{-1}. (**a**) Spermine tetrachloride (10 mM) alone. (**b**) Response after addition of PCMH to 35 μM. (**c**) as for (**b**) but recorded at 0.1 current gain. (**d**) as for (**c**) but upon addition of the substrate *p*-cresol to 3 mM. Redrawn from Ref. 228, with kind permission

PCMH in solution (giving rise to catalytic-type steady-state cyclic voltammograms) *and* through the reversible response that was visible with high (35 μM) enzyme concentration in the *absence of substrate*. The steady-state current amplitudes were found to be proportional to substrate concentration. They increased also, usually in a saturative manner, with the concentration of cation promoter. These ranged from a variety of aminoglycosides to metal aquo-ions such as Mg^{2+}.

The reduction potential value for the reversible process was + 255 mV, in good agreement with potentiometric measurements monitoring oxidation and reduction of the cytochrome in the intact tetramer, but significantly different from the value obtained for the free subunit (+ 180 mV). This observation showed that the electrochemical response stemmed from the enzyme functioning *as a unit* rather than from dissociated cytochrome subunits coupling electron transfer to the FAD-containing entities. Since peak currents were proportional to $\upsilon^{1/2}$ at low scan rates, an effective diffusion coefficient could be evaluated. At 1×10^{-7} $cm^2 s^{-1}$, this was in broad agreement with the "ideal" value predicted by the Stokes–Einstein equation for a molecule of this size. The result is significant since it illustrates very clearly the point made in Sect. 2.2, i.e. that the slower diffusion of macromolecules does not pose a prohibitive problem. By contrast with the other enzymes discussed in this section, PCMH appears not to be tightly adsorbed.

Considerable selectivity was shown among various cations as promoters of the electrochemistry. Neomycin was the most potent reagent, yielding optimal steady-state currents at a concentration of 0.2 mM (whereas 2 mM levels of the comparable aminoglycosides gentamycin and ribostomycin were required). Furthermore, the catalytic response could be achieved also by use of Au electrodes modified by adsorption of the peptides cys-glu-cys (charge − 1), arg-cys-OMe (charge + 2) or lys-cys-thr-cys-cys-ala (+ 1), provided that spermine (2 mM) was present and the electrode had been subjected to a brief polish with alumina *after* pre-adsorption of peptide. While these observations have not as yet been rationalized in terms of protein surface characteristics, they should provide an excellent opportunity for starting to understand the specific interactions governing the electrode interactions of large enzymes.

The clear detection of both reversible active-site and biocatalytic waves represents a "completeness" that establishes the feasibility of applying direct electochemistry to probe the mechanism of action of complex redox enzymes. To take this further, I shall take up the author's prerogative for mentioning studies "currently underway in the laboratory" and mention, briefly, another membrane-bound enzyme, fumarate reductase (FR), isolated from *Escherichia coli*. Structurally, this is closely related to the more familiar succinate dehydrogenase (SDH) which constitutes the major part of Complex II of the mitochondrial respiratory chain. Of the four subunits which make up the membrane-bound system, two may be freed to give a soluble enzyme that is active in fumarate reduction by artificial electron donors [230]. The larger of these, MW approx. 70 000, contains, like SDH, a covalently bound FAD. The smaller, MW approx. 30 000, appears to contain three Fe–S clusters. These are termed centre 1 ([2Fe–2S]), centre 2 ([4Fe–4S]) and centre 3 ([3Fe–4S]). Their respective reduction potentials as determined by potentiometry are − 20 mV, − 320 mV, and − 70 mV [231]. (Although the potential of the FAD has not been determined for FR, the two-electron value for beef heart SDH is − 79 mV at pH 7.0. The radical form is unstable since the two one-electron reductions occur at potentials of − 127 and − 31 mV respectively [232].)

As shown in Fig. 27, this soluble form of FR adsorbs at PGE electrodes giving rise to a voltammetric response having an effective $E^{0'}$ value of = − 40 mV, that appears essentially reversible even at a scan rate of 500 mVs^{-1}. This response may actually be an "envelope" of signals from centres 1, 3, and the FAD.

As with PCMH described above, the reversible response coincides with a biocatalytic wave that is observed if fumarate is added. The overall electrode reaction is given by Eq. (23).

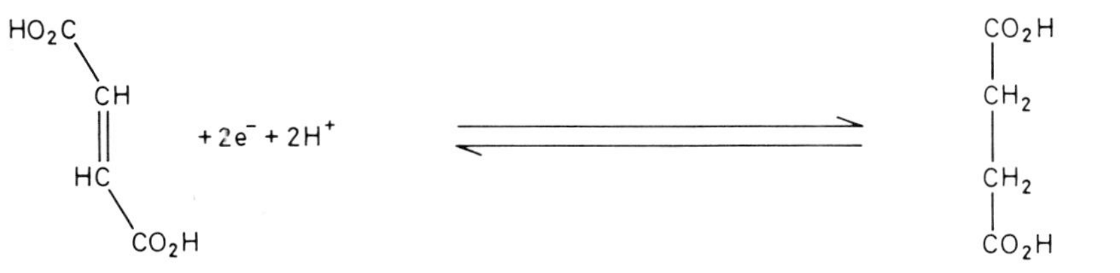

(23)

The efficiency of this "electrocatalytic hydrogenation" is self-evident since it occurs close to the thermodynamic reduction potential calculated for the fumarate/

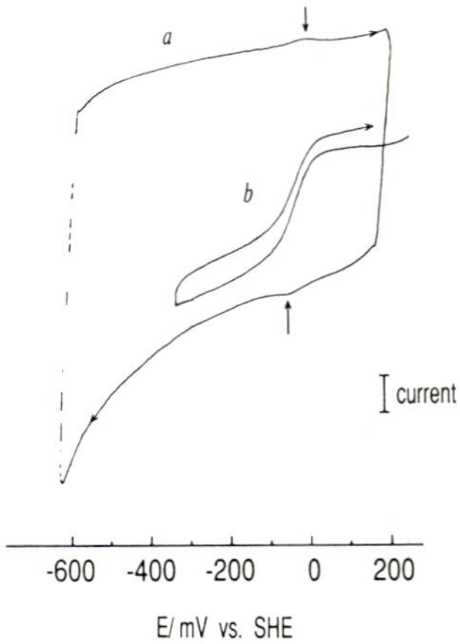

Fig. 27. *a*). Cyclic voltammogram (at a PGE electrode) of a solution of fumarate reductase 13 μM in 20 mM HEPES, 0.1 M NaCl, pH 7.0, containing 2 mM neomycin. Scan rate 500 mVs^{-1}, current scale 2 μA, temperature = 0 °C. The reversible signal (*arrowed*) lies in the potential domain expected for centres 1 ($[2Fe–2S]^{2+/1+}$), 3 ($[3Fe–4S]^{1+/0}$) and FAD/FADH$_2$ *b*). Cyclic voltammogram of a PGE electrode (at which is adsorbed fumarate reductase) immersed in a solution of fumarate (80 μM) in 20 mM HEPES, 0.1 M NaCl, pH 7.0. Scan rate 1 mVs^{-1}, current scale 0.1 μA, temperature 0 °C

succinate couple (+ 30 mV at pH 7.0 [233]). The steady-state current shows Michaelis-Menten-type saturation as the concentration of fumarate is increased.

The role of centre 2, for which the potentiometric reduction potential, at − 320 mV, is too low to be involved directly in electron transport between hydroquinone and fumarate, is not clear [164, 234, 235]. In these electrochemical experiments we have so far detected neither its electrode response nor any change in the steady-state biocatalytic current as the potential is lowered beyond − 320 mV. Cammack and co-workers have proposed [234, 235] that this centre might be active in one-electron reduction and reoxidation of the more reducing FAD *radical* species that must be generated, albeit transiently, during turnover. Their mechanism of reduction of oxidized FAD in SDH is termed the "dual-pathway" model since the two single electrons flowing consecutively out of FAD do so by different routes. The lower potential pathway uses centre 2. An equivalent mechanism, with the direction of electron flow reversed, is applicable for FR.

8 Concluding Remarks and Acknowledgments

As I conclude this article, I hope that I have given the reader some appetite for the field and have shown that direct electrochemistry, far from remaining merely a subject of curiosity, may be applied to interesting and important problems in bioinorganic chemistry and enzymology. Without the requirement for mediators, it is possible to "talk" directly to the active sites of metalloproteins. This "two-way conversation" happens in *real time* and provides a complementary departure from contemporary strategies that are largely dominated by studies on "static" samples: resting states or intermediates quenched during a reaction or catalytic cycle. With this in mind, we have only just begun to see the tip of a very large "iceberg".

It is a great pleasure to acknowledge those individuals whose interest and encouragement have been so forthcoming. Foremost is Allen Hill who introduced me to electrochemistry some eight years ago, and with whose very active research group at Oxford I am proud to have been associated. Among these scientists (to name but a few) Nick Walton, Mark Harmer, Nigel Oliver and Martin Lannon have been the most valued of colleagues. I am grateful to Ancy Thomson, Simon George, Dick Cammack, Geoff Sykes, Alan Bond, and a talented student—Julea Butt, for interesting and continuing collaborations. Further advice and assistance with this article were given by Isao Taniguchi, Katsumi Niki, Jean Haladjian, Hans Freeman, Gary Brayer, Larry Sieker, and Gordon Sanghera. Their help and the editorial comments of John Goodenough are gratefully acknowledged. Finally, I wish to thank The Royal Society for their support over the past six years.

9 References

1. Sawyer DT, Roberts JL Jr (1974) Experimental electrochemistry for chemists, Wiley, New York
2. Bond AM (1980) Modern polarographic methods in analytical chemistry, Dekker, New York
3. Bard AJ, Faulkner LR (1980) Electrochemical methods, fundamentals and applications, Wiley, New York
4. Southampton Electrochemistry Group (1985) Instrumental methods in electrochemistry, Ellis Horwood, Chichester
5. Wilson GS (1978) Meth. Enzymol. 54: 396
6. Dutton PL (1978) Meth. Enzymol. 54: 411
7. Eddowes MJ, Hill HAO (1982) In: Kadish KM (ed.) Electrochemical and spectrochemical studies of biological redox components ACS Adv. Chem. Ser. 201: 173
8. Bowden EF, Hawkridge FM, Blount HN (1985) In: Srinivasan S, Chizmadzhev YA, Bockris JO'M, Conway BE, Yeager E (eds) Comprehensive treatise of electrochemistry, Plenum, New York, p 297
9. Armstrong FA, Hill HAO, Walton NJ (1986) Quart. Rev. Biophys. 18: 261
10. Frew JE, Hill HAO (1988) Eur. J. Biochem. 172: 261
11. Armstrong FA, Hill HAO, Walton NJ (1988) Acc. Chem. Res. 21: 407
12. Scott RA, Mauk AG, Gray HB (1985) J. Chem. Educ. 62: 932
13. Mayo SL, Ellis WR Jr, Crutchley RJ, Gray HB (1986) Science 233: 948
14. Pielak GJ, Concar DW, Moore GR, Williams RJP (1987) Protein Engineering 1: 83
15. Sykes AG (1988) Chem. Brit. 24: 551
16. McLendon G (1988) Acc. Chem. Res. 21: 160
17. Marcus RA (1965) J. Chem. Phys. 43: 679
18. Moore GR, Pettigrew GW (1987) Cytochromes *c*. Springer, Berlin Heidelberg New York (Springer Series in Molecular Biology)
19. Louie GV, Hutcheon WLB, Brayer GD (1988) J. Mol. Biol. 199: 295
20. Salemme FR (1976) J. Mol. Biol. 102: 563
21. Poulos TL, Kraut J (1980) J. Biol. Chem. 255: 10322
22. Lane RF, Hubbard AT (1973) J. Phys. Chem. 77: 1411
23. Lyklema J (1984) Colloids Surf. 10: 33
24. Norde W (1986) Adv. Colloid Interfac. Sci. 25: 267
25. Many relevant articles are to be found in: Brash JL, Horbett TA (eds.) Proteins at Interfaces, ACS Adv. Chem. Ser. 343 (1987)
26. Scheller F, Strnad G (1982) In: Kadish KM (ed) Electrochemical and spectrochemical studies of biological redox components, ACS Adv. Chem. Ser. 201: 219
27. Hubbard AT, Anson FC (1970) Electroanal. Chem., Marcel Dekker, Vol 4, p 129
28. Scheller F, Jänchen M, Prümke H-J (1975) Biopolymers 14: 1553
29. Stankovich MT, Bard AJ (1978) J. Electroanal. Interfac. Electrochem. 85: 189
30. Santhanum KSV, Jespersen N, Bard AJ (1977) J. Amer. Chem. Soc. 99: 274

31. Haser R, Pierrot M, Frey M, Payan F, Astier JP, Bruschi M, Le Gall J (1979) Nature 282: 806
32. Higuchi Y, Kusunoki M, Yasuoka N, Kakudo M, Yagi T (1981) J. Biochem. (Tokyo) 90: 1715
33. Niki K, Yagi T, Inokuchi H, Kimura K (1979) J. Amer. Chem. Soc. 101: 3335
34. Niki K, Yagi T, Inokuchi H (1982) In: Kadish KM (ed) Electrochemical and spectrochemical studies of biological redox components, ACS Adv. Chem. Ser. 201: 199
35. Niki K, Kawasaki Y, Kimura Y, Higuchi Y, Yasuoka N (1987) Langmuir 3: 982
36. Betso SR, Klapper MH, Anderson LB (1972) J. Amer. Chem. Soc. 94: 8197
37. Scheller F, Jänchen M, Lampe J, Prümke H-J, Blanck J, Palecek E (1975) Biochim. Biophys. Acta 412: 157
38. Kuznetsov BA, Shumakovich GP, Mestechkina NM (1977) Bioelectrochem. Bioenerg. 4: 512
39. Haladjian J, Bianco P, Serre PA (1979) Bioelectrochem. Bioenerg. 6: 555
40. Weitzman PDJ, Kennedy IR, Caldwell RA (1971) FEBS Lett. 17: 241
41. Kuznetzov BA, Mestechkina NM, Shumakovich GP (1977) Bioelectrochem. Bioenerg. 4: 1
42. Ikeda T, Toriyama K, Senda M (1979) Bull. Chem. Soc. Japan 52: 1937
43. Kukutani T, Toriyama K, Ikeda T, Senda, M (1980) Bull. Chem. Soc. Japan 53: 947
44. Bianco P, Haladjian J (1977) Biochem. Biophys. Res. Commun. 78: 323
45. Hill CL, Renaud J, Holm RH, Mortenson, LE (1977) J. Amer. Chem. Soc. 99: 2549
46. Bianco P, Haladjian J (1986) J. Electroanal. Interfac. Electrochem. 199: 365
47. van Dijk C, van Leeuwen JW, Veeger C, Schreurs JPGM, Barendrecht E (1982) Bioelectrochem. Bioenerg. 9: 743
48. Eddowes MJ, Hill HAO (1977) J. Chem. Soc. Chem. Commun. 771
49. Eddowes MJ, Hill HAO (1979) J. Amer. Chem. Soc. 101: 4461
50. Nicholson RS, Shain I (1964) Anal. Chem. 36: 706
51. Landrum HL, Salmon RT, Hawkridge FM (1977) J. Amer. Chem. Soc. 99: 3154
52. Crawley CD, Hawkridge FM (1983) J. Electroanal. Interfac. Electrochem. 159: 313
53. Yeh P, Kuwana T (1977) Chem. Lett. 1145
54. Bowden EF, Hawkridge FM, Blount HN (1984) J. Electroanal. Interfac. Electrochem. 161: 355
55. Reed DE, Hawkridge FM (1987) Anal. Chem. 59: 2334
56. Cotton TM, Schultz SG, VanDuyne RP (1980) J. Amer. Chem. Soc. 102: 7960
57. Smulevich G, Spiro TG (1985) J. Phys. Chem. 89: 5168
58. Hildebrandt P, Stockburger M (1986) J. Phys. Chem. 90: 6017
59. Hinnen C, Parsons R, Niki K (1983) J. Electroanal. Interfac. Electrochem. 147: 329
60. Rodrigues CG, Farchione F, Wedd AG, Bond AM (1987) J. Electronal. Interfac. Electrochem. 218: 251
61. Gadsby PMA, Peterson J, Foote N, Greenwood C, Thomson AJ (1987) Biochem. J. 246: 43 and references therein
62. Haladjian J, Pilard R, Bianco P, Serre P-A (1982) Bioelectrochem. Bioenerg. 9: 91
63. Brautigan DL, Ferguson-Miller S, Margoliash E (1978) Methods Enzymol. 53D: 131
64. Bianco P, Haladjian J, Bruschi M (1986) Bioelectrochem. Bioenerg. 15: 57
65. Albery WJ, Eddowes MJ, Hill HAO, Hillman AR (1981) J. Amer. Chem. Soc. 103: 3904
66. Cotton TM, Kaddi D, Iorga D (1983) J. Amer. Chem. Soc. 105: 7462
67. Taniguchi I, Murakami T, Toyosawa K, Yamaguchi H, Yasukouchi K (1982) J. Electroanal Interfac. Electrochem. 131: 397
68. McMahon JJ, Dougherty TP, Riley DJ, Babcock GT, Carter RL (1985) Surface Science 158: 381
69. Taniguchi I, Iseki M, Yamiguchi H, Yasukouchi K (1985) J. Electroanal. Interfac. Electrochem. 186: 299
70. Elliott D, Hamnett A, Lettington OC, Hill HAO, Walton NJ (1986) J. Electroanal. Interfac. Electrochem. 202: 303
71. Taniguchi I, Toyosawa K, Yamaguchi H, Yasukouchi K (1982) J. Chem. Soc. Chem. Commun. 1032
72. Taniguchi I, Iseki M, Yamaguchi H, Yasukouchi K (1984) J. Electroanal. Interfac. Electrochem. 175: 341
73. Allen PM, Hill HAO, Walton NJ (1984) J. Electroanal. Interfac. Electrochem. 178: 69
74. Taniguchi I, Iseki M, Toyosawa K, Yamaguchi H, Yasukouchi K (1984) J. Electroanal. Interfac. Electrochem. 164: 385
75. Haladjian J, Bianco P, Pilard R (1983) Electrochim. Acta 28: 1823
76. Hill HAO, Page DJ, Walton NJ, Whitford DJ (1985) J. Electroanal. Interfac. Electrochem. 187: 315
77. Taniguchi I, Higo N, Umekita K, Yasukouchi K (1986) J. Electroanal. Interfac. Electrochem. 206: 341
78. Hill HAO, Page DJ, Walton NJ (1986) J. Electroanal. Interfac. Electrochem. 208: 395 (1986)
79. DiGleria K, Hill HAO, Lowe VJ, Page DJ (1986) J. Electroanal. Interfac. Electrochem. 213: 333
80. Hill HAO, Page DJ, Walton NJ (1987) J. Electroanal. Interfac. Electrochem. 217: 129

81. Barker PD, Di Gleria K, Hill HAO, Lowe VJ (submitted) Eur. J. Biochem.
82. Haladjian J, Bianco P (1983) Bioelectrochem. Bioenerg. 11: 319
83. Hill HAO, Page DJ, Walton NJ (1987) J. Electroanal. Interfac. Electrochem. 217: 141
84. Niwa K, Furukawa M, Niki K (1988) J. Electroanal. Interfac. Electrochem. 245: 275
85. Bowden EF, Hawkridge FM, Chlebowski JF, Bancroft EE, Thorpe C, Blount HN (1982) J. Amer. Chem. Soc. 104: 7641
86. Koller KB, Hawkridge FM (1985) J. Amer. Chem. Soc. 107: 7412
87. Koller KB, Hawkridge FM (1988) J. Electroanal. Interfac. Electrochem. 239: 291
88. Harmer MA, Hill HAO (1985) J. Electroanal. Interfac. Electrochem. 189: 229
89. Willit JL, Bowden EF (1987) J. Electroanal. Interfac. Electrochem. 221: 265
90. Armstrong FA, Hill HAO, Oliver BN: J. Chem. Soc. Chem. Commun. 1984: 976
91. Armstrong FA, Cox PA, Hill HAO, Lowe VJ, Oliver BN (1987) J. Electroanal. Interfac. Electrochem. 217: 331
92. Panzer RE, Elving PJ (1975) Electrochim. Acta 20: 635
93. Nakamizo M, Tamai K (1984) Carbon 22: 197
94. Kamau GN, Willis WS, Rusling JF (1985) Anal. Chem. 57: 545
95. Arnold DJ, Gerchario KA, Anderson CW (1984) J. Electroanal. Interfac. Electrochem. 172: 379
96. Taniguchi I, Funatsu T, Umekita K, Yamaguchi H, Yasukouchi K (1986) J. Electroanal. Interfac. Electrochem. 199: 455
97. Seiter CHA, Margalit R, Perreault RA (1979) Biochem. Biophys. Res. Commun. 86: 473
98. Armstrong FA, Brown KJ (1987) J. Electroanal. Interfac. Electrochem. 219: 319
99. Armstrong FA, Hill HAO, Oliver BN, Whitford D (1985) J. Amer. Chem. Soc. 107: 1473
100. Armstrong FA, Cox PA, Hill HAO, Oliver BN, Williams AA (1985) J. Chem. Soc. Chem. Commun. 1236
101. van Dijk C, van Eijs T, van Leeuwen JW, Veeger C (1984) FEBS Lett. 166: 76
102. Takabe T, Ishikawa H, Niwa S, Tanaka Y (1984) J. Biochem. (Tokyo) 96: 385
103. Armstrong FA, Hill HAO, Walton NJ (1982) FEBS Lett. 145: 241
104. Armstrong FA, Hill HAO, Oliver BN, Walton NJ (1984) J. Amer. Chem. Soc. 106: 921
105. Armstrong FA, George SJ, Thomson AJ, Yates MG (1988) FEBS Lett. 234: 107
106. Armstrong FA, Driscoll PC, Hill HAO (1985) FEBS Lett. 190: 242
107. Concar DW, Hill HAO, Moore GR, Whitford D, Williams RJP (1986) FEBS Lett. 206: 15
108. Sheludko A (1966) Colloid chemistry, Elsevier p 212
109. Barker PD, Hill HAO (1988) In: King TE, Mason HS, Morrison M (eds) Oxygen and Related Redox Systems
110. Armstrong FA, George SJ, Cammack R, Hatchikian EC, Thomson AJ Biochem. J.
111. George SJ, Armstrong FA, Hatchikian EC, Thomson AJ Biochem. J.
112. Armstrong FA, Butt JN, George SJ, Hatchikian EC, Thomson AJ FEBS Lett. 259, 15 (1989)
113. Bianco P, Manjaqui A, Haladjian J, Bruschi M (1988) J. Electroanal Interfac. Electrochem. 249: 241
114. Bianco P, Haladjian J (1987) J Electroanal. Interfac. Electrochem. 223: 201
115. Nakahara Y, Kimura K, Inokuchi H, Yagi T (1979) Chem. Lett. 877
116. Tam S-C, Williams RJP (1985) Structure and Bonding 63: 103
117. Dhesi R, Cotton TM, Timkovich R (1983) J. Electroanal. Interfac. Electrochem. 154: 129
118. Lewis NS, Wrighton MS (1981) Science (Washington DC) 211: 944
119. Elliott CM, Martin WS (1982) J. Electroanal. Interfac. Electrochem. 137: 377
120. Chao S, Robbins JL, Wrighton MS (1983) J. Amer. Chem. Soc. 105: 181
121. Kulys JJ, Samalius AS, Švirmickas G-JS (1980) FEBS Lett. 114: 7
122. Albery WJ, Bartlett PN, Craston DH (1985) J. Electroanal. Interfac. Electrochem. 194: 223
123. Kulys JJ (1986) Biosensors 2: 3
124. Nicholson RS (1965) Anal. Chem. 37: 1351
125. Armstrong FA, Bond AM, Hill HAO, Psalti ISM, Zoski CG (1989) J. Phys. Chem. 93: 6485
126. Armstrong FA, Bond AM, Hill HAO, Oliver BN, Psalti ISM (1989) J. Amer. Chem Soc. 111: 9185
127. Gueshi T, Tokuda K, Matsuda H (1979) J. Electroanal. Interfac. Electrochem. 101: 29
128. Amatore C, Savéant JM, Tessier D (1983) J. Electroanal. Interfac. Electrochem. 147: 39
129. Sabatini E, Rubinstein I (1987) J. Phys. Chem. 91: 6663
130. Heineman WR, Norris BJ, Goetz JF (1975) Anal. Chem. 47: 79
131. Hawkridge FM, Ke B (1977) Anal. Biochem. 78: 76
132. Watt GD (1979) Anal. Biochem. 99: 399
133. Heineman WR, Anderson CW, Halsall HB, Hurst MM, Johnson JM, Kreishman GP, Norris BJ, Simone MJ, Su C-H (1982) In: Kadish KM (ed) Electrochemical and spectrochemical studies of biological redox components, ACS Adv. Chem. Ser. 201: 1
134. Moore GR, Pettigrew GW, Rogers NK (1986) Proc. Natl. Acad. Sci. USA 83: 4998

135. Mashiko T, Reed CA, Haller KJ, Kastner ME, Scheidt WR (1981) J. Amer. Chem. Soc. 103: 5758
136. Stellwagen E (1978) Nature 275: 73
137. Moore GR (1983) FEBS Lett. 161: 171
138. Osheroff N, Brautigan DL, Margoliash E (1980) Proc. Natl. Acad. Sci. USA 77: 4439
139. Lambeth DO, Campbell KL, Zand R, Palmer G (1973) J. Biol. Chem. 248: 8130
140. Moore GR, Harris DA, Leitch FA, Pettigrew GR (1984) Biochim. Biophys. Acta 764: 331
141. Taniguchi I, Iseki M, Eto T, Toyasawa K, Yamaguchi H, Yasukouchi K (1984) Bioelectrochem. Bioenerg. 13: 373
142. Taniguchi I, Funatsu T, Iseki M, Yamaguchi H, Yasukouchi K (1985) J. Electroanal. Interfac. Electrochem. 193: 295
143. Kreishman GP, Su C-H, Anderson CW, Halsall HB, Heineman WR (1980) In: Blank M (ed) Bioelectrochemistry: Ions, surfaces, membranes, ACS Adv. Chem. Ser. 188: 169
144. Myer YP, Srivastava RB, Kumar S, Raghavendra K (1983) J. Prot. Chem. 2: 13
145. Di Marino M, Marassi R, Santucci R, Brunori M, Ascoli F (1987) Bioelectrochem. Bioenerg. 17: 27
146. Santucci R, Reinhard H, Brunori M (1988) J. Amer. Chem. Soc. 110: 8536
147. Hill HAO, Whitford D (1987) J. Electroanal. Interfac. Electrochem. 235: 153
148. Sorrell TN, Martin PK, Bowden EF (1989) J. Amer. Chem. Soc. 111: 766
149. Bianco P, Haladjian J (1981) Electrochim. Acta 26: 1001
150. Bianco P, Haladjian J (1982) J. Electroanal. Interfac. Electrochem. 137: 367
151. Sokol WF, Evans DH, Niki K, Yagi T (1980) J. Electroanal. Interfac. Electrochem. 108: 107
152. Niki K, Kobayashi Y, Matsuda H (1984) J. Electroanal. Interfac. Electrochem. 178: 333
153. Niki K, Kawasaki, Nishimura N, Higuchi Y, Yasuoka N, Kakudo M (1984) J. Electroanal. Interfac. Electrochem. 168: 275
154. Dolla A, Cambillau C, Bianco P, Haladjian J, Bruschi M (1987) Biochem. Biophys. Res. Commun 147: 818
155. Gray HB, Solomon EI (1981) In: Spiro TG (ed) Copper proteins, Wiley, New York p 1
156. Katoh S (1977) In: Pirson A, Zimmerman MH (eds) Encyclopaedia of plant physiology, vol 5, Wiley, New York, p 247
157. Guss JM, Harrowell PR, Murata M, Norris VA, Freeman HC (1986) J. Mol. Biol. 192: 361
158. Moore JM, Case DA, Chazin WJ, Gippert GP, Havel TF, Powls R, Wright PE (1988) Science (Washington DC) 240: 314
159. Segal MG, Sykes AG (1978) J. Amer. Chem. Soc. 100: 4584
160. Sinclair-Day JD, Sykes AG (1986) J. Chem. Soc. Dalton Trans: 2069
161. Katoh S, Shiratori I, Takamiya A (1962) J. Biochem. (Tokyo) 51: 32
162. Kojiro CL, Markley JL (1983) FEBS Lett. 162: 52
163. Lappin AG, Lewis CA, Ingledew WJ (1985) Inorg. Chem. 24: 1446
164. Cammack R (1987) In: Matsubara H (ed) Iron sulfur protein research, Japanese Sci. Soc. Press, Tokyo/Springer, Berlin Heidelberg New York
165. Adman ET, Sieker LC, Jensen LH (1973) J. Biol. Chem. 248: 3987
166. Fukuyama K, Nagahara Y, Tsukihara T, Katsabe Y (1988) J. Mol. Biol. 199: 183
167. DePamphilis BV, Averill BA, Herskovitz T, Que L Jr, Holm RH (1974) J. Amer. Chem. Soc. 96: 4159
168. Kassner RJ, Yang W (1977) J. Amer. Chem. Soc. 99: 4351
169. Carter CW (1977) In: Lovenberg W (ed) Iron-sulfur proteins, Academic, New York p 157
170. Stout CD (1989) J. Mol. Biol. 205: 545
171. Stout GH, Turley S, Sieker LC, Jensen LH (1988) Proc. Natl. Acad. Sci USA 85: 1020
172. Kissinger CR, Adman ET, Sieker LC, Jensen LH (1988) J. Amer. Chem. Soc. 110: 8721
173. Beinert H, Thomson AJ (1983) Arch. Biochem. Biophys. 222: 333
174. Emptage MH, Dreyer J-L, Kennedy MC, Beinert H (1983) J. Biol. Chem. 258: 11106
175. Beinert H (1986) Biochem. Soc. Trans. 14: 527
176. Sweeney WV, Rabinowitz JC, Yoch DC (1975) J. Biol. Chem. 250: 7842
177. Stephens PJ, Morgan TV, Stout CD, Burgess BK (1986) In: Xavier AV (ed) Frontiers of bioinorganic chemistry, VCH, Weinheim, p 637
178. George SJ, Richards AJM, Thomson AJ, Yates MG (1984) Biochem. J. 224: 247
179. Johnson MK, Bennett DE, Fee JA, Sweeney WV (1987) Biochim. Biophys. Acta 911: 81
180. Thomson AJ, Robinson AE, Johnson MK, Cammack R, Rao KK, Hall DO (1981) Biochim. Biophys. Acta 637: 423
181. Armstrong FA, Hill HAO, Walton NJ (1982) FEBS Lett. 150: 214
182. Dus K, De Klerk H, Sletton K, Bartsch RG (1967) Biochim. Biophys. Acta 140: 291
183. Nocera DG, Winkler JR, Yocum KM, Bordignon E, Gray HB (1984) J. Amer. Chem. Soc. 106: 5145
184. Isied SS, Kuehn C, Worosila G (1984) J. Amer. Chem. Soc. 106: 1722

185. Bechtold R, Kuehn C, Lepre C, Isied SS (1986) Nature 322: 286
186. Jackman MP, McGinnis J, Powls R, Salmon GA, Sykes AG (1988) J. Amer. Chem. Soc. 110: 5880
187. Jackman MP, Lim M-C, Sykes AG, Salmon GA (1988) J. Chem. Soc. Dalton Trans. 2843
188. Armstrong FA, Butt JN, McGinnis J, Sykes AG (manuscript in preparation)
189. Emptage MH, Kent TA, Kennedy MC, Beinert H, Münck E (1983) Proc. Natl. Acad. Sci. USA 80: 4674
190. Moura I, Moura JJG, Münck E, Papaefthymiou V, LeGall J (1986) J. Amer. Chem. Soc. 108: 349
191. Surerus KK, Münck E, Moura I, Moura JJG, LeGall J (1987) J. Amer. Chem. Soc. 109: 3805
192. Job RC, Bruice TC (1975) Proc. Natl. Acad. Sci. USA 72: 2478
193. Bovier-Lapierre G, Bruschi M, Bonicel J, Hatchikian EC (1987) Biochim. Biophys. Acta 913: 20
194. Crichton R (1987) Bioelectrochem. Bioenerg. 18: 105
195. Watt GD, Frankel RB, Papaefthymiou GC (1985) Proc. Natl. Acad. Sci. USA 82: 3640
196. Olafson RW (1988) Bioelectrochem. Bioenerg. 19: 111
197. Hill HAO, Walton NJ, Higgins IJ (1981) FEBS Lett. 126: 282
198. Hill HAO, Walton NJ (1982) J. Amer. Chem. Soc. 104: 6515
199. Wikström M, Krab K, Saraste M (1981) Cytochrome *c* oxidase; A synthesis, Academic, London
200. Cohen DJ, Hawkridge FM, Blount HN, Hartzell CR (1984) In: Allen MJ, Usherwood PRN (eds) Charge and field effects in biosystems, Abacus, Tunbridge Wells, p 19
201. Barker PD, Coleman JOD, Hill HAO, Walton NJ, Whitford D (1986) Biochem. Soc. Trans. 14: 130
202. Long RC, Hawkridge FM, Hartzell CR (1986) J. Electroanal. Interfac. Electrochem. 198: 89
203. Long RC, Hawkridge FM, Chlebowski JF, Hartzell CR (1988) J. Electroanal. Interfac. Electrochem. 256: 111
204. Coleman JOD, Hill HAO, Walton NJ, Whatley FR (1983) FEBS Lett. 154: 319
205. Hill HAO, Walton NJ, Whitford D (1985) J. Electroanal. Interfac. Electrochem. 187: 109
206. Hill HAO, Walton NJ, Whitford D, Coleman JOD (1985) J. Inorganic Biochem. 23: 303
207. Frew JE, Harmer MA, Hill HAO, Libor SI (1986) J. Electroanal. Interfac. Electrochem. 201: 1
208. Cass AEG, Davis G, Hill HAO, Nancarrow DJ (1985) Biochim. Biophys. Acta 828: 51
209. Hill HAO, Oliver BN, Page DJ, Hopper DJ (1985) J. Chem. Soc. Chem. Commun: 1469
210. Berezin IV, Bogdanovskaya VA,Varfolomeev SD, Tarasevich MR, Yarapolov AI (1978) Dokl. Akad. Nauk SSSR. 240: 619
211. Tarasevich MR, Yaropolov AI, Bogdanovskaya VA, Varfolomev SD (1979) Bioelectrochem. Bioenerg 6: 393
212. Reinhammar B, Malmström BG (1981) In: Spiro TG (ed) Copper proteins, Wiley, New York, p 109
213. Reinhammar BRM (1972) Biochim. Biophys. Acta 275: 245
214. Lee C-W, Gray HB, Anson FC, Malmström BG (1984) J. Electroanal. Interfac. Electrochem. 172: 289
215. Yonetani T (1976) In: Boyer PD (ed) The enzymes, vol 13, Academic, p 345
216. Finzel BC, Poulos TL, Kraut J (1984) J. Biol. Chem. 259: 13027
217. Mochan E, Nicholls P (1971) Biochem. J. 121: 69
218. Hazzard JT, Poulos TL, Tollin G (1987) Biochemistry 26: 2836
219. Assefa H, Bowden EF (1986) Biochem. Biophys. Res. Commun. 139: 1003
220. Armstrong FA, Lannon AM (1987) J. Amer. Chem. Soc. 109: 7211
221. Armstrong FA, Bond AM, Hamnett A, Hill HAO, Lannon AM, Lettington OC, Zoski CG (submitted for publication)
222. Razumas VJ, Gudavičius AV, Kulys JJ (1983) J. Electroanal. Interfac. Electrochem. 151: 311
223. Razumas VJ, Gudavičius AV, Kulys JJ (1986) J. Electroanal. Interfac. Electrochem. 198: 81
224. Yaropolov AI, Malovik V, Varfolomeev SD, Berezin IV (1979) Dokl. Acad. Nauk SSSR 249: 1399
225. Razumas VJ, Jasaitis JJ, Kulys JJ (1984) Bioelectrochem. Bioenerg. 12: 297
226. Ikeda T, Fushimi F, Miki K, Senda M (1988) Agric. Biol. Chem. 52: 2655
227. Ikeda T, Miki K, Fushimi F, Senda M (1987) Agric. Biol. Chem. 51: 747
228. Guo LH, Hill HAO, Hopper DJ, Lawrance GA, Sanghera GS (1989) J. Electroanal. Interfac. Electrochem. 266: 379
229. McIntire W, Hopper DJ, Singer TP (1985) Biochem. J. 228: 325
230. Dickie P, Weiner JH (1979) Can. J. Biochem. 57: 813
231. Cammack R, Patil DS, Weiner JH (1986) Biochim. Biophys. Acta 870: 545
232. Ohnishi T, King TS, Salerno JC, Blum H, Bowyer JR, Maida T (1981) J. Biol. Chem. 256: 5577
233. Clark WM (1960) Oxidation-reduction potentials of Organic Systems. Williams and Wilkins, p 507
234. Cammack R, Crowe BA, Cook ND (1986) Biochem. Soc. Trans. 14: 1207
235. Cammack R, Maguire JJ, Ackrell BAC (1987) in "Cytochrome Systems Molecular biology and bioenergetics (Papa S, Chance B, Ernster L, eds.) Plenum. p 485
236. Paddock RM, Bowden EF (1989) J. Electroanal. Interfac. Electrochem. 260: 487

Author Index Volumes 1–72